FUNCTIONAL DECOMPOSITION WITH APPLICATION TO FPGA SYNTHESIS

Functional Decomposition with Application to FPGA Synthesis

by

Christoph Scholl

Institute of Computer Science,
Albert-Ludwigs-University,
Freiburg im Breisgau, Germany

KLUWER ACADEMIC PUBLISHERS
BOSTON / DORDRECHT / LONDON

A C.I.P. Catalogue record for this book is available from the Library of Congress.

ISBN 978-1-4419-4929-5

Published by Kluwer Academic Publishers,
P.O. Box 17, 3300 AA Dordrecht, The Netherlands.

Sold and distributed in North, Central and South America
by Kluwer Academic Publishers,
101 Philip Drive, Norwell, MA 02061, U.S.A.

In all other countries, sold and distributed
by Kluwer Academic Publishers,
P.O. Box 322, 3300 AH Dordrecht, The Netherlands.

Printed on acid-free paper

To Barbara.

Contents

Preface

During the last few years Field Programmable Gate Arrays (FPGAs) have become increasingly important. Thanks to recent breakthroughs in technology, FPGAs offer millions of system gates at low cost and considerable speed.

Functional decomposition has emerged as an essential technique in automatic logic synthesis for FPGAs. Functional decomposition as a technique to find realizations for Boolean functions was already introduced in the late fifties and early sixties by Ashenhurst (Ashenhurst, 1959), Curtis (Curtis, 1961), Roth and Karp (Roth and Karp, 1962; Karp, 1963). In recent years, however, it has attracted a great deal of renewed attention, for several reasons. First, it is especially well suited for the synthesis of lookup-table based FPGAs. Also, the increased capacities of today's computers as well as the development of new methods have made the method applicable to larger-scale problems. Modern techniques for functional decomposition profit from the success of Reduced Ordered Binary Decision Diagrams (ROBDDs) (Bryant, 1986), data structures that provide compact representations for many Boolean functions occurring in practical applications. We have now seen the development of algorithms for functional decomposition which work directly based on ROBDDs, so that the decomposition algorithm works based on compact representations and not on function tables or decomposition matrices as in previous approaches.

The book presents, in a consistent manner, a comprehensive presentation of a multitude of results stemming from the author's as well as various researchers' work in the field. Apart from the basic method, it also covers functional decomposition for incompletely specified functions, decomposition for multi–output functions and non–disjoint decomposition.

In essence the exposition is self–contained; a basic knowledge of mathematics and of some fields of computer science such as complexity theory and graph theory will be helpful for an understanding.

CHRISTOPH SCHOLL

Freiburg, August 2001

Acknowledgments

I would like to thank all my colleagues and friends both from the University of Saarland, Saarbrücken, and from the Albert–Ludwigs–University, Freiburg, who contributed to a pleasant working climate.

Special thanks to Dr. Harry Hengster and Dr. Uwe Sparmann for many helpful discussions.

I acknowledge the work of Stephan Melchior, who made contributions to the theory of symmetries of incompletely specified functions.

I would like to express my thanks to Prof. Dr. Paul Molitor for supporting my research, for cooperating on many subjects in this monograph, for engaging in helpful discussions, and for encouraging me to write this book.

I am also grateful to Prof. Dr. Günter Hotz for numerous suggestions, for his advice, and his confident support.

I would like to thank Prof. Dr. Bernd Becker for giving me the opportunity to finish this book by sparing me any interruptions and for his continuous support.

I owe gratitude to Prof. Dr. Paul Molitor and Mary Masters, who have read and improved the draft of this monograph.

Finally, I would like to thank Cindy Lufting, James Finlay, and Mark de Jongh of Kluwer Academic Publishers for their patience and help in preparing the final manuscript.

Introduction

Integrated Circuits. In recent years *Integrated Circuits* have extended to a rapidly increasing number of areas of everyday life. Not only have microprocessors, as components of Personal Computers, spread to an unforeseen extent in our information and communication society, but also Integrated Circuits (microprocessors and ASICs = Application Specific Integrated Circuits), which come as integral parts of *Embedded Systems* in automotive applications and various household appliances, are on the march. Electronic anti-theft devices, anti-lock braking systems, communication equipment, compact disc players, and chip cards, etc., are only a few examples of application of Integrated Circuits.

ASICs and FPGAs. Apart from standard components such as microprocessors, these systems frequently require Integrated Circuits which are tailored to the application. *Application Specific Integrated Circuits* are designed to perform a specific function in one system. Whereas standard components are designed in a *Full–Custom Design* style, where every part of the circuit is especially optimized for the purpose it serves in the design, ASICs are mostly designed in a *Semi–Custom Design* style by assembling pre–designed and pre–characterized subcircuits. The reason for these different design styles lies in the fact that standard components are typically produced in much higher volumes than ASICs, so that for standard parts more time can be spent on reducing area and increasing performance of the chip. In contrast, development costs and times tend to be a much more dominant factor for ASICs.

Field Programmable Gate Arrays (FPGAs) are devices which allow even smaller development costs and times than ASICs. FPGAs consist of logic blocks and interconnections which can be *programmed in the field*. Thus FPGA devices are manufactured regardless of the functionality they finally have to realize, and they obtain their functionality by programming. No manufacturing is needed after the design of the circuit to be realized by an FPGA. So in some sense FPGAs combine advantages of standard components and ASICs: They are produced in high volumes leading to relatively low costs for single devices *and* their flexibility leads to short design times, which has a great effect on their time–to–market.

Only a few years ago FPGAs were still rather expensive, slow and small. But thanks to recent breakthroughs in technology, FPGAs offer millions of system gates at reduced cost and at considerable speed, so that they have become serious competitors of ASICs.

The role of Design Automation.   The systems which fit into a single chip today are so large that it is obvious that they can neither be specified nor optimized without using *Design Automation*. This is true for full–custom designs as well as for ASIC and FPGA designs. For instance, the Pentium 4 processor, which was launched in November 2000, includes 42 million transistors. Modern FPGAs offer millions of system gates, too.

It is clear that designing such large systems is a big challenge and it cannot be accomplished by manual design. It is simply impossible to design such complicated circuits manually, without any form of design automation. On the one hand the complexity of the design task has to be kept under control and on the other hand the design has to be completed in a design time which is as short as possible to reduce time–to–market and thus enable product success on the market. So automatic computer–aided design (CAD) tools must come into play to shorten the design cycle by automating part of it.

This book is devoted to the automation of logic synthesis, which is a crucial step in the design flow.

Design flow.   The task which has to be solved in circuit design consists of computing a correct *implementation* from a *specification* of the circuit. The design flow can be divided into several steps:

- In the *high–level* or *architectural* design phase, a decomposition of the design into larger functional blocks is performed (either manually or automatically). Such a top–level block diagram can be expressed in a Hardware Description Language (HDL) or by graphical means.

- Based on the specification which is provided by the architectural design phase, logic synthesis generates a gate–level description. The result of logic synthesis is a netlist using basic gates or cells.

- Validation or verification checks whether the implementation of the chip is equivalent to its specification.

- Test generation computes test stimuli which are used later on to identify faulty chips after fabrication. During test generation, hardware may be inserted to improve the 'testability' of the final chip.

- Physical layout synthesis generates a physical description of the final chip. Cells are placed and wired. The result of layout synthesis are masks, which are used to fabricate the chip on silicon. In the case of FPGA synthesis, the result of physical layout synthesis is a program which can be used to configure the FPGA chip to realize the computed logic cells and interconnection wires.

Typically, a verification or validation step is done after each step in the design flow to check whether the result is correct.

Logic synthesis can be divided into *combinational* logic synthesis and *sequential* logic synthesis. The goal of sequential logic synthesis is to compute realizations for *sequential* circuits, i.e., for circuits which contain memory elements. The result of sequential synthesis is a Boolean network of combinational logic blocks and memory elements. The combinational blocks are then synthesized by *combinational logic synthesis*. Combinational logic synthesis computes a gate realization of *Boolean functions* $f : \{0, 1\}^n \to \{0, 1\}^m$. In this book we assume that the separation of the circuit into memory elements and combinational blocks is already done and we deal only with combinational logic synthesis.

Combinational logic synthesis. Of course, there is a large variety of possibilities for computing a gate realization of a Boolean function $f : \{0, 1\}^n \to \{0, 1\}^m$. Optimization goals in computing *good* gate realizations are the following:

- *Area* occupied by the logic gates and interconnection wires.

- *Delay* of the resulting circuit.

- *Power* consumed by the circuit.

- Degree of *testability*, which is an approximation of the probability of detecting faults which are present in a fabricated chip.

Combinational logic synthesis is traditionally separated into *technology independent optimization* and *technology mapping*. Technology independent optimization derives an 'optimal' structure independent of the gates available in a particular technology. Technology mapping maps the optimized network produced by technology independent optimization to particular gates from a pre–designed library, called a *technology library*. Technology mapping only introduces local changes to the netlist produced by technology independent optimization.

First approaches to an automatic logic synthesis were restricted to *two–level logic synthesis*. In two–level logic synthesis the search space for good realizations is restricted to sums–of–products.[1] Two–level solutions can easily be realized by Programmable Logic Arrays (PLAs) and there are matured heuristic methods such as ESPRESSO (Brayton et al., 1984), which are able to find good sums-of-products. However there are many Boolean functions occurring in practical applications where even the optimal two–level representation is huge. So for large instances *multi–level logic synthesis* is preferred in most cases.

Apart from methods based on local transformations (e.g. (Darringer et al., 1981; Darringer et al., 1984)) the 'classical method' for multi–level logic synthesis is based on methods of two–level logic synthesis combined with methods for an algebraic factorization of sum-of-product representations (implemented in *misII* (Brayton et al., 1987) and *sis* (Sentovich et al., 1992)).

Logic synthesis for FPGAs.   Earlier methods for FPGA synthesis used the results of technology independent optimization by algebraic factorization (Brayton et al., 1987; Sentovich et al., 1992) and they viewed FPGA synthesis methods only as a technology mapping step to map the results of technology independent optimization to logic blocks of FPGAs.

(Francis et al., 1990) and (Francis et al., 1991) present a technology mapping algorithm which is based on the results of technology independent optimization by *misII* (Brayton et al., 1987). After the technology independent optimization, the circuit is mapped to lookup tables. Lookup tables (LUTs) are the basic blocks of so–called LUT–based FPGAs and each LUT is able to realize an arbitrary function $lut : \{0,1\}^b \rightarrow \{0,1\}$ up to a certain number b of inputs.[2] *chortle* (Francis et al., 1990) and *chortle–crf* (Francis et al., 1991) start with an input Boolean network which consists of nodes implementing sums-of-products. *chortle* divides the Boolean network into a forest of trees and determines an optimal mapping of each tree by dynamic programming. *chortle–crf* employs a bin–packing approach to decompose sums-of-products of nodes in the Boolean network into lookup tables. Both methods have the common property that they are only structural approaches, i.e., they are only able to combine existing gates

---

[1]For the definition of sums–of–products see page 9.
[2]More details can be found in Section 1.4.

into lookup tables (in a clever way), but they are not able to restructure the circuit.

In (Murgai et al., 1990) and (Murgai et al., 1991) the technology mapping is also based on a Boolean network produced by *misII*. In these papers, a multitude of different methods is used to decompose the Boolean network into lookup tables with a fixed number *b* of inputs and to minimize the number of lookup tables after computing an initial solution. As in (Francis et al., 1990; Francis et al., 1991), structural methods are used in (Murgai et al., 1990; Murgai et al., 1991) to decompose sums-of-products which are too large to fit into one lookup table, but (Murgai et al., 1990; Murgai et al., 1991) also use Roth–Karp decomposition of sums-of-products (Roth and Karp, 1962) and Shannon decompositions (Shannon, 1938) of sums-of-products, which are *not purely structural* methods (even if these decompositions are applied only to single nodes in the Boolean network). One method for node minimization after computing an initial solution is reduced to a binate covering problem (Rudell, 1989); another method consists of locally combining several lookup tables into one, if possible.

(Karplus, 1991) first constructs 'if–then–else dags' as a decomposition of nodes in the Boolean network produced by *misII* and then uses a covering procedure to map them into lookup tables.

For a detailed description of the approaches to FPGA synthesis mentioned above we refer the reader to (Brown et al., 1992) and (Murgai et al., 1995). In this book we concentrate on logic synthesis for FPGAs based on *functional decomposition*.

Functional decomposition. FPGA synthesis by *functional decomposition* is viewed as a *combination of technology independent logic synthesis and technology mapping*. Functional decomposition is a *Boolean* method which goes beyond algebraic factorizations.

The basic idea of functional decomposition is illustrated by the following figure:

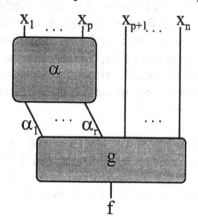

To decompose a Boolean function $f : \{0,1\}^n \to \{0,1\}$ with input variables $x_1, \ldots, x_n$, the input variables are partitioned into two disjoint sets $\{x_1, \ldots, x_p\}$ (the set of 'bound variables') and $\{x_{p+1}, \ldots, x_n\}$ (the set of 'free variables'). There are r 'decomposition functions' $\alpha_1, \ldots, \alpha_r$, which compute an 'intermediate result' based on the variables of the bound set. The function g is called 'composition function' and g computes the result of f based on the intermediate result computed by the decomposition functions and on the free variables.

The applicability of functional decomposition to the synthesis of lookup table based FPGAs is obvious: If the FPGA is able to realize functions with up to b variables by lookup tables and if p is smaller than or equal to b, then each decomposition function α_i can be realized by one lookup table. If the number of inputs $r + n - p$ of g is also equal to or smaller than b, then also one lookup table suffices to realize g. If the number of inputs of a block is larger than b, then functional decomposition has to be applied recursively.

If decompositions are used for logic synthesis, it is expedient to look for decompositions where the number of decomposition functions is as small as possible. This has the following advantages:

- If the number of decomposition functions is smaller, then we have to realize fewer (decomposition) functions, when the method is applied recursively. Thus we hope to minimize the total cost of the overall circuit.

- If the number of decomposition functions is smaller, the number of inputs of the composition function g is likewise smaller. The complexity of g is hopefully smaller, if g has a smaller number of inputs. If we decompose the composition function recursively until we obtain blocks with only b inputs, then the number of recursive steps is also (hopefully) smaller for a composition function with a smaller number of inputs.

- Based on the netlist, which is produced by logic synthesis, a layout of the circuit realizing the given Boolean function is computed. In many cases we can observe circuits in which the layout area is not affected primarily by the number of logic gates or cells, but rather the number of global wires which have to be implemented to connect logic gates. A decomposition with a small number of decomposition functions suggests a realization with a small number of global wires.

Functional decomposition is a rather old technique, considered already by Ashenhurst (Ashenhurst, 1959), Curtis (Curtis, 1961), Roth and Karp (Roth and Karp, 1962; Karp, 1963) in the late fifties and early sixties. In recent years functional decomposition has attracted a great deal of renewed interest due to several reasons:

- The increased capacities of today's computers as well as the development of new methods have made the method applicable to larger–scale problems.

- As already mentioned, functional decomposition is especially well suited for the synthesis of lookup-table based FPGAs. During the last few years FPGAs have become increasingly important.

- The success of functional decomposition during the last few years is owing to (among other things) the success of **Reduced Ordered Binary Decision Diagrams** (ROBDDs) (Bryant, 1986), which are data structures that provide compact representations for many Boolean functions occurring in practical applications. Algorithms for functional decomposition have been developed, which work directly based on ROBDDs, so that the decomposition algorithm works based on compact representations and not on function tables as in earlier approaches.

Functional decomposition is most widely used for the synthesis of FPGAs, especially for look-up table based FPGAs. Although functional decomposition using decompositions with a minimum number of decomposition functions is not *restricted* to FPGA synthesis, it is *especially suited* for the synthesis of look-up table based FPGAs.

This book reviews the developments of the last few years concerning functional decomposition (for FPGA synthesis) and consistently presents a multitude of results stemming from the author's as well as various researchers' work in the field.

Organization of the book. The first chapter gives basic definitions and notations for Boolean functions, which are used in the following chapters. Basic notations for circuits and Boolean expressions are also given in the first two sections. Readers who are familiar with these subjects can skip the first part and use it only for reference when they are in doubt about the exact definitions and notations in the book. Then, Binary Decision Diagrams (Bryant, 1986) for representing completely and incompletely specified Boolean functions are briefly reviewed. The general structure of FPGAs and some examples of typical FPGA devices are given in the next section, and finally, the chapter is concluded by some theoretical considerations about the complexity of Boolean functions to clarify what algorithms for logic synthesis can accomplish and what they cannot be expected to do.

Since ROBDDs are the basic data structure to represent Boolean functions during functional decomposition and since minimized ROBDDs represent a good starting point for decomposition, the second chapter addresses minimization methods for ROBDDs, both for ROBDDs representing completely specified functions as well as ROBDDs representing incompletely specified functions. ROBDDs are very sensitive to the variable ordering, i.e., depending on the variable ordering, the sizes of the representations may vary from linear to exponential (measured in the number of inputs). We review existing methods for optimizing the variable ordering and we study the role of symmetries in improving on existing reordering algorithms. For incompletely specified functions we have the additional option of assigning values to *don't cares* to obtain small ROBDD representations. We describe two methods to exploit don't cares: The first is based on 'communication minimization' and the second is based on symmetries of incompletely specified functions. Both methods are used in an overall strategy to exploit don't cares and to optimize the variable order of the ROBDD at the same time.

In Chapter 3 we consider the basic method for decomposing a single–output Boolean function $f : \{0,1\}^n \to \{0,1\}$. First, the method is explained by means of *decomposition matrices* and some illustrative examples for decomposed circuits are given. Then, we explain how the methods can be applied to ROBDD representations without needing to compute function tables or decomposition matrices. Two fundamental problems in decomposition are addressed in the rest of the chapter: the problem of finding a good selection for the 'bound set' in decomposition and the problem of making use of some degree of freedom during the selection of decomposition functions (also called the 'encoding problem'). We present different solutions to the encoding problem, which are intended to reduce the complexity of the composition function or the complexity of the decomposition functions. In addition, arguments are given which give reasons for the restriction to a special class of decomposition functions, the so–called *strict* decomposition functions.

Chapter 4 deals with functional decomposition of multi–output functions $f : \{0,1\}^n \to \{0,1\}^m$ with $m > 1$. The efficiency of good realizations of multi–output functions is often based on the fact that 'logic can be shared between different outputs'. This means that the single–output functions $f_1, \ldots, f_m$ of $f : \{0,1\}^n \to \{0,1\}^m$ are not realized separately, but rather subcircuits are identified which *several* single–output functions may take advantage of. To compute shared logic in the decomposition approach, we make use of some degree of freedom in the selection of decomposition functions: Decomposition functions are selected in such a way that many of them can be used for the decomposition of as many output functions as possible. An algorithm is presented to compute 'common decomposition functions'. The approach is illustrated by examples, and the efficiency of the approach is evaluated by experiments. Alternatives to this approach of selecting common decomposition functions of several output functions are discussed, too.

After concentrating on the decomposition of completely specified functions in Chapters 3 and 4, we present an extension of the methods to incompletely specified functions in Chapter 5. The decomposition of incompletely specified functions is important, since many logic synthesis problems occurring in practical applications correspond to incompletely specified functions. Moreover, incompletely specified functions occur during the recursive decomposition process in a natural way — even if the original function to be decomposed is completely specified.

Chapter 6 gives a brief review of non–disjoint decompositions where the set of bound variables and the set of free variables are not disjoint. It is shown where it possibly makes sense to use non–disjoint decompositions, and it is also shown how non–disjoint decompositions can be integrated in the algorithmic framework already given.

Finally, Chapter 7 closes with some remarks on processing very large circuits in combination with functional decomposition techniques.

Chapter 1

REALIZATIONS OF BOOLEAN FUNCTIONS

The first chapter gives basic definitions and notations for Boolean functions. These notations will be used in the following chapters. The main representations for Boolean functions introduced in this chapter are Ω–circuits (which are simply circuits based on a cell library Ω) and ROBDDs. FPGAs (Field Programmable Gate Arrays) are special classes of Ω–circuits.

Section 1.5 discusses what algorithms for logic synthesis can accomplish and what they cannot be expected to do, based on what we know about the complexity of Boolean functions from literature.

The problem in logic synthesis is to find a circuit which realizes a given Boolean function at minimal cost. Cost measures such as area, depth or delay are defined in this chapter, too.

DEFINITION 1.1 (COMPLETELY SPECIFIED BOOLEAN FUNCTIONS)
$f : \{0,1\}^n \to \{0,1\}^m$ $(n, m \in I\!N)$ is a (completely specified) Boolean function *with n inputs and m outputs. The set of all (completely specified) Boolean functions with n inputs and m outputs is denoted by* $B_{n,m} := \{f : \{0,1\}^n \to \{0,1\}^m\}$.

NOTATION 1.1 $B_n := B_{n,1}$

DEFINITION 1.2 (INCOMPLETELY SPECIFIED BOOLEAN FUNCTIONS)
$f : D \to \{0,1\}^m$ $(D \subseteq \{0,1\}^n,\ n, m \in I\!N)$ is an *incompletely specified Boolean function with n inputs, m outputs and domain D. D is also denoted by $D(f)$. $BP_{n,m} := \{f : D \to \{0,1\}^m \mid D \subseteq \{0,1\}^n\}$ is the set of all incompletely specified Boolean functions with n inputs and m outputs.*

1

NOTATION 1.2 $BP_n := BP_{n,1}$

NOTATION 1.3 *For $D \subseteq \{0,1\}^n$ let $BP_{n,m}(D) = \{f : D \to \{0,1\}^m\}$ be the set of all incompletely specified Boolean functions with m outputs and domain D.*

NOTATION 1.4 $BP_n(D) := BP_{n,1}(D)$

DEFINITION 1.3 (ON–SET, OFF–SET, DC–SET)
Let $f : D \to \{0,1\}$, $D \subseteq \{0,1\}^n$. Then

- $DC(f) = \{0,1\}^n \setminus D$ *is the* **don't care–set** *(DC–set) of f,*

- $ON(f) = \{x \in D \mid f(x) = 1\}$ *is the* **ON–set** *of f,*

- $OFF(f) = \{x \in D \mid f(x) = 0\}$ *is the* **OFF–set** *of f.*

DEFINITION 1.4 (EXTENSION)
An incompletely specified function $f : D_f \to \{0,1\}^m$ ($D_f \subseteq \{0,1\}^n$) is an extension of an incompletely specified function $g : D_g \to \{0,1\}^m$ ($D_g \subseteq \{0,1\}^n$), if and only if $D_g \subseteq D_f$ and $g(\epsilon) = f(\epsilon)$ for all $\epsilon \in D_g$. If $D_f = \{0,1\}^n$, then f is called a completely specified extension *of g.*

To define new Boolean functions based on given Boolean functions, functions from B_1 and B_2 are used as operations on $BP_n(D)$. If $g, h \in BP_n(D)$ and f is an operation from B_2, then we usually write $g \ f \ h$ in infix notation instead of $f(g,h)$, i.e., for all $x \in D$ the notation $(g \ f \ h)(x)$ will denote $f(g(x), h(x))$ in the following.

We use the following common notations for functions from B_1 and B_2:

NOTATION 1.5 $not \in B_1$, $and, nand, or, nor, exor, equiv \in B_2$.

(a) $not(x) = 1$ *iff* $x = 0$. *not is called* negation *of x. Notation: $\overline{x}$ or $\neg x$.*

(b) $and(x,y) = 1$ *iff* $x = y = 1$. *and is called* conjunction *of x and y. Notation: $x \wedge y$, $x \cdot y$ or xy.*

(c) $nand(x,y) = 0$ *iff* $x = y = 1$.

(d) $or(x,y) = 0$ *iff* $x = y = 0$. *or is called* disjunction *of x and y. Notation: $x \vee y$ or $x + y$.*

(e) $nor(x,y) = 1$ *iff* $x = y = 0$.

(f) $exor(x, y) = 1$ *iff* $x \neq y$. *exor is called* exclusive-or function *of* x *and* y. *Notation:* $x \oplus y$.

(g) $equiv(x, y) = 1$ *iff* $x = y$. *Notation:* $x \equiv y$.

(h) **0** *is the* constant 0 function.

(i) **1** *is the* constant 1 function.

REMARK 1.1 *Let* $\pi_i^n \in B_n$ *be the ith projection function, i.e.,* $\pi_i^n(x_1, \ldots, x_n)$ $= x_i$. *If the meaning will be clear from the context, we will replace* π_i^n *simply by* x_i.

DEFINITION 1.5 (COFACTOR OF f)
If $f \in B_n$, *then the function* $f_{x_{i_1}^{\epsilon_{i_1}} \ldots x_{i_k}^{\epsilon_{i_k}}}$ *defined by*

$$\forall (x_1, \ldots, x_n) \in \{0, 1\}^n : f_{x_{i_1}^{\epsilon_{i_1}} \ldots x_{i_k}^{\epsilon_{i_k}}}(x_1, \ldots, x_n) = f(y_1, \ldots, y_n)$$

with

$$y_j = \begin{cases} \epsilon_j, & \text{if } j \in \{i_1, \ldots, i_k\} \\ x_j, & \text{if } j \notin \{i_1, \ldots, i_k\} \end{cases}$$

is denoted as cofactor *of* f *with respect to* $x_{i_1}^{\epsilon_{i_1}} \ldots x_{i_k}^{\epsilon_{i_k}}$.

1.1. Circuits

The most general model to represent Boolean functions are circuits over a (fixed) library Ω. To realize Boolean functions, a certain set of (simpler) Boolean functions is available to express more complex functions. The set of all functions which are already available is collected in a *cell library*:

DEFINITION 1.6 (CELL LIBRARY)
A cell library is a finite subset $\Omega \subseteq \bigcup_{n \in \mathbb{N}} B_n$.

In the following we often use the cell libraries $\Omega = B_2, \Omega = STD = \{0, 1, and,$ $nand, or, nor, not\}$ or $\Omega = B_2 \setminus \{exor, equiv\} =: R_2$.

Usually realizations for functions from a cell library are called *gates*.

The two following definitions specify how Boolean functions can be represented based on functions from a cell library. Boolean functions are represented by Ω-circuits, which can be immediately interpreted as netlist of circuits.

DEFINITION 1.7 (CIRCUIT)
Let Ω *be a cell library. Let* $T = \Omega \cup \{EPAD, APAD\}$.
An Ω-*circuit* S *with* n *inputs and* m *outputs is a tuple* $(G = (V, E), type, pe, pa)$ *with:*

- *G is a directed, acyclic, node oriented[1] graph. V is the set of nodes (or cells), E the set of edges (or signals of the circuit).*
  Source and target of an edge are given by functions $So : E \to V$ and $Ta : E \to V$.
  The indegree of a node v is defined by $indeg : V \to \mathbb{N}$, $indeg(v) = |\{e \in E \mid Ta(e) = v\}|$.
  Accordingly the outdegree of v is $outdeg : V \to \mathbb{N}$, $outdeg(v) = |\{e \in E \mid So(e) = v\}|$.
  The node orientations of G are incompletely specified one–to–one functions $I, O : V \times \mathbb{N} \rightsquigarrow E$. If $1 \le i \le indeg(v)$, then $I(v, i)$ is specified as $I(v, i) = e$ with $Ta(e) = v$ and e is the ith input of node v; if $1 \le i \le outdeg(v)$, then $O(v, i)$ is specified as $O(v, i) = e$ with $So(e) = v$ and e is the ith output of v.

- *$type : V \to T$ is a function which assigns a 'type' (function from the cell library or $EPAD$ or $APAD$) to each node.*
  It holds: $|type^{-1}(EPAD)| = n$, $|type^{-1}(APAD)| = m$.
  If $type(v) = t$, then t is called the type of v and v is called a t–node.
  If $type(v) \in \Omega$, then $type(v) \in B_{indeg(v)}$. If $type(v) = EPAD$, then $indeg(v) = 0$, if $type(v) = APAD$, then $indeg(v) = 1$, $outdeg(v) = 0$.

- *pe, pa are one-to-one functions, which give the 'orders' for the $EPAD$– and $APAD$–nodes: $pe : \{1, \ldots, n\} \to \{v \in V \mid type(v) = EPAD\}$, $pa : \{1, \ldots, m\} \to \{v \in V \mid type(v) = APAD\}$. $pe(i)$ ($pa(i)$) is called the ith primary input (output).*

The following definition gives the correlation between an Ω–circuit and the Boolean function realized by the Ω–circuit.

DEFINITION 1.8 (FUNCTION DEFINED BY Ω–CIRCUIT)
Let $S = (G, type, pe, pa)$ be an Ω–circuit with n inputs and m outputs. Let $e \in E$ be an edge with $e = O(v, j)$.

- *If $type(v) = EPAD$ and $v = pe(i)$, then the function computed by edge e is defined by*

$$f_e : \{0, 1\}^n \to \{0, 1\}, \quad f_e(x_1, \ldots, x_n) = x_i.$$

- *If $type(v) = g \in \Omega$, $e_k = I(v, k)$ for $k = 1, \ldots, indeg(v)$, then the function computed by edge e is defined by*

$$f_e : \{0, 1\}^n \to \{0, 1\}, \quad f_e(\mathbf{x}) = g(f_{e_1}(\mathbf{x}), \ldots, f_{e_{indeg(v)}}(\mathbf{x}))$$

---

[1] A graph $G = (V, E)$ is called *node oriented*, if for all nodes $v \in V$ there is an order for the incoming edges and an order for the outgoing edges.

For $1 \leq i \leq m$ let $y_i = I(pa(i), 1)$, f_{y_i} *the functions computed by* y_i*, respectively. The function* $f_S : \{0,1\}^n \to \{0,1\}^m$ *realized by* S *is defined by*

$$f_S(\mathbf{x}) = (f_{y_1}(\mathbf{x}), \ldots, f_{y_m}(\mathbf{x})).$$

Moreover a circuit S is not only viewed as a realization of a single completely specified function, but also a realization for *all* functions which result from f_S by restricting the domain of f_S:

DEFINITION 1.9 (S REALIZES g) *Let* $f_S : \{0,1\}^n \to \{0,1\}^m$ *be the completely specified Boolean function realized by the* Ω*–circuit* S *and* g *an incompletely specified function.* S is a realization of g, *if* f_S *is an extension of* g.

EXAMPLE 1.1 Figure 1.1 illustrates Definition 1.7 for a circuit realizing a full adder, i.e., a function

$$fa : \{0,1\}^3 \to \{0,1\}^2, \ (x_1, x_2, x_3) \mapsto (c, s) \ \text{with} \ x_1 + x_2 + x_3 = s + 2c.$$

We chose the cell library $\Omega = \{\oplus, \cdot, \vee\}$.

Usually we will modify the graphical representation of the circuit to a simpler form such as in Figure 1.2.

To compare different realizations of the same Boolean function we define cost functions for Ω–circuits. To do so, we extend cell libraries by cell areas and delays.

DEFINITION 1.10 (EXTENDED CELL LIBRARY) *An extended cell library is a triple* $(\Omega, A_\Omega, T_\Omega)$*, which consists of a cell library* Ω*, and two functions* $A_\Omega : \Omega \to \mathbb{R}_0^+$ *and* $T_\Omega : \Omega \to \mathbb{R}_0^+$*. Here* A_Ω *assigns a cell area to each function from* Ω*, and* T_Ω *assigns a delay to each function from* Ω*.*

For the definition of cost functions for Ω–circuits we need an additional notation:

NOTATION 1.6 *Let* $S = (G = (V, E), type, pe, pa)$ *be an* Ω*–circuit. The* set of paths *in* S *is defined as*

$$Paths(S) := \{(v_1, \ldots, v_l) \ | \ \forall 1 \leq i \leq l \ \ v_i \in V,$$
$$type(v_1) = EPAD, type(v_l) = APAD,$$
$$\forall 1 \leq i \leq l - 1 \ (v_i, v_{i+1}) \in E\}.$$

Figure 1.1.   Circuit realizing a full adder.

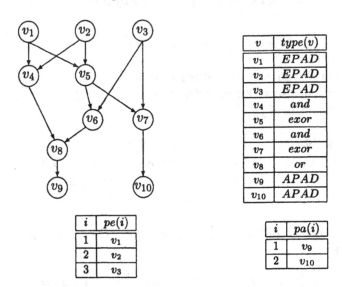

v	$type(v)$
v_1	$EPAD$
v_2	$EPAD$
v_3	$EPAD$
v_4	and
v_5	$exor$
v_6	and
v_7	$exor$
v_8	or
v_9	$APAD$
v_{10}	$APAD$

i	$pe(i)$
1	v_1
2	v_2
3	v_3

i	$pa(i)$
1	v_9
2	v_{10}

Figure 1.2.   Simplified representation of the circuit in Figure 1.1.

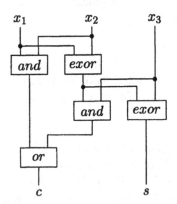

DEFINITION 1.11 (COST FUNCTIONS)

1. The Ω–complexity of an Ω–circuit $S = (G = (V, E), type, pe, pa)$ is

$$C_\Omega(S) = |V \setminus \{v \in V \mid type(v) \in \{EPAD, APAD\}\}|,$$

i.e., the Ω–complexity is determined by the number of cells in the circuit which are not $EPAD$– or $APAD$–cells.

The Ω–complexity of a Boolean function $f \in BP_{n,m}(D)$ is

$$C_\Omega(f) = \min\{C_\Omega(S) \mid f_S|_D = f\}.^2$$

A realization S of f with $C_\Omega(S) = C_\Omega(f)$ is C_Ω–optimal or simply Ω–optimal.

2. The cell area of an Ω–circuit $S = (G = (V, E), type, pe, pa)$ is defined as

$$A_\Omega(S) = \sum_{\substack{v \in V \\ type(v) \notin \{EPAD, APAD\}}} A_\Omega(type(v)),$$

i.e., the cell area is the sum of the gate areas of all cells in the circuit. The cell area of a Boolean function $f \in BP_{n,m}(D)$ is defined accordingly:

$$A_\Omega(f) = \min\{A_\Omega(S) \mid f_S|_D = f\}.$$

A realization S of f with $A_\Omega(S) = A_\Omega(f)$ is A_Ω–optimal.

3. The depth of an Ω–circuit $S = (G = (V, E), type, pe, pa)$ is defined as

$$D_\Omega(S) = \max\{l \mid (v_1, \ldots, v_l) \in Paths(S)\} - 2,$$

i.e., $D_\Omega(S)$ is the maximum number of cells of a directed path in G, where $EPAD$– and $APAD$–cells are not counted. The depth of a Boolean function $f \in BP_{n,m}(D)$ is defined accordingly:

$$D_\Omega(f) = \min\{D_\Omega(S) \mid f_S|_D = f\}.$$

A realization S of f with $D_\Omega(S) = D_\Omega(f)$ is D_Ω–optimal.

4. The delay of an Ω–circuit $S = (G = (V, E), type, pe, pa)$ is defined as

$$T_\Omega(S) = \max\{\sum_{i=2}^{l-1} T_\Omega(type(v_i)) \mid (v_1, \ldots, v_l) \in Paths(S)\},$$

i.e., $T_\Omega(S)$ is the maximum sum of all delays on a directed path from an $EPAD$–node to an $APAD$–node. The delay of a Boolean function $f \in BP_{n,m}(D)$ is

$$T_\Omega(f) = \min\{T_\Omega(S) \mid f_S|_D = f\}.$$

A realization S of f with $T_\Omega(S) = T_\Omega(f)$ is T_Ω–optimal.

---

[2] The notation $g|_D$ is defined for each (completely or) incompletely specified function $g \in BP_{n,m}(D')$ with $D' \supseteq D$ and means the function $g|_D \in BP_{n,m}(D)$ with $g|_D(\epsilon) = g(\epsilon)$ for all $\epsilon \in D$.

REMARK 1.2 *To model the delay of a physical circuit more exactly, delays can also be assigned to wires in a circuit, i.e., to edges in an Ω–circuit. Moreover the delay of a node can depend not only on its cell function, but also on the outdegree of the node.*

When it is clear from the context what the underlying cell library is, we also write $C(S)$ and $C(f)$ instead of $C_\Omega(S)$ and $C_\Omega(f)$.

1.2.   Boolean Expressions

Boolean Expressions are another means of representing Boolean functions in B_n.

Boolean Expressions have a direct correspondence to Ω–circuits over the cell library $\{0, 1, and, or, not\}$ and in the restricted form of 'sums-of-products' they are especially used to describe two-level realizations like PLAs (Programmable Logic Arrays).

DEFINITION 1.12 (BOOLEAN EXPRESSION OVER X)
Let $X = \{x_1, \ldots, x_n\}$ be a set of n variables and $E = \{0, 1, (,), \cdot, +, \neg\} \cup X$. The set $\mathcal{E}(X)$ of all Boolean Expressions over X is the smallest set L of strings over E with the following properties:

- *$\{0, 1\} \cup X \subseteq L$,*

- *$w_1, \ldots, w_k \in L \Longrightarrow (w_1 \cdot \ldots \cdot w_k) \in L$ and $(w_1 + \ldots + w_k) \in L$,*

- *$w \in L \Longrightarrow (\neg w) \in L$,*

Notation: Instead of $(\neg w)$ we also write $\overline{w}$.

DEFINITION 1.13 (FUNCTION DEFINED BY BOOLEAN EXPRESSION)
The Boolean function $f_w \in B_n$ defined by the Boolean Expression w over $X = \{x_1, \ldots, x_n\}$ is defined by

- *$f_w = 0$, if $w = 0$,*

- *$f_w = 1$, if $w = 1$,*

- *$f_w = \pi_i^n$, if $w = x_i$,*

- *$f_w = f_{w_1} \cdot \ldots \cdot f_{w_k}$, if $w = (w_1 \cdot \ldots \cdot w_k)$,*

- *$f_w = f_{w_1} + \ldots + f_{w_k}$, if $w = (w_1 + \ldots + w_k)$,*

- *$f_w = not(f_v)$, if $w = (\neg v)$.*

A sum-of-products is a special case of a Boolean expression:

DEFINITION 1.14 (SUM-OF-PRODUCTS)
A product $m \in \mathcal{E}(\{x_1, \ldots x_n\})$ *is a Boolean expression* $m = (x_{i_1}^{\epsilon_1} \cdot \ldots \cdot x_{i_j}^{\epsilon_j})$ *with* $1 \leq j \leq n$, *and* $\epsilon_k \in \{0, 1\}$ *for* $k = 1, \ldots, j$, $x_{i_k}^1 = x_{i_k}$ *and* $x_{i_k}^0 = \overline{x_{i_k}}$.

$p \in \mathcal{E}(\{x_1, \ldots x_n\})$ *is a* sum-of-products, *iff*

- $p = 0$ *or*

- $p = 1$ *or*

- $p = (m_1 + \ldots + m_k)$, *where* $\forall l \in \{1, \ldots, k\}$ m_l *is a product.*

The definition of the cost of a sum-of-products p is motivated by the R_2–complexity of the R_2–circuit, which can be constructed from p in a straightforward manner:

DEFINITION 1.15 (COST OF SUM-OF-PRODUCTS)
The costs $C(p)$ *of a sum-of-products* p *are*

- $C(p) = 0$, *if* $p = 0$ *or* $p = 1$,

- $C(p) = k - 1 + \sum_{i=1}^{k} C(m_i)$, *if* $p = (m_1 + \ldots + m_k)$, *where for* $1 \leq l \leq k$ *the cost of a product* $m_l = (x_{i_1}^{\epsilon_1} \cdot \ldots \cdot x_{i_j}^{\epsilon_j})$ *is* $C(m_l) = j - 1$.

Two-level representations (sums-of-products) are used in many designs to represent Boolean functions. On the one hand they can be realized by Programmable Logic Arrays (PLAs) and on the other hand there are efficient heuristic algorithms like ESPRESSO (Brayton et al., 1984), which are able to compute good sum-of-products representations.

However in many cases it is not advisable to confine the search for good representations of Boolean functions to two-level realizations, since for many Boolean functions even the best sum-of-products representation is substantially more expensive than multi-level representations. Extreme cases are the parity functions (*equiv* and *exor* with n inputs) or the binary adder, which can be realized by (multi–level) C_{R_2}–circuits with linear cost. However all sum-of-products representations for these functions have exponential cost.

1.3.   Binary Decision Diagrams (BDDs)

1.3.1   Binary Decision Diagrams to represent completely specified functions

Binary Decision Diagrams (BDDs) as a data structure for representation of Boolean functions were first introduced by Lee (Lee, 1959) and further popularized by Akers (Akers, 1978) and Moret (Moret, 1982). In the restricted form of ROBDDs (Reduced Ordered Binary Decision Diagrams) they gained widespread application, because ROBDDs are a canonical representation and allow efficient manipulations (Bryant, 1986). Some fields of application are logic design verification, logic synthesis, test generation, and fault simulation (Malik et al., 1988; Bryant, 1992). It will be made clear in Chapter 3 that ROBDDs are a fundamental data structure to perform logic synthesis by decomposition.

Many Boolean functions which occur in practical applications have a compact ROBDD representation, such that they can be used for internal representations of Boolean functions in logic synthesis tools. Moreover many algorithms using ROBDDs have run times which are polynomial in the size of the ROBDDs. The size of ROBDDs cannot 'explode' using binary Boolean operations or negation. Whereas other representations of Boolean functions such as sums-of-products can show an exponential increase of size only by performing one negation operation, negation leaves the size of an ROBDD unchanged. In the case of binary operations the size of the result can be at most as large as the product of the sizes of the two operand ROBDDs.

Binary Decision Diagrams rely on the representation of a Boolean function by the Shannon–expansion (Shannon, 1938)

$$
\begin{aligned}
f(x_1, \ldots, x_n) \;=\; & \overline{x_i} \cdot f(x_1, \ldots, x_{i-1}, 0, x_{i+1}, \ldots, x_n) \\
& + x_i \cdot f(x_1, \ldots, x_{i-1}, 1, x_{i+1}, \ldots, x_n).
\end{aligned}
$$

However it was not until Bryant (Bryant, 1986) introduced a restriction to a certain subclass of BDDs, the *Ordered* Binary Decision Diagrams, that it was possible to perform efficient operations on these representations of Boolean functions.

Ordered Binary Decision Diagrams (OBDDs) are defined as follows:

DEFINITION 1.16 (ORDERED BINARY DECISION DIAGRAM (OBDD))
Let $X = \{x_1, \ldots, x_n\}$ be a set of variables, index : $\{1, \ldots, n\} \to \{1, \ldots, n\}$ a permutation of $\{1, \ldots, n\}$. index induces a linear order $<_{index}$ on X: $x_{index(i)} <_{index} x_{index(j)}$ if and only if $i < j$. $<_{index}$ is called variable

order. *An* Ordered Binary Decision Diagram (OBDD) *F with variable set X and variable order $<_{index}$ is a pair (G,m) consisting of a directed graph $G = (V, E)$ and a labeling function $m : V \to (X \cup \{0,1\})$.*
G has exactly one source $q \in V$. Each node $v \in V$ which is not a sink has exactly 2 sons: the 0–son v^0 and the 1–son v^1.
The labeling function m assigns a label $m(v) \in X$ to each node $v \in V$ which is not a sink. In addition for all sinks s we have $m(s) = 0$ or $m(s) = 1$. For each node $v \in V$ whose son v^ϵ ($\epsilon \in \{0,1\}$) is not a sink, we have: $m(v) <_{index} m(v^\epsilon)$.

An *Ordered* Binary Decision Diagram is defined in such a way that for each path from the source to a sink, each variable can occur at most once. Furthermore the variables occur in the order $x_{index(1)}, \ldots, x_{index(n)}$.

NOTATION 1.7 *The size of an* OBDD *$F = ((V, E), m)$ is defined as the number $|V|$ of nodes.*

An OBDD with variable set $X = \{x_1, \ldots, x_n\}$ (and variable order $<_{index}$) defines a Boolean function $f \in B_n$:

DEFINITION 1.17 (FUNCTION DEFINED BY AN OBDD)
Let $F = (G, m)$ be an Ordered Binary Decision Diagram with variable set $X = \{x_1, \ldots, x_n\}$ and variable order $<_{index}$. Let $v \in V$ be a node of G.

- *If $m(v) = 0$, then the function f_v defined by v is equal to constant **0**.*
  $f_v(x_1, \ldots, x_n) = 0 \quad \forall(x_1, \ldots, x_n) \in \{0,1\}^n$.

- *If $m(v) = 1$, then the function f_v defined by v is equal to constant **1**.*
  $f_v(x_1, \ldots, x_n) = 1 \quad \forall(x_1, \ldots, x_n) \in \{0,1\}^n$.

- *Let $m(v) = x_i$, v^0 the 0–son of v, v^1 the 1–son. Let $f_{v^0} \in B_n$ be the function defined by v^0, $f_{v^1} \in B_n$ the function defined by v^1. Then the function $f_v \in B_n$ defined by v is equal to*

$$f_v(x_1, \ldots, x_n) = \overline{x_i} \cdot f_{v^0}(x_1, \ldots, x_n) + x_i \cdot f_{v^1}(x_1, \ldots, x_n)$$
$$\forall(x_1, \ldots, x_n) \in \{0,1\}^n.$$

Let $q \in V$ be the only source of G. The function $f_F \in B_n$ defined by $F = (G, m)$ is $f_F = f_q$.

EXAMPLE 1.2 Figure 1.3 shows two OBDDs for function $f(x_1, \ldots, x_4) = x_2 + x_1 x_3 x_4$ with variable order $x_2 <_{index} x_1 <_{index} x_3 <_{index} x_4$.

Figure 1.3. 2 OBDDs for $f(x_1, \ldots, x_4) = x_2 + x_1 x_3 x_4$ with variable order $x_2 <_{index}$
$x_1 <_{index} x_3 <_{index} x_4$.

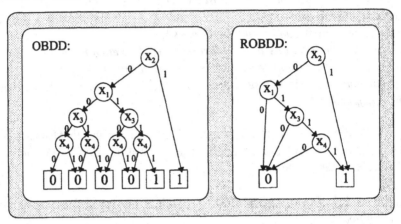

According to Definition 1.17, an OBDD can be evaluated by following a computation path through the OBDD defined by an assignment to the input variables:

DEFINITION 1.18
*Let $F = (G = (V, E), m)$ be an OBDD with variable set $X = \{x_1, \ldots, x_n\}$ and
variable order $<_{index}$, $\epsilon_{index} = (\epsilon_{index(1)}, \ldots, \epsilon_{index(k)}) \in \{0, 1\}^k$, $k \leq n$.
The node v of G reached by ϵ_{index} is defined by the following algorithm:*

1. *Start at source q of G. $v := q$. $i := 1$.*

2. *While $i \leq k$ do:*

    (a) *If $m(v) = x_{index(i)}$, go to $\epsilon_{index(i)}$–son of v, i.e., $v := v^{\epsilon_{index(i)}}$.*

    (b) *increment i by 1.*

If $\epsilon = (\epsilon_1, \ldots, \epsilon_n)$, then the node reached by $(\epsilon_{index(1)}, \ldots, \epsilon_{index(n)})$ is a sink
s of G. It holds:

$$f_F(\epsilon_1, \ldots, \epsilon_n) = m(s).$$

Thus f_F is evaluated for input ϵ beginning at the source of G and following a
path towards a sink according to the input assignment ϵ.

REMARK 1.3 *Let v be the node in OBDD $F = (G, m)$, which is reached
by $\epsilon_{index} = (\epsilon_{index(1)}, \ldots, \epsilon_{index(k)})$ with $k \leq n$. Then the subgraph of G
containing all nodes which can be reached starting at v (together with their*

labels) can be interpreted as an OBDD. *It is easy to see that the function defined by this* OBDD *is exactly the cofactor* $f_{x_{index(1)}^{\epsilon_{index(1)}} \cdots x_{index(k)}^{\epsilon_{index(k)}}}$.

Based on an OBDD we obtain a *Reduced* Ordered Binary Decision Diagram (ROBDD) by removing 'redundant' nodes and by removing subgraphs from the OBDD which occur more than once. Thus the size of the OBDD is reduced without changing the function defined by it.

DEFINITION 1.19 (REDUCED ORDERED DECISION DIAGRAM)
An OBDD $F = (G = (V, E), m)$ *is* **reduced**, *iff*

- *there is no node whose 0–son and 1–son are identical and*

- *there are no sinks with the same mark and no pairs of nodes* $v_1 \neq v_2 \in V$ *with* $v_1^0 = v_2^0$ *and* $v_1^1 = v_2^1$.

LEMMA 1.1 (BRYANT '86) *An* OBDD $F = (G = (V, E), m)$ *can be transformed into an* ROBDD $F' = (G', m')$ *realizing the same function in time* $O(|V| \cdot \log(|V|))$.

It can be proved for ROBDDs (with a fixed variable order) that they are a canonical representation of Boolean functions.

THEOREM 1.1 (BRYANT '86) *For every Boolean function* $f \in B_n$ *and every fixed variable order there is (up to isomorphism) a unique* ROBDD *which defines* f *(and all other* OBDDs *which define* f *have more nodes).*

The following remark results easily from Remark 1.3 and Theorem 1.1.

REMARK 1.4 *If* $F = (G, m)$ *is an* ROBDD *with variable set* $\{x_1, \ldots, x_n\}$ *and variable order* $<_{index}$, *which defines function* f, *and if two cofactors*

$$f_{x_{index(1)}^{\epsilon_{index(1)}} \cdots x_{index(k)}^{\epsilon_{index(k)}}} \text{ and } f_{x_{index(1)}^{\delta_{index(1)}} \cdots x_{index(k)}^{\delta_{index(k)}}}$$

are equal, then the nodes of G *reached by* $(\epsilon_{index(1)}, \ldots, \epsilon_{index(k)})$ *and* $(\delta_{index(1)}, \ldots, \delta_{index(k)})$ *are identical.*

This fact comes into play later in Chapter 3, when decompositions of Boolean functions are derived from ROBDD–representations.

Information about the complexity of some basic operations for ROBDDs can be found in Table 1.1 (see also (Bryant, 1986)).

Table 1.1.   Complexity of some basic operations for ROBDDS

input	output	time
ROBDD $F = ((V, E), m)$ for $f \in B_n$	ROBDD for $\bar{f}$	$O(|V|)$
ROBDD $F_1 = ((V_1, E_1), m_1)$ for f_1,	ROBDD for $f_1 <op> f_2$,	$O(|V_1| \cdot |V_2|)$
ROBDD $F_2 = ((V_2, E_2), m_2)$ for f_2	$op \in \{and, nand, or, exor, \ldots\}$	
ROBDD $F = ((V, E), m)$ for $f \in B_n$	ROBDD for cofactor $f_{x_i^{\epsilon_i}}, \epsilon_i \in \{0, 1\}$	$O(|V| \log(|V|))$
ROBDD $F = ((V, E), m)$ for $f \in B_n$	arbitrary element of $ON(f)$	$O(n)$
ROBDD $F = ((V, E), m)$ for $f \in B_n$	$ON(f)$	$O(n \cdot |ON(f)|)$
ROBDD $F = ((V, E), m)$ for $f \in B_n$	$|ON(f)|$	$O(|V|)$

1.3.2   Binary Decision Diagrams to represent incompletely specified functions

ROBDDs as defined in the previous section represent completely specified Boolean functions. An obvious technique for the representation of incompletely specified Boolean functions is to introduce a third sink (apart from the sinks labeled 0 and 1), which is labeled 'dc'. All paths in the ROBDD which correspond to vectors in the don't care set lead to this new sink. The resulting reduced and ordered Decision Diagram is called DCBDD in the following. The first Decision Diagram in Figure 1.4 shows such a representation for function $f \in BP_3(D)$ with $D = \{(0, 0, 0), (0, 1, 1), (1, 0, 1), (1, 1, 0), (1, 1, 1)\}$,

$$f(\epsilon) = \begin{cases} 1, & \text{if } \epsilon = (1, 1, 1) \\ 0, & \text{if } \epsilon \in D \setminus \{(1, 1, 1)\}. \end{cases}$$

As an alternative, an incompletely specified Boolean function f can also be represented by ROBDDs for the characteristic functions f_{on}, f_{off} and f_{dc} for the sets $ON(f)$, $OFF(f)$ and $DC(f)$, respectively.[3] Since for each incompletely specified Boolean function f $DC(f) = \{0, 1\}^n \setminus (ON(f) \cup OFF(f))$, it is sufficient to represent two out of three sets $ON(f)$, $OFF(f)$ and $DC(f)$. To combine the representations of f_{on} and f_{off} into one ROBDD, Chang, Cheng and Marek–Sadowska (Chang et al., 1994) introduce a new variable z apart from the input variables $\{x_1, \ldots, x_n\}$ of f and they define

$$ext(f) = \bar{z} \cdot f_{off} + z \cdot f_{on}. \tag{1.1}$$

---

[3]$\forall (x_1, \ldots, x_n) \in \{0, 1\}^n$:
$f_{on}(x_1, \ldots, x_n) = 1$ iff $(x_1, \ldots, x_n) \in ON(f)$,
$f_{off}(x_1, \ldots, x_n) = 1$ iff $(x_1, \ldots, x_n) \in OFF(f)$,
$f_{dc}(x_1, \ldots, x_n) = 1$ iff $(x_1, \ldots, x_n) \in DC(f)$.

Figure 1.4. Representations of the incompletely specified function f with $f_{on}(x_1, x_2, x_3) =$
$x_1 \cdot x_2 \cdot x_3$, $f_{off}(x_1, x_2, x_3) = \overline{x_1 \oplus x_2 \oplus x_3}$

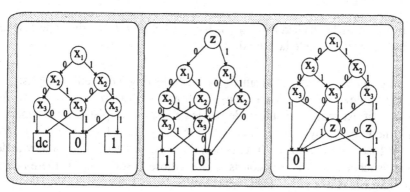

Then the incompletely specified function f is represented by the ROBDD for
$ext(f)$. The second ROBDD in Figure 1.4 shows this representation for the
same example as on the left hand side under the assumption that the additional
variable z is the first variable in the variable order. The ROBDD on the right
hand side shows the same function again, when the variable z is the last one in
the variable order. The following lemma shows that there is no major difference
between this representation with z as the last variable in the variable order and
the DCBDD (for illustration see Figure 1.4):

LEMMA 1.2 *Let* $f \in BP_n(D)$. *The following holds for the DCBDD F_1 for f
with variable order $<_{index}$ and the ROBDD F_2 for ext(f) with the same variable
order (with the only difference that there is an additional variable z occurring
as the last variable in the order):*

$$size(F_2) - 2 \le size(F_1) \le size(F_2),$$

where $size(F_1)$ *is the number of nodes in* F_1, $size(F_2)$ *the number of nodes in*
F_2.

PROOF:
The following equivalences for $\epsilon_{index} \in \{0, 1\}^n$ result directly from Equation
(1.1):

- In F_1 the node reached by ϵ_{index} is labeled 0. $\Longleftrightarrow$ In F_2 the node reached
  by ϵ_{index} is the node labeled z, whose 0–son is labeled 1 and whose 1–son
  is labeled 0.

- In F_1 the node reached by ϵ_{index} is labeled 1. $\iff$ In F_2 the node reached by ϵ_{index} is the node labeled z, whose 0–son is labeled 0 and whose 1–son is labeled 1.

- In F_1 the node reached by ϵ_{index} is labeled dc. $\iff$ In F_2 the node reached by ϵ_{index} is the node labeled 0.

Thus F_2 can be constructed from F_1 by the following local replacements: Insert two new nodes v_0 with $m(v_0) = 0$ and v_1 with $m(v_1) = 1$ into F_1. Replace (if existent) in F_1 the node labeled dc by v_0. Replace (if existent) the *old* 0–node of F_1 by a new node, which is labeled z, whose 0–son is v_1 and whose 1–son is v_0. Replace (if existent) the *old* 1–node of F_1 by a new node, which is labeled z, whose 0–son is v_0 and whose 1–son is v_1. If there are no incoming edges of v_0 (v_1), then remove v_0 (v_1). $\qquad\square$

1.4. Field Programmable Gate Arrays (FPGAs)

Field Programmable Gate Arrays (FPGAs) are special devices which allow the implementation of Boolean functions. They are *programmable*, i.e., the function realized by an FPGA device can be changed using a special programming hardware after manufacturing the device.[4]

In general, an FPGA consists of a matrix of *logical blocks* for implementation of the required functions. The general structure of an FPGA is illustrated in Figure 1.5. There are three types of resources: logic blocks, I/O blocks and interconnection wires. I/O blocks connect the interconnection wires to the pins of the FPGA package. The interconnection wires are arranged as horizontal and vertical routing channels between rows and columns of logic blocks. The routing channels contain wires and programmable switches, which allow the logic blocks to be interconnected in many ways. The programmable switches are used to establish connections between horizontal and vertical routing wires.

The architecture of an FPGA is characterized not only by the arrangement of routing channels, but also by the structure of the logic blocks. Each logic block typically has a small number of inputs and a small number of outputs. The most commonly used logic block is a *lookup table* (LUT). A lookup table with b inputs is able to realize arbitrary functions $lut : \{0,1\}^b \to \{0,1\}$ up to a certain number b of inputs. A lookup table can be realized using 2^b static RAM cells. When the system is started, the lookup table can be loaded by an arbitrary function with b variables.

---

[4]Here we give on overview of the architecture of FPGAs. More details (also on the physical implementation of FPGAs) can be found in (Brown et al., 1992) or (Brown and Vranesic, 2000).

Figure 1.5. General structure of an FPGA device.

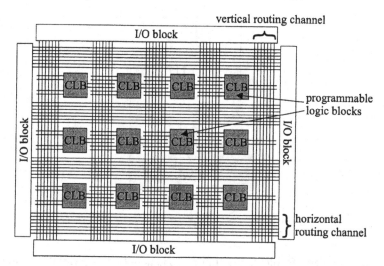

Figure 1.6. A 3–input lookup table.

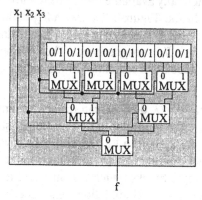

In Figure 1.6 a lookup table with 3 inputs is illustrated. It contains $2^3 = 8$ storage elements, which can be programmed to be 0 or 1, respectively. The storage cells can realize the function table of any Boolean function with 3 inputs. A tree of multiplexer cells (MUX–cells) is used to evaluate the function for input x_1, x_2, x_3. A multiplexer realizes a function $mux : \{0,1\}^3 \rightarrow \{0,1\}$ defined by $mux(s, x_0, x_1) = \bar{s} \cdot x_0 + s \cdot x_1$.

Figure 1.7.   FPGA logic block with additional flip–flop.

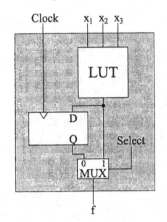

If we have to realize a Boolean function $f \in B_{n,m}$ with $n > b$ variables using an FPGA device with LUTs of b inputs, then f has to be *decomposed* into blocks of functions each having at most b inputs. Chapters 3, 4, 5 and 6 present concepts which can be used to solve this task.

In many cases of commercially available FPGAs the logic blocks have additional circuitry in each logic block. Figure 1.7 shows a logic block which contains a flip–flop[5] in addition to the LUT. Not only the LUT, but also the input *select* of the multiplexer in Figure 1.7 is programmed. Thus during programming time the output of the logic block is fixed either to the output of the flip–flop or to the output of the LUT.

Commercially available FPGA devices can differ in many ways: They have different architectures of the routing resources, different structures of the logic blocks, different sizes etc.. Modern FPGAs also may have additional features like RAM blocks, which can be used to store information, or special configuration facilities of the logic blocks which are supposed to support an efficient implementation of some arithmetic functions, e.g. an efficient carry computation for addition.

As examples of typical FPGA devices we list some commercially available devices together with their main features. Of course, this list gives only a few examples and is far from being complete:

■ The following three devices are provided by *Altera* (www.altera.com):

---

[5]A flip–flop is used to store values applied at the D input under control of a *clock input* CK.

Altera FLEX 10K This series of devices is based on 'logic elements' containing one 4–input LUT and an additional flip–flop. FLEX 10K chips are available in different sizes offering up to $12,160$ logic elements and $250,000$ equivalent gates.[6] FLEX 10K chips contain not only logic elements, but also RAM blocks. A detailed description of the FLEX 10K device is given in Appendix A.

Altera APEX 20K In the APEX 20K device the basic block is also a logic element with one 4–input LUT and one flip–flop. APEX 20K can have up to $51,840$ logic elements and up to $1,500,000$ equivalent gates (including RAM blocks).

Altera APEX II The basic block of APEX II is a logic element with one 4–input LUT and one flip–flop, again. APEX II can have up to $89,280$ logic elements and up to $4,000,000$ equivalent gates (including RAM blocks).

- The following devices are provided by *Actel* (www.actel.com):

Actel ACT 1 The basic logic module of the ACT 1 is the ACT 1 logic module shown in Figure 1.8. The ACT 1 logic module is not a lookup table. Actel Act 1 belongs to a different class of FPGAs, the so–called multiplexer based FPGAs. The ACT 1 module has 8 inputs, but it cannot implement all Boolean functions with 8 inputs. The ACT 1 device can have up to 547 ACT 1 logic modules and up to 2000 equivalent gates.

Actel ACT 2 The ACT 2 is based on C modules and M modules as basic logic blocks. The C module is shown in Figure 1.9. Similarly to the ACT 1 module it is also a multiplexer based cell with 8 inputs. The M module is equal to the C module, but has an additional flip–flop. The ACT 2 device has up to 1232 logic modules (about half of them C modules and half of them M modules) and up to 8000 equivalent gates.

Actel ACT 3 The ACT 3 device is also based on C modules and M modules. It has up to 1377 logic modules (about half of them M modules) and up to $10,000$ equivalent gates.

Actel SX The Actel SX device is based on C cells (slightly modified C modules) and R cells, which contain one flip–flop. It has up to 2880 logic modules and up to $32,000$ equivalent gates.

- The following devices are provided by *Xilinx* (www.xilinx.com):

---

[6]*Equivalent gates* is a commonly used measure and means the total number of two–input *nand* gates that would be needed to build the circuit. This measure is used to give a rough estimate of the complexity of the circuit.

Xilinx XC3000 The logic block of the XC3000 family consists of one *Configurable Logic Block* (CLB). A CLB can implement either one 5–input LUT or two 4–input LUTs. If the CLB implements two 4–input LUTs, then the total number of inputs of the two 4–input LUTs must not be larger than 5. Each CLB additionally contains two flip–flops. There are devices in the XC3000 family with up to 484 CLBs and up to 7500 equivalent gates.

Xilinx XC4000 The logic block of the XC4000 is also called a CLB. However, the CLB differs from the XC3000 CLB: For the XC4000 series each CLB consists of two separated 4–input LUTs (with 8 inputs in total). So the CLB can realize two 4–input functions. However there is additional circuitry, which allows implementation of functions with more inputs: An additional 3–input LUT is connected to one or two of the 4–input LUTs (or to none of them) and to an additional input variable. Thus the CLB can realize a function with up to 9 variables in this mode. Since not all 9–input functions can be realized by the CLB, it does not implement a 9–input LUT. Devices of the XC4000 family can have up to 3136 CLBs and up to 180, 000 equivalent gates (including RAM).

Xilinx XC5200 The XC5200 CLB consists of 4 logic cells. Each logic cell can implement one separated 4–input LUT (with inputs independent of the other logic cells) and contains one flip–flop. Optionally two adjacent logic cells in a CLB can be used as one 5–input LUT. Devices of this family can have up to 484 CLBs and up to 23, 000 equivalent gates.

Xilinx Virtex II Each CLB contains 4 'slices'. Each slice contains two 4–input LUTs and two flip–flops. There are additional multiplexers in each slice, which make it possible to implement any 5–, 6–, 7– or 8–input function and also certain 9–input functions (giving up the lookup table concept in this case).

1.5.   Complexity of realizations of Boolean functions

This section has essentially two goals: On the one hand it is intended to give basic information about the complexity of Boolean functions and on the other hand it should clarify what algorithms for logic synthesis can accomplish and what they cannot be expected to do.

Already in 1949 Shannon succeeded in proving the following theorem by simple counting arguments (Shannon, 1949):

Figure 1.8. ACT 1 logic module.

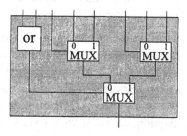

Figure 1.9. *C* module of ACT 2 and ACT 3 device.

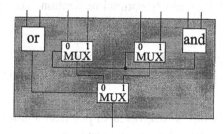

THEOREM 1.2 (SHANNON '49) *For sufficiently large n almost all [7] functions $f \in B_n$ have a complexity*

$$C_{B_2}(f) \geq \frac{2^n}{n}.$$

On the other hand a theorem by Lupanov (Lupanov, 1958) says the following:

THEOREM 1.3 (LUPANOV '58) *For each Boolean function $f \in B_n$ there is a realization L_f (called Lupanov's k–s–realization) with*

$$C_{B_2}(L_f) \leq \frac{2^n}{n} + o\left(\frac{2^n}{n}\right).$$

---

[7]The notion 'almost all functions $f \in B_n$ have property P' stands for the assertion that

$$|\{f \in B_n \mid f \text{ does } not \text{ have property } P \}| \, / \, |B_n| \to 0 \text{ as } n \to \infty.$$

We can conclude from these two theorems that almost all functions $f \in B_n$ have almost the same complexity as the hardest function in B_n. This was called the *Shannon effect* by Lupanov (Lupanov, 1970).

Similar results also hold for multi-output functions (functions in $B_{n,m}$). It can be shown that almost all functions with n inputs and m outputs have a complexity $> m\frac{2^n}{n}$. However this is only true if m is not too large (Scholl and Molitor, 1993). To realize a Boolean function from $B_{n,m}$, it is easy to find a representation with cost $m\frac{2^n}{n} + o(m\frac{2^n}{n})$: Each output function is realized separately by its k–s–realization. In so far we also have a 'Shannon effect' for multi-output functions.

Considering these results, it might seem at first glance that dealing further with the subject would be superfluous: For almost all functions in B_n Lupanov's k–s–realization provides a nearly optimal realization. However the following points give reason to look for other synthesis methods:

- The theorems mentioned above give *asymptotic* results. For small n the complexity of Lupanov's k–s–realization can be substantially larger than $\frac{2^n}{n}$.

- The Boolean functions, which occur in practical applications, are *not* randomly chosen functions, but are specified by human users and arise from certain algorithmic problems. Thus they show certain *regularities* and *structural properties* in most cases. Therefore they can be realized using much less than $\frac{2^n}{n}$ gates in most cases.

Algorithms for logic synthesis have the task of detecting such structural properties and of making use of them to find good realizations for Boolean functions.

One example of a special structural property is the nontrivial decomposability of a Boolean function. Chapters 3, 4 and 5 deal with a detailed description of a synthesis method relying on the exploitation of nontrivial decompositions.

If m is not too large and if m arbitrary Boolean functions $f_1, \ldots, f_m \in B_n$ are randomly selected, the probability is not high that the realization of a function f_j can profit from larger parts of the realization of function f_i ($j \neq i$). However the situation is different for practical examples: Frequently the different output functions of multi-output Boolean functions show a similar structure. Efficient realizations make use of this fact by utilizing the same subcircuits in the realization of several output functions. In many cases the efficiency of good realizations of Boolean functions is based especially on the reuse of identical subcircuits. Solutions to the problem of discovering subcircuits for shared use are presented in Chapter 4 as well.

Chapter 2

MINIMIZATION OF BDDS

Chapter 3 will show that the minimization of ROBDDs plays an important role for the realization of Boolean functions by decomposition. Before applying decomposition, the ROBDD representation of the function is minimized to obtain a good starting point for optimization. Therefore in this chapter we deal with several methods for minimizing the ROBDD representation of Boolean functions, including variable reordering and don't care exploitation to minimize ROBDDs for incompletely specified Boolean functions.

2.1. BDD minimization for completely specified functions

2.1.1 The role of variable orders

Reduced Ordered Decision Diagrams (ROBDDs) are canonical representations of Boolean functions, if the variable order is fixed (Theorem 1.1). But this does not mean that ROBDDs do not change when the variable order is changed.

In fact the size of ROBDDs for many Boolean functions depends heavily on the variable order. Changing the variable order can lead us from ROBDDs with linear size to ROBDDs with exponential size (and vice versa). An example of a class of such functions was given by Bryant in (Bryant, 1986): It is the class of all functions $g_n \in B_{2n}$ with $g_n(x_1, \ldots, x_{2n}) = x_1 \cdot x_2 + \ldots + x_{2n-1} \cdot x_{2n}$. The ROBDD with variable order $<_{index}$, $x_1 <_{index} x_2 <_{index} \cdots <_{index} x_{2n}$, has size $2n + 2$, whereas the ROBDD with variable order $<_{index'}$, $x_1 <_{index'} x_3 <_{index'} \cdots <_{index'} x_{2n-1} <_{index'} x_2 <_{index'} x_4 <_{index'} \cdots <_{index'} x_{2n}$, needs 2^{n+1} nodes. Figure 2.1 shows ROBDDs for g_3 with variable orders $<_{index}$ and $<_{index'}$.

Figure 2.1. ROBDDs for $x_1 \cdot x_2 + x_3 \cdot x_4 + x_5 \cdot x_6$. For the ROBDD on the left hand side $x_1 <_{index} \cdots <_{index} x_6$ is used, and for the ROBDD on the right hand side $x_1 <_{index'} x_3 <_{index'} x_5 <_{index'} x_2 <_{index'} x_4 <_{index'} x_6$.

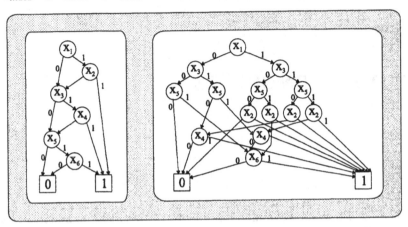

2.1.2   ROBDD size and communication complexity

Given a fixed variable order, the size of an ROBDD for a function f respecting this variable order can be determined based on the size of certain sets of cofactors of f. Here arguments known from communication complexity are used.

Let $f \in B_n$ be a Boolean function, F an ROBDD with variable set $\{x_1, \ldots, x_n\}$ and variable order $<_{index}$. Then a node v of F labeled $x_{index(p+1)}$ is a cofactor of f with respect to $x_{index(1)}^{\epsilon_{index(1)}} \cdots x_{index(p)}^{\epsilon_{index(p)}}$ for some $\epsilon_{index} = (\epsilon_{index(1)}, \cdots, \epsilon_{index(p)}) \in \{0,1\}^p$, such that v is reached by ϵ_{index} (cf. Remark 1.3, page 13); moreover

$$\left(f_{x_{index(1)} \cdots x_{index(p)}^{\epsilon_{index(1)} \cdots \epsilon_{index(p)}}} \right)_{x_{index(p+1)}} \neq \left(f_{x_{index(1)} \cdots x_{index(p)}^{\epsilon_{index(1)} \cdots \epsilon_{index(p)}}} \right)_{\overline{x_{index(p+1)}}},$$

i.e., $f_{x_{index(1)} \cdots x_{index(p)}^{\epsilon_{index(1)} \cdots \epsilon_{index(p)}}}$ 'depends essentially' on $x_{index(p+1)}$, since otherwise $f_{x_{index(1)} \cdots x_{index(p)}^{\epsilon_{index(1)} \cdots \epsilon_{index(p)}}}$ would not be reduced.

DEFINITION 2.1 *A function $g \in B_n$ depends essentially on variable x_i, if and only if*

$$g_{x_i} \neq g_{\overline{x_i}}.$$

Then we say 'x_i is in the support of g'.

On the other hand

- every cofactor $f_{x_{index(1)}^{\epsilon_{index(1)}} \cdots x_{index(p)}^{\epsilon_{index(p)}}}$ of f is represented by a node in F (labeled $x_{index(p+1)}$ only if $f_{x_{index(1)}^{\epsilon_{index(1)}} \cdots x_{index(p)}^{\epsilon_{index(p)}}}$ depends essentially on $x_{index(p+1)}$), and

- if two cofactors $f_{x_{index(1)}^{\epsilon_{index(1)}} \cdots x_{index(p)}^{\epsilon_{index(p)}}}$ and $f_{x_{index(1)}^{\delta_{index(1)}} \cdots x_{index(p)}^{\delta_{index(p)}}}$ are equal, then the nodes reached by $(\epsilon_{index(1)}, \ldots, \epsilon_{index(p)})$ and $(\delta_{index(1)}, \ldots, \delta_{index(p)})$ are identical.

Consequently, the number of nodes in F which are labeled $x_{index(p+1)}$ is equal to the *number of different cofactors* of f with respect to variables $x_{index(1)}, \ldots,$ $x_{index(p)}$, which *essentially depend on* $x_{index(p+1)}$.

Decomposition matrices are a means to illustrate the number of different cofactors with respect to a variable set $X^{(1)}$:

DEFINITION 2.2 (DECOMPOSITION MATRIX)
The decomposition matrix *of an (incompletely specified) Boolean function* $f \in BP_n(D)$ *with input variables* $x_1, \ldots, x_p, x_{p+1}, \ldots, x_n$ *with respect to the subset* $X^{(1)} = \{x_1, \ldots, x_p\}$ *is a* $(2^p \times 2^{n-p})$*-matrix* $Z(X^{(1)}, f)$*, which is defined by:* $\forall\, 0 \le i \le 2^p - 1$ *and* $\forall\, 0 \le j \le 2^{n-p} - 1$

$$Z(X^{(1)}, f)_{ij} = \begin{cases} f(bin_p(i), bin_{n-p}(j)), & \text{if } (bin_p(i), bin_{n-p}(j)) \in D \\ \star, & \text{if } (bin_p(i), bin_{n-p}(j)) \notin D \end{cases}$$

(Here $bin_p(i) \in \{0,1\}^p$ *corresponds to the binary representation of* i*,* $bin_{n-p}(j) \in \{0,1\}^{n-p}$ *to the binary representation of* j*.)*

It is easy to see that for $X^{(1)} = \{x_{index(1)}, \ldots, x_{index(p)}\}$ and $\epsilon_{index} = (\epsilon_{index(1)}, \ldots, \epsilon_{index(p)}) \in \{0,1\}^p$ the row of $Z(X^{(1)}, f)$ with index $int(\epsilon_{index})$ [1] represents a function table of the cofactor $f_{x_{index(1)}^{\epsilon_{index(1)}} \cdots x_{index(p)}^{\epsilon_{index(p)}}}$ of f.[2] Thus the number of different cofactors of f with respect to variables $x_{index(1)}, \ldots, x_{index(p)}$ is equal to the number of different row patterns of $Z(X^{(1)}, f)$.

NOTATION 2.1 *Let* $f \in B_n$ *be a function with input variables* $x_1, \ldots, x_p,$ $x_{p+1}, \ldots, x_n$. *Let* $Z(X^{(1)}, f)$ *be the decomposition matrix of* f *with respect to variable set* $X^{(1)} := \{x_1, \ldots, x_p\}$. *Then the number of different row patterns of* $Z(X^{(1)}, f)$ *is denoted by* $nrp(X^{(1)}, f)$.

---

[1] $int(\epsilon_{index})$ is the integer value of binary number ϵ_{index}.
[2] Since a cofactor of $f \in B_n$ with respect to $x_1^{\epsilon_1} \ldots x_p^{\epsilon_p}$ does not depend essentially on $x_1, \ldots, x_p$, it is often viewed as a function in B_{n-p} in the following.

Figure 2.2.   Decomposition matrix for g_3 with respect to $\{x_1, x_2, x_3\}$.

	x_4	0	0	0	0	1	1	1	1
	x_5	0	0	1	1	0	0	1	1
	x_6	0	1	0	1	0	1	0	1
$x_1 x_2 x_3$									
0 0 0		0	0	0	1	0	0	0	1
0 0 1		0	0	0	1	1	1	1	1
0 1 0		0	0	0	1	0	0	0	1
0 1 1		0	0	0	1	1	1	1	1
1 0 0		0	0	0	1	0	0	0	1
1 0 1		0	0	0	1	1	1	1	1
1 1 0		1	1	1	1	1	1	1	1
1 1 1		1	1	1	1	1	1	1	1

NOTATION 2.2 *If $Z(X^{(1)}, f)$ is the decomposition matrix of a function f with respect to the variable set $X^{(1)}$, then equality of row patterns of $Z(X^{(1)}, f)$ provides an equivalence relation on $\{0, 1\}^{|X^{(1)}|}$, if one defines: $\epsilon^{(1)} \equiv \epsilon^{(2)}$ iff the row patterns of rows with indices $int(\epsilon^{(1)})$ and $int(\epsilon^{(2)})$ are equal[3]. The number of equivalence classes of this relation $\equiv$ is equal to $nrp(X^{(1)}, f)$. The set of equivalence classes $\{K_1, \ldots, K_{nrp(X^{(1)}, f)}\}$ of $\equiv$ is denoted as $\{0, 1\}^{|X^{(1)}|}/_{\equiv}$ in the following.*

EXAMPLE 2.1 Figure 2.2 shows the decomposition matrix with respect to $\{x_1, x_2, x_3\}$ of function g_3 from Section 2.1.1 (for ROBDDs representing g_3 see Figure 2.1). The number of different row patterns is equal to 3 and consequently the number of different cofactors of g_3 with respect to $\{x_1, x_2, x_3\}$ is 3. Only one out of these three cofactors depends on variable x_4. Therefore in the ROBDD for g_3 with variable order $x_1 <_{index} \cdots <_{index} x_6$ (see left hand side of Figure 2.1) the number of nodes labeled x_4 is one.

The variable order $x_1 <_{index'} x_3 <_{index'} x_5 <_{index'} x_2 <_{index'} x_4 <_{index'} x_6$, which is inferior to $<_{index}$, is associated with a decomposition matrix with more row patterns. Figure 2.3 shows the decomposition matrix with respect to $\{x_1, x_3, x_5\}$. It has 8 different row patterns, i.e., there are 8 different cofactors with respect to $\{x_1, x_3, x_5\}$. 4 out of these 8 cofactors depend essentially on x_2 and thus the ROBDD with variable order $<_{index'}$ (see right hand side of Figure 2.1) has 4 nodes labeled x_2.

---

[3] $int(\epsilon) = \sum_{j=1}^{|X^{(1)}|} \epsilon_j \cdot 2^{|X^{(1)}|-j}$.

Figure 2.3. Decomposition matrix for g_3 with respect to $\{x_1, x_3, x_5\}$.

| $x_1x_3x_5$ / x_2 | 0 | 0 | 0 | 0 | 1 | 1 | 1 | 1 |
| | x_4 0 | 0 | 1 | 1 | 0 | 0 | 1 | 1 |
	x_6 0	1	0	1	0	1	0	1
0 0 0	0	0	0	0	0	0	0	0
0 0 1	0	1	0	1	0	1	0	1
0 1 0	0	0	1	1	0	0	1	1
0 1 1	0	1	1	1	0	1	1	1
1 0 0	0	0	0	0	1	1	1	1
1 0 1	0	1	0	1	1	1	1	1
1 1 0	0	0	1	1	1	1	1	1
1 1 1	0	1	1	1	1	1	1	1

Figure 2.4. Decomposition matrix with respect to $\{x_1, x_2, x_3\}$ and ROBDD of an example function in B_5. There are exactly 4 different row patterns and thus exactly 4 different 'linking nodes' (shaded gray) below the cut line below the variables x_1, x_2, x_3 in the ROBDD.

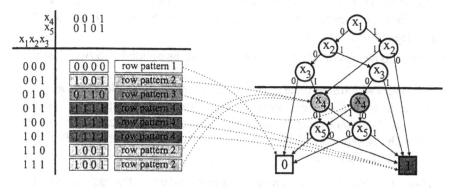

Figure 2.4 illustrates the relation between different row patterns in the decomposition matrix and ROBDD nodes, which are roots of sub-ROBDDs representing the respective cofactors. These nodes with one-to-one correspondence to row patterns are called 'linking nodes' in the following. The name is explained by the fact that these nodes are located in the ROBDD immediately below a imaginary cut line after the first p variables (the variables in variable set $X^{(1)}$). Linking nodes are exactly those nodes which are located below this cut line, but are connected to at least one node located above the cut line.

The number of nodes in the ROBDD F of $f \in B_n$ with variable order $<_{index}$ is equal to

$$
\sum_{p=0}^{n-1} \left| \left\{ f_{x_{index(1)} \ldots x_{index(p)}}^{\epsilon_{index(1)} \ldots \epsilon_{index(p)}} \mid \epsilon_{index} \in \{0,1\}^p, \right. \right.
$$
$$
\left. \left. f_{x_{index(1)} \ldots x_{index(p)}^{\epsilon_{index(1)} \ldots \epsilon_{index(p)}} x_{index(p+1)}^0}^{} \neq f_{x_{index(1)} \ldots x_{index(p)}^{\epsilon_{index(1)} \ldots \epsilon_{index(p)}} x_{index(p+1)}^1}^{} \right\} \right|
$$
$$
+ \left| \left\{ f_{x_{index(1)} \ldots x_{index(n)}}^{\epsilon_{index(1)} \ldots \epsilon_{index(n)}} \mid \epsilon_{index} \in \{0,1\}^n \right\} \right|.
$$

2.1.3 Variable reordering

The fact that the size of ROBDDs heavily depends on the variable order used confronts the user of ROBDDs with the problem of finding appropriate variable orders for given Boolean functions in order to represent these functions as compactly as possible. Frequently a good variable order can easily be found manually through an understanding of the structure of the given Boolean function. However it is desirable that the system finds such an order automatically without leaving this task to the user. Since the problem of finding a variable order for a Boolean function which minimizes the ROBDD size turned out to be NP–complete (Bollig et al., 1994; Tani et al., 1993; Bollig and Wegener, 1996), we use heuristics to determine a good variable order for Boolean functions which depend on a large number of inputs.

Heuristics to determine good variable orders can be divided into two classes:

1. Heuristics which determine a good 'initial order' before building the ROBDD. These heuristics derive a variable order from a given specification of a Boolean function by analyzing the structure of the specification. Examples for such heuristics are given in (Malik et al., 1988; Fujita et al., 1988; Fujita et al., 1991; Fujii et al., 1993).

2. Heuristics which improve on given variable orders on the basis of ROBDDs already built, e.g. (Ishiura et al., 1991; Rudell, 1993; Felt et al., 1993; Bollig et al., 1995; Drechsler et al., 1995). These heuristics can additionally be used for *dynamic* reordering, i.e., not only *after* computing some ROBDD, but also *during* its computation: If an ROBDD is computed from 'smaller' ROBDDs by Boolean operations and an intermediate result grows too large during this computation, reordering heuristics can be applied to minimize the sizes of the intermediate ROBDDs already computed, so that the following computation makes use of this new variable order.

The heuristic reordering method which has emerged as the most successful algorithm so far is the *sifting* algorithm by Rudell (Rudell, 1993). It profits from the fact that exchanging two variables x_i and x_j, which are adjacent in the variable order $<_{index}$, is a local operation, i.e., during the exchange of such a pair of variables in the variable order only those nodes are affected which are labeled x_i and x_j, whereas all other nodes remain unchanged. Moreover, the exchange can be done efficiently. The sifting algorithm begins with a variable x_i from $\{x_1, \ldots, x_n\}$ and moves this variable by exchanging adjacent variables to all possible n positions while keeping the other variables fixed. Afterwards the variable is shifted to the position in the order for which the resulting ROBDD size was minimal. The procedure is then repeated with the next variable until the position of all variables is determined after $O(n^2)$ exchanges of variables.

2.1.4 Symmetries of completely specified functions

Although the results of the sifting algorithm are very good in most cases, experiments performed by Panda and Somenzi (Panda and Somenzi, 1995) revealed a weakness of the sifting algorithm: They frequently observed groups of variables which are always adjacent to each other in good variable orders. Variables from such groups tend to remain unchanged in their position when the sifting algorithm is applied. Suppose there is a strong 'attraction' between two adjacent variables x_i and x_j. Then x_i tends to returning to its position adjacent to x_j, when it is 'sifted', and the same is true for x_j. As a result the sifting algorithm is not able to optimize the *relative* positions of such groups of variables attracting each other.

One example of such an attraction between variables are *symmetric* variables. For this reason Möller / Molitor / Drechsler (Möller et al., 1994) and Panda / Somenzi / Plessier (Panda et al., 1994) introduced 'symmetric sifting', which moves not single variables, but *groups* of variables in which a Boolean function is symmetric. Here also the relative position of these groups is optimized.

The impact of symmetries of Boolean functions on their ROBDD representations motivates us to have a closer look at symmetry properties. In Chapter 3 we will see that symmetry is also an important structural property in logic synthesis, especially in synthesis by decomposition.

2.1.4.1 Definitions and basic properties

In the following, let $X = \{x_1, \ldots, x_n\}$ be the set of variables of a Boolean function.

We give the definition of symmetry in two variables, in a set of variables, and in a partition of the set of input variables of a completely specified Boolean function:

DEFINITION 2.3 *A completely specified Boolean function* $f : \{0,1\}^n \to \{0,1\}$ *is* symmetric in a pair of input variables (x_i, x_j) *if and only if*

$$f(\epsilon_1, \ldots, \epsilon_i, \ldots, \epsilon_j, \ldots, \epsilon_n) = f(\epsilon_1, \ldots, \epsilon_j, \ldots, \epsilon_i, \ldots, \epsilon_n)$$

$\forall \epsilon \in \{0,1\}^n$. f *is* symmetric in a subset λ *of* X *iff* f *is symmetric in* x_i *and* x_j $\forall x_i, x_j \in \lambda$. f *is* symmetric in a partition $P = \{\lambda_1, \ldots, \lambda_k\}$ *of the set of input variables iff* f *is symmetric in* λ_i $\forall 1 \leq i \leq k$. f *is called* totally symmetric *iff* f *is symmetric in* X.

If f is symmetric in a subset λ of the set of input variables, then we say that 'the variables in λ form a symmetric group'.

Symmetry as defined in Definition 2.3 is a special case of the so–called G–symmetry (Hotz, 1974). G–symmetry is defined based on a set $G \subseteq \mathbf{P}_n$, where $\mathbf{P}_n$ is a special set of permutations of $\{0,1\}^n$, which is generated by 'variable exchanges' and 'variable negations':

DEFINITION 2.4 *For* $1 \leq i, k \leq n$ *let*

$$\sigma_{ik} : \{0,1\}^n \to \{0,1\}^n,$$
$$\sigma_{ik}(\epsilon_1, \ldots, \epsilon_i, \ldots, \epsilon_k, \ldots, \epsilon_n) = (\epsilon_1, \ldots, \epsilon_k, \ldots, \epsilon_i, \ldots, \epsilon_n)$$

and

$$\nu_i : \{0,1\}^n \to \{0,1\}^n, \nu_i(\epsilon_1, \ldots, \epsilon_i, \ldots, \epsilon_n) = (\epsilon_1, \ldots, \overline{\epsilon_i}, \ldots, \epsilon_n).$$

Then

$$\mathbf{P}_n := \{\tau_1 \circ \ldots \circ \tau_l \mid l \geq 0, \forall j \; \tau_j \in \{\sigma_{ik} \mid 1 \leq i, k \leq n\} \cup \{\nu_i \mid 1 \leq i \leq n\}\},$$

i.e., $\mathbf{P}_n$ *is the group generated by* $\{\sigma_{ik} \mid 1 \leq i, k \leq n\} \cup \{\nu_i \mid 1 \leq i \leq n\}$.

DEFINITION 2.5 *Let* $G \subseteq \mathbf{P}_n$. *A completely specified Boolean function* $f : \{0,1\}^n \to \{0,1\}$ *is called* G–symmetric, *if and only if for all* $\tau \in G$ *and all* $\epsilon \in \{0,1\}^n$

$$f(\tau(\epsilon)) = f(\epsilon).$$

REMARK 2.1 *Symmetry in variables* x_i, x_j *in the sense of Definition 2.3 is the same as G–symmetry with* $G = \{\sigma_{ij}\}$.

Other types of symmetry such as equivalence symmetry (Edwards and Hurst, 1978) can be expressed as G–symmetry, too.

DEFINITION 2.6 *A completely specified Boolean function* $f : \{0,1\}^n \rightarrow \{0,1\}$ *is equivalence symmetric in a pair of input variables* (x_i, x_k) *if and only if* $\forall \epsilon_l \in \{0,1\}, l \in \{1,\ldots,n\} \setminus \{i,k\}$

$$f(\epsilon_1,\ldots,\epsilon_{i-1},0,\epsilon_{i+1},\ldots,\epsilon_{k-1},0,\epsilon_{k+1},\ldots,\epsilon_n) =$$
$$f(\epsilon_1,\ldots,\epsilon_{i-1},1,\epsilon_{i+1},\ldots,\epsilon_{k-1},1,\epsilon_{k+1},\ldots,\epsilon_n).$$

REMARK 2.2 *Equivalence symmetry in variables* x_i, x_j *is the same as G–symmetry with* $G = \{\nu_i \circ \sigma_{ij} \circ \nu_i\}$.

Symmetry in variables x_i, x_j according to Definition 2.3 is sometimes also called *nonequivalence symmetry* to distinguish between symmetry and equivalence symmetry.

DEFINITION 2.7 *If* $f \in B_n$ *is both nonequivalence symmetric and equivalence symmetric in* (x_i, x_j) *(i.e., G–symmetric with* $G = \{\sigma_{ij}, \nu_i \circ \sigma_{ij} \circ \nu_i\}$), *then* f *is called* multiply symmetric *in* (x_i, x_j).

2.1.4.2 Symmetry variable orders

Symmetric sifting (Möller et al., 1994; Panda et al., 1994) produces variable orders, where variables in symmetric groups are located side by side. Variable orders of this kind are called 'symmetry orders'.

DEFINITION 2.8 *Let* f *be a completely specified Boolean function, which is symmetric in the variable partition* $P = \{\lambda_1,\ldots,\lambda_k\}$. *A variable order* $<_{index}$ *is called a* symmetry variable order *if for each symmetry set* $\lambda_i \in P$ *there exists* j *such that* $\{x_{index(j)}, x_{index(j+1)},\ldots,x_{index(j+|\lambda_i|-1)}\} = \lambda_i$.

Of course, the size of a symmetry ordered ROBDD for a function f which is symmetric in λ_i does not change, when only the position of variables in λ_i is changed, because f does not change, when symmetric variables from λ_i are exchanged. For that reason symmetric sifting does not need to exchange variables in symmetric groups.

Moreover symmetry is a structural property of Boolean functions which leads to small ROBDDs sizes, when symmetry orders are used. For totally symmetric functions $f \in B_n$, e.g., it is well–known that the size of the ROBDD is bounded by $O(n^2)$. This is due to the observation that for functions symmetric in (x_i, x_j) the equation $f_{x_i \bar{x}_j} = f_{\bar{x}_i x_j}$ holds. If the function f represented by an ROBDD F is symmetric in the first two variables $x_{index(1)}$ and $x_{index(2)}$ of the variable order and if f depends essentially on $x_{index(1)}$ and $x_{index(2)}$, then the left son of the right son of the root of F is equal to the right son of the left son of the root. Thus, ROBDDs representing totally symmetric functions grow within each level at most by one node. This is demonstrated by the following diagram:

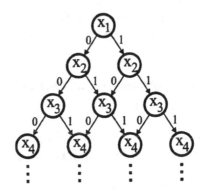

In general, in a symmetry ordered ROBDD there exist a lot of sub-ROBDDs where all variables in the upper part form a symmetry set. If k is the size of such a symmetry set, the upper parts of these sub-ROBDDs consisting of all nodes labeled by variables from the symmetry set have $O(k^2)$ nodes.

Furthermore, the value of a function that is symmetric in some variables $\{x_{i_1}, \ldots, x_{i_q}\}$ does not depend on the exact assignment of these variables but only on their weight $\sum_{j=1}^{q} x_{i_j}$. If one uses symmetry ordered ROBDDs, this weight is computed in neighboring levels and no information about partial weights has to be kept over several non-symmetric levels – and keeping information may cause large ROBDD sizes. Symmetry variable orders often avoid this drawback.

Whereas there are functions where symmetric variable orders do not provide optimal variable orders (Möller et al., 1994; Panda and Somenzi, 1995), the results of intensive experimentation proved that locating symmetric variables side by side is a successful heuristic method for variable ordering (Möller et al., 1994; Panda et al., 1994; Panda and Somenzi, 1995; Scholl et al., 1999).

2.1.4.3   Detection of symmetries

To make use of symmetries in variable ordering (and also in other logic synthesis problems) we need a method to detect symmetries in variable sets which are as large as possible (or variable partitions which are as small as possible).

Fortunately, symmetry in pairs of variables is transitive. If f is symmetric in the variable pairs (x_i, x_j) and (x_j, x_k), it is also symmetric in (x_i, x_k) and thus in $\{x_i, x_j, x_k\}$. If we define for a function $f \in B_n$ the relation $\sim_{sym}$ on $X = \{x_1, \ldots, x_n\}$ by

$$x_i \sim_{sym} x_j \quad \Longleftrightarrow \quad f \text{ is symmetric in } (x_i, x_j),$$

then $\sim_{sym}$ forms an equivalence relation on X. Thus, there is a unique minimum partition P of X (namely the set of the equivalence classes of this relation) such that f is symmetric in P. The equivalence classes of $\sim_{sym}$ are the maximal variable sets in which f is symmetric. If the partition into equivalence classes for $\sim_{sym}$ is $P_{\sim_{sym}} = \{\mu_1, \ldots, \mu_k\}$, then f is symmetric in a variable set λ, if and only if $\lambda \subseteq \mu_i$ for some $1 \leq i \leq k$. $P_{\sim_{sym}}$ is the minimum partition in which f is symmetric and $P_{\sim_{sym}}$ gives information about the maximal sets μ_i, in which f is symmetric.

In the rest of the book we use symmetry graphs to illustrate symmetries of Boolean functions. The symmetry graph $G^f_{sym} = (X, E)$ of a Boolean function $f \in B_n$ is a undirected graph with node set X (the set of input variables of f) and edges $\{x_i, x_j\} \in E$ iff f is symmetric in (x_i, x_j). For completely specified Boolean functions f G^f_{sym} has a special structure: The connected components of the graph form cliques as symmetry in two variables forms an equivalence relation.

Thus the problem of computing a minimum partition of f such that f is symmetric in P can be reduced to testing for symmetry in all pairs of input variables. $\binom{n}{2} = \frac{1}{2}(n^2 - n)$ symmetry tests are sufficient.

If f is represented by an ROBDD G_f, this can be done by computing the cofactors $f_{x_i \overline{x_j}}$ and $f_{\overline{x_i} x_j}$ and by testing whether the two cofactors are equal. The check can be done in time $O(size(G_f) \cdot \log(size(G_f)))$ (see Chapter 1). In (Möller et al., 1993) a method was presented to accelerate symmetry detection. It does not need to compute cofactors for all $\binom{n}{2} = \frac{1}{2}(n^2 - n)$ pairs of variables. The method relies on simple checks based on the structure of ROBDD G_f and excludes pairs of variables, in which f cannot be symmetric. Only the variable pairs which remain after these checks are checked for symmetry.

2.1.4.4 Detection of nonequivalence symmetries and equivalence symmetries

In this section we study how the situation changes when equivalence symmetry is also considered (apart from nonequivalence symmetry). Unfortunately pairwise equivalence symmetry does not provide an equivalence relation on X. This is shown by the following counterexample:

EXAMPLE 2.2

x_1	x_2	x_3	$f(x_1, x_2, x_3)$
0	0	0	0
0	0	1	0
0	1	0	1
0	1	1	0
1	0	0	0
1	0	1	1
1	1	0	0
1	1	1	0

It is easy to see that f is equivalence symmetric both in (x_1, x_2) and (x_2, x_3), but f is *not* equivalence symmetric in (x_1, x_3) because of $f(0,0,0) \neq f(1,0,1)$. However f is nonequivalence symmetric in (x_1, x_3).

The following lemma shows that this holds in general:

LEMMA 2.1 *Let $f \in B_n$. If f is equivalence symmetric in the variable pairs (x_i, x_j) and (x_j, x_k) (x_i, x_j, x_k pairwise different), then f is nonequivalence symmetric in (x_i, x_k).*

PROOF: We have to prove $\forall \epsilon_l \in \{0, 1\}$, $l \in \{1, \ldots, n\} \setminus \{i, k\}$

$$f(\epsilon_1, \ldots, \epsilon_{i-1}, 0, \epsilon_{i+1}, \ldots, \epsilon_{k-1}, 1, \epsilon_{k+1}, \ldots, \epsilon_n) =$$
$$f(\epsilon_1, \ldots, \epsilon_{i-1}, 1, \epsilon_{i+1}, \ldots, \epsilon_{k-1}, 0, \epsilon_{k+1}, \ldots, \epsilon_n)$$

Case 1: $\epsilon_j = 0$

$$f(\ldots, \underbrace{0}_{\epsilon_i}, \ldots, \underbrace{0}_{\epsilon_j}, \ldots, \underbrace{1}_{\epsilon_k}, \ldots)$$

$$= f(\ldots, \underbrace{1}_{\epsilon_i}, \ldots, \underbrace{1}_{\epsilon_j}, \ldots, \underbrace{1}_{\epsilon_k}, \ldots)$$

(since f equivalence symm. in (x_i, x_j))

$$= f(\ldots, \underbrace{1}_{\epsilon_i}, \ldots, \underbrace{0}_{\epsilon_j}, \ldots, \underbrace{0}_{\epsilon_k}, \ldots)$$

(since f equivalence symm. in (x_j, x_k))

Case 2: $\epsilon_j = 1$

$$f(\ldots, \underbrace{0}_{\epsilon_i}, \ldots, \underbrace{1}_{\epsilon_j}, \ldots, \underbrace{1}_{\epsilon_k}, \ldots)$$

$$= f(\ldots, \underbrace{0}_{\epsilon_i}, \ldots, \underbrace{0}_{\epsilon_j}, \ldots, \underbrace{0}_{\epsilon_k}, \ldots)$$

(since f equivalence symm. in (x_j, x_k))

$$= f(\ldots, \underbrace{1}_{\epsilon_i}, \ldots, \underbrace{1}_{\epsilon_j}, \ldots, \underbrace{0}_{\epsilon_k}, \ldots)$$

(since f equivalence symm. in (x_i, x_j))

$\square$

The following can be shown in an analogous manner:

LEMMA 2.2 *Let* $f \in B_n$. *If* f *is equivalence symmetric in* (x_i, x_j) *and nonequivalence symmetric in* (x_j, x_k), *then* f *is equivalence symmetric in* (x_i, x_k).

If f is equivalence symmetric in (x_i, x_j), then the function f', which results from negating x_i (or x_j), is nonequivalence symmetric in (x_i, x_j). If in Example 2.2 x_1 and x_3 are negated (or x_2 is negated), then the resulting function is (nonequivalence) symmetric in (x_1, x_2) and (x_2, x_3) and thus totally symmetric.

Now our goal is to find a selection of input variables, such that negating these variables leads to a function which is (nonequivalence) symmetric in a partition which is as small as possible.

In the following we show that there is a unique minimum partition with this property (in the sense that every other partition with this property is a refinement of it) and we show how to find such a minimum partition (together with an appropriate negation of variables).

DEFINITION 2.9 *Let* $P = (p_1, \ldots, p_n) \in \{0, 1\}^n$ *and* $f \in B_n$. *The function* $f_P \in B_n$ *defined by* $f_P(x_1, \ldots, x_n) = f(x_1^{p_1}, \ldots, x_n^{p_n})$ $\forall (x_1, \ldots, x_n) \in \{0, 1\}^n$ *is called the Boolean function generated from* f *using polarity vector* P. P *is called* polarity vector.

Thus we are looking for an 'optimal' polarity vector and its associated variable partition (i.e., we are looking for a polarity vector, such that the function generated using this polarity vector is symmetric in a minimum partition). To this end, the notion of 'extended symmetry' is defined by:

DEFINITION 2.10 *A function* $f \in B_n$ *is* extended symmetric *in the variable pair* (x_i, x_j) *iff* f *is (nonequivalence) symmetric or equivalence symmetric in* (x_i, x_j).

LEMMA 2.3 *Let $f \in B_n$. The relation $\sim_{esym}$ on $X = \{x_1, \ldots, x_n\}$ defined by*

$$x_i \sim_{esym} x_j \quad \Longleftrightarrow \quad f \text{ is extended symmetric in } (x_i, x_j)$$

is a equivalence relation on X.

PROOF: The proof follows from the fact that $\sim_{sym}$ is transitive and from Lemmas 2.1 and 2.2. □

The symmetry graph $G^f_{esym} = (X, E)$ for extended symmetry is defined like the symmetry graph G^f_{sym} for (nonequivalence) symmetry. Additionally there is a type for each edge which is 'nonequivalence symmetric', if f is *not* equivalence symmetric in the respective variable pair, 'equivalence symmetric', if f is *not* nonequivalence symmetric in the pair and 'multiply symmetric' otherwise.

THEOREM 2.1 *Let $f \in B_n$. There is a polarity vector P, such that f_P is (nonequivalence) symmetric in partition $P_{\sim_{esym}}$, which is formed by the equivalence classes of relation $\sim_{esym}$ of f.*

PROOF: We build polarity vector P, such that f_P is (nonequivalence) symmetric in $P_{\sim_{esym}}$.

Let $\mu_k = \{x_{i_1}, \ldots, x_{i_{l_k}}\} \in P_{\sim_{esym}}$. Then choose $p_{i_1} = \epsilon$ arbitrarily. If f is (nonequivalence) symmetric in (x_{i_1}, x_{i_2}), then choose $p_{i_2} = \epsilon$, if f is equivalence symmetric in (x_{i_1}, x_{i_2}), then choose $p_{i_2} = \bar{\epsilon}$. Continue the procedure with (x_{i_2}, x_{i_3}) etc. until $p_{i_{l_k}}$ is determined.

Obviously, f_P is (nonequivalence) symmetric in μ_k. □

If the polarity vector according to the construction in the proof is chosen, then the symmetry graph $G^{f_P}_{sym}$ for f_P has exactly the same edges as G^f_{esym} for f (with respect to extended symmetry). If another polarity vector P' is chosen, $G^{f_{P'}}_{sym}$ may possibly have fewer edges than G^f_{esym}. There is no edge between two nodes in $G^{f_{P'}}_{sym}$, if the corresponding edge in G^f_{esym} has type '(nonequivalence) symmetric', but the two nodes have different polarities in P', or if the corresponding edge in G^f_{esym} has type 'equivalence symmetric', but the two nodes have the same polarity in P'. Thus $f_{P'}$ has to be symmetric in a 'refinement' of $P_{\sim_{esym}}$.[4] Figure 2.5 illustrates the situation showing a graph G^f_{esym} of a function $f \in B_9$ and the resulting symmetry graph $G^{f_{P'}}_{sym}$ for $f_{P'}$ with $P' = (1, 0, 0, 1, 1, 1, 1, 1, 0)$.

---

[4] A partition $Q = \{Q_1, \ldots, Q_l\}$ is called a *refinement* of a partition $P = \{P_1, \ldots, P_k\}$ if and only if $\cup_{i=1}^k P_i = \cup_{i=1}^l Q_i$ and for all Q_i ($1 \leq i \leq l$) there is a $j \in \{1, \ldots, k\}$ with $Q_i \subseteq P_j$.

Figure 2.5. Left hand side: G^f_{esym} of a function $f \in B_9$. Solid lines stand for 'nonequivalence symmetric', dashed lines for 'equivalence symmetric' and 'double edges' (both solid and dashed lines) for 'multiply symmetric'. Right hand side: Resulting symmetry graph $G^{f_{P'}}_{sym}$ for $f_{P'}$ with $P' = (1, 0, 0, 1, 1, 1, 1, 1, 0)$. Negated variables are represented by bold face nodes.

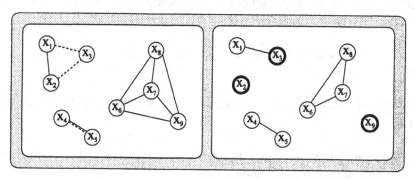

Figure 2.6. All possible subgraphs with 3 nodes of G^f_{esym} (except for renaming variables). Solid lines stand for 'nonequivalence symmetric', dashed lines for for 'equivalence symmetric' and 'double edges' (both solid and dashed lines) for 'multiply symmetric'.

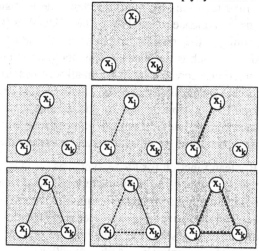

Finally, we will have a closer look at the structure of symmetry graphs for extended symmetry (for application in decomposition, Chapter 3):

In Figure 2.6 all possible subgraphs of the symmetry graph containing three nodes are shown (except for renaming variables).

The fact that no other subgraphs can occur follows from the transitivity of $\sim_{sym}$, Lemma 2.1, Lemma 2.2 and the following lemma:

LEMMA 2.4 *If* f *is multiply symmetric in* (x_i, x_j) *and nonequivalence symmetric (equivalence symmetric) in* (x_j, x_k), *then* f *is also multiply symmetric in* (x_i, x_k) *and in* (x_j, x_k).

PROOF: The proof follows from transitivity of $\sim_{sym}$ and Lemmas 2.1 and 2.2.   □

Consequently only two kinds of cliques in the symmetry graph G^f_{esym} can occur:

1. Cliques with nonequivalence and equivalence symmetric pairs of variables, but without multiply symmetric pairs.

2. Cliques *only* with multiply symmetric pairs.

2.2.   BDD minimization for incompletely specified functions

Incompletely specified functions play an important role in many applications such as checking the equivalence of two Finite State Machines (FSMs) (Coudert et al., 1989b), minimizing the transition relation of an FSM or logic synthesis for FPGA realizations, which is the focus of this book. In many cases incompletely specified functions represent the functionality of subcircuits which are embedded into a larger system, such that certain input vectors cannot be applied to the subcircuit resulting in 'don't cares' for these input vectors. If changing the output of a subcircuit will not change the output of the overall system for certain input vectors, then these input vectors can be used as don't cares, too.

To decompose a Boolean function for FPGA synthesis (see Chapter 5) we start with an ROBDD representation of an incompletely specified function. In a first step the ROBDD representation is minimized exploiting don't cares, i.e., we are looking for an extension of the incompletely specified function whose ROBDD has a minimal number of nodes. In the following we present some ideas for ROBDD minimization of incompletely specified functions.

2.2.1   The 'communication minimization' approach

As described in Chapter 1 an incompletely specified function $f \in BP_n(D)$ can be represented by two out of three ROBDDs for f_{on}, f_{off} and f_{dc}, which are the characteristic functions for the sets $ON(f)$, $OFF(f)$ and $DC(f)$), respectively, or by an ROBDD for $ext(f) = \overline{z} \cdot f_{off} + z \cdot f_{on}$.

Of course, the ROBDD for f_{on} represents a (completely specified) extension of f. But the special don't care assignment for f_{on} ($f_{on}(\epsilon) = 0$ for all $\epsilon \notin D$) will not lead to optimal ROBDD sizes in general. In most cases we will need a more sophisticated don't care assignment method.

Sauerhoff and Wegener (Sauerhoff and Wegener, 1996) proved that solving the minimization problem exactly (for a fixed variable order) is an NP-complete problem.

For practical application we use the observation from Section 2.1.2, which establishes a relation between the number of linking nodes immediately below a cut line after p variables in the ROBDD and the number of different cofactors with respect to the first p variables in the variable order. Like Chang, Cheng, Marek–Sadowska (Chang et al., 1994) and Shiple et al. (Shiple et al., 1994) we use don't cares to minimize the number of linking nodes below cut lines in a given ROBDD. Certainly the minimization can be performed for different cut lines in the ROBDD. At first we assume that we are using a fixed variable order $<_{index}$ and we consider a fixed cut line after the first p variables.

2.2.1.1 The problem CM

As in Section 2.1.2 we use decomposition matrices for illustration. We know from Section 2.1.2 that for a completely specified function f the number of different cofactors with respect to variables $x_{index(1)}, \ldots, x_{index(p)}$ equals the number of different row patterns of the decomposition matrix $Z(X^{(1)}, f)$ with $X^{(1)} = \{x_{index(1)}, \ldots, x_{index(p)}\}$, which is also equal to the number of linking nodes immediately below a cut line after the first p variables in the ROBDD for f with variable order $<_{index}$. This means if we want to find a completely specified extension f of an incompletely specified function g, whose ROBDD with variable order $<_{index}$ has a minimum number of nodes immediately below a cut line after the first p variables, then we have to look for an assignment to the don't care values which results in a decomposition matrix $Z(X^{(1)}, f)$ for f with minimum number of row patterns.

EXAMPLE 2.3
Figure 2.7 shows the decomposition matrix $Z(\{x_1, x_2, x_3\}, isp)$ for an incompletely specified function isp. Note that entries $*$ represent don't cares. Obviously our problem corresponds to a replacement of all entries $*$ by 0 and 1, such that the number of row patterns in $Z(\{x_1, x_2, x_3\}, isp)$ is minimized.

Again, the rows of decomposition matrices for incompletely specified functions can be interpreted as function tables of cofactors:

Figure 2.7.   Decomposition matrix $Z(\{x_1, x_2, x_3\}, isp)$ of a function isp

		x_4	0	0	0	0	1	1	1	1
		x_5	0	0	1	1	0	0	1	1
		x_6	0	1	0	1	0	1	0	1
$x_1 x_2 x_3$										
0 0 0			*	1	1	*	*	0	0	*
0 0 1			1	0	0	1	*	*	*	*
0 1 0			1	0	0	1	*	*	*	*
0 1 1			*	1	1	*	*	0	0	*
1 0 0			*	*	*	*	0	1	1	0
1 0 1			0	*	*	0	1	*	*	1
1 1 0			0	*	*	0	1	*	*	1
1 1 1			*	*	*	*	0	1	1	0

NOTATION 2.3 *Let $f \in BP_n(D)$. The cofactor of f with respect to $x_1^{\epsilon_1} \ldots x_p^{\epsilon_p}$ is a function $f_{x_1^{\epsilon_1} \ldots x_p^{\epsilon_p}}$ in $BP_n(D')$, where*

$$D' = \{(y_1, \ldots, y_{n-p}) \in \{0,1\}^{n-p} \mid (\epsilon_1, \ldots, \epsilon_p, y_1, \ldots, y_{n-p}) \in D\}$$

and

$$f_{x_1^{\epsilon_1} \ldots x_p^{\epsilon_p}}(y_1, \ldots, y_{n-p}) = f(\epsilon_1, \ldots, \epsilon_p, y_1, \ldots, y_{n-p}) \; \forall (y_1, \ldots, y_{n-p}) \in D'.$$

In order to find don't care assignments with a minimum number of row patterns in a decomposition matrix $Z(X^{(1)}, f)$, we define the notion of *compatible rows* of $Z(X^{(1)}, f)$ (cf. (Roth and Karp, 1962)). Two rows are called compatible, if and only if there is a replacement of *'s (representing don't cares) by 0's and 1's, which makes the row patterns equal.

DEFINITION 2.11 (COMPATIBLE ROWS)
Let $f \in BP_n(D)$ be a function with input variables $x_1, \ldots, x_n$. Let $Z(X^{(1)}, f)$ be the decomposition matrix of f with respect to $X^{(1)} = \{x_1, \ldots, x_p\}$. For $\epsilon^{(1)} = (\epsilon_1^{(1)}, \ldots, \epsilon_p^{(1)}) \in \{0,1\}^p$, $\epsilon^{(2)} = (\epsilon_1^{(2)}, \ldots, \epsilon_p^{(2)}) \in \{0,1\}^p$, $i = int(\epsilon^{(1)})$ and $j = int(\epsilon^{(2)})$, the rows of $Z(X^{(1)}, f)$ with indices i and j are called compatible $(\epsilon^{(1)} \sim \epsilon^{(2)})$, if and only if there is no column index $0 \le k \le 2^{n-p} - 1$ with

$$(Z(X^{(1)}, f)_{ik} = 0 \text{ and } Z(X^{(1)}, f)_{jk} = 1) \qquad \text{or}$$
$$(Z(X^{(1)}, f)_{ik} = 1 \text{ and } Z(X^{(1)}, f)_{jk} = 0).$$

(In other words:

$$\epsilon^{(1)} \sim \epsilon^{(2)} \quad \Longleftrightarrow$$

$\nexists \delta \in \{0,1\}^{n-p}$ with $(\epsilon^{(1)}, \delta), (\epsilon^{(2)}, \delta) \in D$ and $f(\epsilon^{(1)}, \delta) \neq f(\epsilon^{(2)}, \delta)$.)

If '$\sim$' were an equivalence relation on $\{0,1\}^p$, then our task of finding an assignment to don't cares to minimize the number of row patterns would be easy. Unfortunately '$\sim$' is *not* an equivalence relation on $\{0,1\}^p$. This is shown by the decomposition matrix

row 0	*	1
row 1	0	*
row 2	0	0
row 3	0	0

Rows 0 and 1 are compatible, rows 1 and 2 are compatible, but rows 0 and 2 are *not* compatible.

Similarly to the number of different row patterns $nrp(X^{(1)}, f)$ we use the following notation:

NOTATION 2.4 *Let $f \in BP_n(D)$ be a Boolean function with input variables $x_1, \ldots, x_n$. Let $Z(X^{(1)}, f)$ be the decomposition matrix of f with respect to variable set $X^{(1)} := \{x_1, \ldots, x_p\}$. Then the minimum number of sets in which $\{0,1\}^p$ can be partitioned, such that every pair of elements of the same set is compatible with respect to $\sim$, is denoted by $ncc(X^{(1)}, f)$.*

Suppose the elements of $\{0,1\}^p$ are partitioned into different sets $PK_1, \ldots,$ $PK_{ncc(X^{(1)}, f)}$ in such a way that the relation $\epsilon^{(1)} \sim \epsilon^{(2)}$ holds for all pairs $\epsilon^{(1)}$ and $\epsilon^{(2)} \in PK_i$ ($1 \leq i \leq ncc(X^{(1)}, f)$). Thus for every pair of rows in $Z(X^{(1)}, f)$ with indices in PK_i it cannot happen that there is a column in which one row has entry 1 and the other row has entry 0. This implies that the don't cares of f ('*' in the decomposition matrix) can be replaced by 0's and 1's, such that all rows belonging to some set PK_i will become equal. In that way a completely specified extension f' of f is obtained with the property that in the decomposition matrix for f' (with respect to $X^{(1)}$) two rows are equal iff the corresponding indices belong to the same set PK_i and consequently $nrp(X^{(1)}, f') = ncc(X^{(1)}, f)$ holds.

However, the problem of computing $ncc(X^{(1)}, f)$ for an incompletely specified function is substantially harder than determining the number of different row patterns $nrp(X^{(1)}, f)$. It turns out that this problem is NP–hard. Thus we do not expect to find an algorithm computing $ncc(X^{(1)}, f)$ with run time polynomial in the size of the decomposition matrix.

The following problem has to be solved:

Problem CM (Communication Minimization)

Given: Incompletely specified function $f \in BP_n(D)$ with input variables $x_1, \ldots, x_n$ and decomposition matrix $Z(X^{(1)}, f)$ with respect to $X^{(1)} = \{x_1, \ldots, x_p\}$.

Find: $ncc(X^{(1)}, f)$.

The following theorem holds (Scholl and Molitor, 1993):

THEOREM 2.2 *Problem* CM *is* NP–*hard.*

PROOF: Polynomial time transformation from the NP–complete problem 'Partition into Cliques' to CM, see Appendix B. $\qquad\square$

2.2.1.2 Solution to problem CM

The following procedure can be used to compute $ncc(X^{(1)}, f)$ (or more precisely to compute an upper bound on $ncc(X^{(1)}, f)$) and to find an appropriate don't care assignment:

- Compute the decomposition matrix $Z(X^{(1)}, f)$.

- Compute the compatibility relation $\sim$ on $\{0, 1\}^p$.

- Interpret $\sim$ as the set E of edges of an undirected graph G with node set $\{0, 1\}^p$. There is an edge $\{\epsilon_1, \epsilon_2\}$ for $\epsilon_1, \epsilon_2 \in \{0, 1\}^p$, if and only if $\epsilon_1 \sim \epsilon_2$.

- Compute $ncc(X^{(1)}, f)$ as the minimum number K, such that $\{0, 1\}^p$ can be partitioned into sets $V_1, \ldots, V_K$, where the subgraphs of all nodes in V_i are complete graphs, respectively. Then the elements in V_i are pairwise compatible with respect to $\sim$. The computation can be done using heuristics to solve the well-known problem 'Partition into Cliques' for graph G (see (Garey and Johnson, 1979)).

In general heuristics for 'Partition into Cliques' will provide only an upper bound on the number of cliques and thus on $ncc(X^{(1)}, f)$.

2.2.1.3 Solution to problem CM for ROBDD representations

Our motivation for dealing with problem CM was the search for a (completely specified) extension of an incompletely specified function f, such that in the ROBDD (for fixed variable order $<_{index}$) of the resulting extension the number of linking nodes immediately below a cut line after the first p variables is minimized. When we are starting with a compact ROBDD representation for the incompletely specified function, we are certainly not interested in computing the corresponding decomposition matrix $Z(X^{(1)}, f)$ for f with $X^{(1)} = \{x_{index(1)}, \ldots, x_{index(p)}\}$, since the size of the decomposition matrix is exponential in n.

Fortunately, we can reduce the search space for computation of a minimum partition $\{PK_1, \ldots, PK_{ncc(X^{(1)}, f)}\}$ of $\{0, 1\}^p$, such that elements of PK_i are pairwise compatible. Using this reduced search space we can avoid computing the decomposition matrix $Z(X^{(1)}, f)$: Let $\{0, 1\}^p/_\equiv = \{K_1, \ldots, K_{nrp(X^{(1)}, f)}\}$ be the partition into equivalence classes which is induced by *equal row patterns* of $Z(X^{(1)}, f)$ ($\epsilon^{(1)} \equiv \epsilon^{(2)}$ iff the row patterns of rows $int(\epsilon^{(1)})$ and $int(\epsilon^{(2)})$ are exactly the same, including entries '∗', cf. Notation 2.2). It is clear that $\{PK_1, \ldots, PK_{ncc(X^{(1)}, f)}\}$ can be chosen in such a way that each set PK_i is the union of certain sets K_j belonging to equal row patterns, i.e., $\{PK_1, \ldots, PK_{ncc(X^{(1)}, f)}\}$ can be chosen in such a way that all indices of rows with equal row patterns in $Z(X^{(1)}, f)$ belong to the same compatibility class PK_i.

EXAMPLE 2.3 (CONTINUED):
For function isp in Figure 2.7 the graph of the compatibility relation on rows of $Z(\{x_1, \ldots, x_3\}, isp)$ looks like

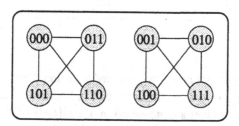

If the relation is restricted to $\{K_1, \ldots, K_{nrp(\{x_1, \ldots, x_3\}, f)}\}$, the following graph results:

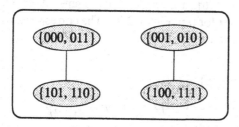

A restriction of the search to cliques in this reduced graph will not affect the quality of the result.

Figure 2.8.   ROBDD for $ext(isp) = \overline{z} \cdot isp_{off} + z \cdot isp_{on}$.

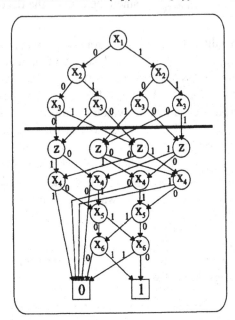

Now assume that the incompletely specified function f is given as an ROBDD for $ext(f) = \overline{z} \cdot f_{off} + z \cdot f_{on}$ as described in Section 1.3.2. For an example see Figure 2.8, which gives a representation of $ext(isp) = \overline{z} \cdot isp_{off} + z \cdot isp_{on}$ for our example function isp. The decomposition matrix for $ext(isp)$ can be found in Figure 2.9.

The decomposition matrix for $ext(isp)$ in Figure 2.9 easily results from the decomposition matrix for isp (Figure 2.7): The corresponding row of $ext(isp)$, i.e., the function table for

$$ext(isp)_{x_1^{\epsilon_1}...x_p^{\epsilon_p}} =$$
$$= (\overline{z} \cdot isp_{off} + z \cdot isp_{on})_{x_1^{\epsilon_1}...x_p^{\epsilon_p}}$$
$$= (\overline{z} \cdot isp_{off\,x_1^{\epsilon_1}...x_p^{\epsilon_p}} + z \cdot isp_{on\,x_1^{\epsilon_1}...x_p^{\epsilon_p}}),$$

is obtained from a row of matrix isp (i.e., the function table of a cofactor $isp_{x_1^{\epsilon_1}...x_p^{\epsilon_p}}$) by putting the function tables of $isp_{off\,x_1^{\epsilon_1}...x_p^{\epsilon_p}}$ and $isp_{on\,x_1^{\epsilon_1}...x_p^{\epsilon_p}}$ side by side (see Figure 2.10).

Figure 2.9. Decomposition matrix $Z(\{x_1, \ldots, x_3\}, ext(isp))$ for $ext(isp) = \overline{z} \cdot isp_{off} + z \cdot isp_{on}$.

z	0	0	0	0	0	0	0	0	1	1	1	1	1	1	1	1
x_4	0	0	0	0	1	1	1	1	0	0	0	0	1	1	1	1
x_5	0	0	1	1	0	0	1	1	0	0	1	1	0	0	1	1
x_6	0	1	0	1	0	1	0	1	0	1	0	1	0	1	0	1
$x_1 x_2 x_3$																
0 0 0	0	0	0	0	0	1	1	0	0	1	1	0	0	0	0	0
0 0 1	0	1	1	0	0	0	0	0	1	0	0	1	0	0	0	0
0 1 0	0	1	1	0	0	0	0	0	1	0	0	1	0	0	0	0
0 1 1	0	0	0	0	0	1	1	0	0	1	1	0	0	0	0	0
1 0 0	0	0	0	0	1	0	0	1	0	0	0	0	0	1	1	0
1 0 1	1	0	0	1	0	0	0	0	0	0	0	0	1	0	0	1
1 1 0	1	0	0	1	0	0	0	0	0	0	0	0	1	0	0	1
1 1 1	0	0	0	0	1	0	0	1	0	0	0	0	0	1	1	0

Figure 2.10. Computation of decomposition matrix row of $ext(f)$ from the corresponding row in the matrix of f.

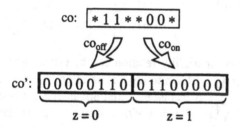

Since a cofactor $f_{x_{index(1)}^{\epsilon_1} \ldots x_{index(p)}^{\epsilon_p}}$ of an (incompletely specified) function f can be described in a unique manner by

$$(f_{x_{index(1)}^{\epsilon_1} \ldots x_{index(p)}^{\epsilon_p}})_{off} \text{ and } (f_{x_{index(1)}^{\epsilon_1} \ldots x_{index(p)}^{\epsilon_p}})_{on},$$

there is a one-to-one correspondence between cofactors $f_{x_{index(1)}^{\epsilon_1} \ldots x_{index(p)}^{\epsilon_p}}$ of f and cofactors $ext(f)_{x_{index(1)}^{\epsilon_1} \ldots x_{index(p)}^{\epsilon_p}}$ of $ext(f)$. Since $ext(f)$ is a completely specified function, again there is a one-to-one correspondence between cofactors $ext(f)_{x_{index(1)}^{\epsilon_1} \ldots x_{index(p)}^{\epsilon_p}}$ of $ext(f)$ and 'linking nodes' (below a cut line after $x_{index(1)}, \ldots, x_{index(p)}$) in an ROBDD for $ext(f)$ with variable order $<_{index}$. (Here we assume that the additional variable z of $ext(f)$ is located after variables $x_{index(1)}, \ldots, x_{index(p)}$ in the variable order.)

EXAMPLE 2.4 For function isp in Figure 2.8 there are exactly 4 'linking nodes' and accordingly, exactly 4 different row patterns in the decomposition matrix of $ext(isp)$ (Figure 2.9) and 4 different row patterns in the decomposition matrix of isp (Figure 2.7). Thus there are 4 classes of the partition of $\{0,1\}^p$, which is induced by equal row patterns, namely $K_1 = \{(000),(011)\}$, $K_2 = \{(101),(110)\}$, $K_3 = \{(001),(010)\}$ and $K_4 = \{(100),(111)\}$.

After identifying the $nrp(\{x_1,\ldots,x_p\},f)$ different linking nodes the 'compatibility graph' of relation '$\sim$' on equivalence classes $K_1,\ldots,K_{nrp(\{x_1,\ldots,x_p\},f)}$ is built. ($K_i \sim K_j$, if for $\epsilon \in K_i$ and $\delta \in K_j$ the relation $\epsilon \sim \delta$ holds, i.e., in the decomposition matrix the row with index ϵ has no 1 in a column, where the row with index δ has a 0 or vice versa.) The following equivalences hold:

$$
\begin{aligned}
&K_i \sim K_j \\
\Longleftrightarrow\quad &\text{for } \epsilon \in K_i, \delta \in K_j: \quad (f_{x_1^{\epsilon_1}\ldots x_p^{\epsilon_p}})_{on} \wedge (f_{x_1^{\delta_1}\ldots x_p^{\delta_p}})_{off} = 0 \text{ and} \\
&\qquad\qquad\qquad\qquad\qquad (f_{x_1^{\epsilon_1}\ldots x_p^{\epsilon_p}})_{off} \wedge (f_{x_1^{\delta_1}\ldots x_p^{\delta_p}})_{on} = 0. \\
\Longleftrightarrow\quad &\text{for } \epsilon \in K_i, \delta \in K_j: \quad ext(f)_{x_1^{\epsilon_1}\ldots x_p^{\epsilon_p} z} \wedge ext(f)_{x_1^{\delta_1}\ldots x_p^{\delta_p} \bar{z}} = 0 \text{ and} \\
&\qquad\qquad\qquad\qquad\qquad ext(f)_{x_1^{\epsilon_1}\ldots x_p^{\epsilon_p} \bar{z}} \wedge ext(f)_{x_1^{\delta_1}\ldots x_p^{\delta_p} z} = 0.
\end{aligned}
$$

It is clear that the compatibility relation can be established directly based on the ROBDD for $ext(f)$. Computing the relation $\sim$ on equivalence classes $K_1,\ldots,K_{nrp(\{x_1,\ldots,x_p\},f)}$ has an essential advantage compared to computing the relation $\sim$ on $\{0,1\}^p$: Starting with a compact representation of an incompletely specified function f by an ROBDD for $ext(f)$ a compact representation of the compatibility relation is obtained as well.

Then the problem CM can be solved by computing a solution for 'Partition into Cliques' applied to the compatibility graph of '$\sim$'. A partition $\{PK_1,\ldots,PK_{ncc(X^{(1)},f)}\}$ of $\{0,1\}^p$ is obtained, where all classes PK_i are unions of certain sets K_j.[5]

EXAMPLE 2.3 (CONTINUED):
We have $PK_1 = K_1 \cup K_2 = \{(000),(011),(101),(110)\}$, $PK_2 = K_3 \cup K_4 = \{(001),(010),(100),(111)\}$, $ncc(\{x_1,\ldots,x_3\},isp) = 2$.

---

[5]If *heuristics* are used to solve 'Partition into Cliques', possibly more than $ncc(X^{(1)},f)$ different classes are computed, of course.

After solving an instance of 'Partition into Cliques' we still have the task of finding an appropriate assignment of zeros and ones to the don't cares. The assignment has to lead to an extension f' with the following property: For arbitrary elements $\epsilon^{(i_j)} \in K_{i_j}$ with $PK_i = \bigcup_{j=1}^{l_i} K_{i_j}$ the cofactors $f'_{x_{index(1)}^{\epsilon_1^{(i_1)}} \cdots x_{index(p)}^{\epsilon_p^{(i_1)}}}, \ldots,$

$f'_{x_{index(1)}^{\epsilon_1^{(i_{l_i})}} \cdots x_{index(p)}^{\epsilon_p^{(i_{l_i})}}}$ are equal, such that finally $nrp(X^{(1)}), f') = ncc(X^{(1)}), f)$.

EXAMPLE 2.3 (CONTINUED):

In Example 2.3 all rows of the decomposition matrix in Figure 2.7 having the forms

$$\boxed{\star\,1\,1\,\star\,\star\,0\,0\,\star}$$

and

$$\boxed{0\,\star\,\star\,0\,1\,\star\,\star\,1}$$

have to be replaced by their common extension

$$\boxed{0\,1\,1\,0\,1\,0\,0\,1}$$

All rows of forms

$$\boxed{1\,0\,0\,1\,\star\,\star\,\star\,\star}$$

and

$$\boxed{\star\,\star\,\star\,\star\,0\,1\,1\,0}$$

have to be replaced by

$$\boxed{1\,0\,0\,1\,0\,1\,1\,0}.$$

In the example above incompletely specified cofactors were replaced by their 'common extension'. The following notion defines exactly how 'common extensions' of several cofactors are computed.

NOTATION 2.5 *Let* $co_1 \in BP_{n-p}(D_1), \ldots, co_l \in BP_{n-p}(D_l)$ *with the property that there is no pair* $i, j \in \{1, \ldots, l\}$, *such that there is* $\epsilon \in D_i \cap D_j$ *with* $co_i(\epsilon) \neq co_j(\epsilon)$. *Then the function* $com\_ext(\{co_1, \ldots, co_l\}) : \bigcup_{i=1}^{l} D_i \to$

$\{0,1\}$, *which is defined for each* $\epsilon \in \cup_{i=1}^{l} D_i$ *by* $com\_ext(\{co_1, \ldots, co_l\})(\epsilon) = co_j(\epsilon)$ *for arbitrary* co_j *with* $\epsilon \in D_j$, *is the* common extension *of* $co_1, \ldots, co_l$.

The $ext(.)$–representation for common extensions can easily be computed based on the $ext(.)$–representations of the cofactors involved:

REMARK 2.3 *Let* $co_1 \in BP_{n-p}(D_1), \ldots, co_l \in BP_{n-p}(D_l)$ *with the property that there is no pair* $i,j \in \{1, \ldots, l\}$, *such that there is* $\epsilon \in D_i \cap D_j$ *with* $co_i(\epsilon) \neq co_j(\epsilon)$. *Let* $ext(co_i) = \overline{z} \cdot (co_i)_{off} + z \cdot (co_i)_{on}$ $(1 \leq i \leq l)$. *Then*

$$ext(com\_ext(\{co_1, \ldots, co_l\})) = \overline{z} \cdot \bigvee_{i=1}^{l} (co_i)_{off} + z \cdot \bigvee_{i=1}^{l} (co_i)_{on}.$$

Let again $\epsilon^{(i_j)}$ be arbitrary elements of classes K_{i_j} with $PK_i = \bigcup_{j=1}^{l_i} K_{i_j}$. The don't care assignment based on the ROBDD of $ext(f)$ is done as follows: For all $1 \leq i \leq ncc(X^{(1)}, f)$ the linking nodes which are reached by $\epsilon^{(i_1)}, \ldots, \epsilon^{(i_{l_i})}$ are replaced (together with the subgraphs rooted by these nodes) by the ROBDD for $ext(com\_ext(\{f_{x_{index(1)}^{\epsilon_1^{(i_1)}} \cdots x_{index(p)}^{\epsilon_p^{(i_1)}}}, \ldots, f_{x_{index(1)}^{\epsilon_1^{(i_{l_i})}} \cdots x_{index(p)}^{\epsilon_p^{(i_{l_i})}}}\}))$, respectively. The ROBDD for $ext(com\_ext(\{f_{x_{index(1)}^{\epsilon_1^{(i_1)}} \cdots x_{index(p)}^{\epsilon_p^{(i_1)}}}, \ldots, f_{x_{index(1)}^{\epsilon_1^{(i_{l_i})}} \cdots x_{index(p)}^{\epsilon_p^{(i_{l_i})}}}\}))$ is computed as stated in Remark 2.3 above based on

$$ext(f)_{x_{index(1)}^{\epsilon_1^{(i_1)}} \cdots x_{index(p)}^{\epsilon_p^{(i_1)}}}, \ldots, ext(f)_{x_{index(1)}^{\epsilon_1^{(i_{l_i})}} \cdots x_{index(p)}^{\epsilon_p^{(i_{l_i})}}}.$$

(We assume that the additional variable z of $ext(f)$ is located after variables $x_{index(1)}, \ldots, x_{index(p)}$ in the variable order.)

EXAMPLE 2.3 (CONTINUED):
In the ROBDD of Example 2.3 (see Figure 2.8) the first and the second linking node (together with the respective subgraphs) must be replaced by the ROBDD for

$$ext(com\_ext(\{isp_{x_1^0 x_2^0 x_3^0}, isp_{x_1^1 x_2^0 x_3^1}\})) =$$
$$\overline{z} \cdot \overline{exor(x_4, x_5, x_6)} + z \cdot exor(x_4, x_5, x_6)$$

and the third and fourth linking node (together with the respective subgraphs) must be replaced by the ROBDD for

$$ext(com\_ext(\{isp_{x_1^0 x_2^0 x_3^1}, isp_{x_1^1 x_2^0 x_3^0}\})) =$$
$$\overline{z} \cdot exor(x_4, x_5, x_6) + z \cdot \overline{exor(x_4, x_5, x_6)}.$$

Figure 2.11. On the left hand side the ROBDD of function $ext(isp')$ after don't care assignment is shown, on the right hand side the ROBDD for isp'_{on}.

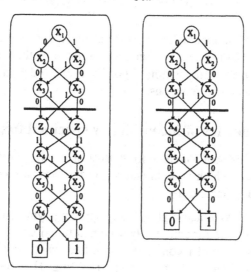

The ROBDD for the resulting function $ext(isp')$ is shown on the left hand side of Figure 2.11. The ROBDD on the right hand side represents $isp'_{on} = exor(x_1, \ldots, x_6)$.

After don't care assignment an ROBDD for an extension $ext(f')$ results with $nrp(X^{(1)}, f') = ncc(X^{(1)}, f)$ linking nodes. Since the common extensions used are not necessarily completely specified, the result $ext(f')$ can represent an *incompletely specified* function again. If a completely specified extension of f with $ncc(X^{(1)}, f)$ linking nodes is needed, then the remaining don't cares can be assigned the same value, for instance.

2.2.1.4 Using CM solutions for ROBDD minimization

The method from the previous section minimizes the number of linking nodes below a cut line after the first p variables $x_{index(1)}, \ldots, x_{index(p)}$ in the variable order $<_{index}$. This approach is used in (Chang et al., 1994) to find, for a fixed variable order, extensions of incompletely specified functions whose ROBDD sizes are as small as possible. As already mentioned, after the don't care assignment we obtain again an incompletely specified function. This suggests further exploitations of don't cares — now to minimize the number of linking nodes also below *other* cut lines. Chang and Marek–Sadowska use the described method to minimize the number of linking nodes for a series of cuts: The first cut is after variable $x_{index(1)}$, the second after $x_{index(2)}$ and so on up to variable $x_{index(n-1)}$.

A disadvantage of this approach lies in the fact that it uses a fixed variable order for minimization. In the next section we describe a method which combines this approach with the computation of good variable orders using symmetries of incompletely specified functions.

2.2.2   Symmetries of incompletely specified functions

In this section we present a method for ROBDD minimization for incompletely specified functions, which is — in contrast to the previous section — not restricted to a fixed variable order.

As determining the symmetric groups and applying symmetric sifting results in good variable orders for completely specified functions, it also seems to be a good idea in the case of incompletely specified functions to first determine symmetric groups and then to apply symmetric sifting. However, the symmetric groups of incompletely specified functions are not uniquely defined as will be demonstrated by some counterexamples. Therefore we have to ask for good partitions of the Boolean variables into symmetric groups with respect to ROBDD minimization.

As a first step we present a theory of symmetries of incompletely specified functions, which was developed in (Scholl, 1996; Scholl et al., 1997; Scholl et al., 1999).

Symmetries of incompletely specified functions will play an important role not only for ROBDD minimization, but also for decompositions of incompletely specified functions (see Chapter 5).

The definition of symmetry of an incompletely specified Boolean function f is reduced to the definition of symmetry of completely specified extensions of f (see Definition 1.4 on page 2).

DEFINITION 2.12 *An incompletely specified Boolean function* $f : D \to \{0, 1\}$ *is symmetric in a pair of input variables* (x_i, x_j) *(in a subset* λ *of* X */ in a partition* $P = \{\lambda_1, \ldots, \lambda_k\}$ *of* X*) iff there is a completely specified extension* f' *of* f *which is symmetric in* (x_i, x_j) *(in* λ */ in* P*).*

REMARK 2.4 *In the same way equivalence symmetry and multiple symmetry of incompletely specified Boolean function is reduced to equivalence symmetry and multiple symmetry of completely specified extensions.*

2.2.2.1   Nonequivalence symmetry of incompletely specified functions

Now we discuss methods for the detection of symmetries of incompletely specified Boolean functions represented by ROBDDs. Based on the natural generalization of the symmetry definition from completely specified functions to incompletely specified functions, we point out difficulties in the computation of symmetries for incompletely specified functions and demonstrate an approach to overcome these difficulties. This approach is based on the concept of 'strong symmetries' and a heuristic solution to find a minimum sized partition of the variables of an incompletely specified function into symmetry groups.

Difficulties with symmetry of incompletely specified functions.   In order to minimize the ROBDD size for an incompletely specified Boolean function f, we are looking for a minimum partition (or for maximal variable sets) such that f is symmetric in this partition (or these sets). Unfortunately there are some difficulties in the computation of such partitions: First of all, symmetry of f in two variables does not form an equivalence relation on X in the case of *incompletely* specified Boolean functions (see also (Dietmeyer and Schneider, 1967) or (Kim and Dietmeyer, 1991)):

EXAMPLE 2.5   The following function shows that symmetry in two variables does not lead to an equivalence relation on the variable set in the case of incompletely specified Boolean functions:

$$f(\epsilon) = \begin{cases} 1 & \text{for } \epsilon = (1,0,0) \\ dc & \text{for } \epsilon = (0,1,0) \\ 0 & \text{for } \epsilon = (0,0,1) \\ 0 & \text{otherwise} \end{cases}$$

It is easy to see that f is symmetric in x_1 and x_2 (for the corresponding completely specified extension f' of f it holds $f'(0,1,0) = 1$). Moreover, f is symmetric in x_2 and x_3 (for the corresponding completely specified extension f' of f it holds $f'(0,1,0) = 0$). However, f obviously is *not* symmetric in x_1 and x_3 (because of $f(1,0,0) \neq f(0,0,1)$).

Since symmetry in pairs of variables does not form an equivalence relation, it will be much more difficult to deduce symmetries in larger variable sets from symmetries in pairs of variables than in the case of completely specified Boolean functions.

Again, we use symmetry graphs to illustrate symmetries of Boolean functions. In contrast to completely specified Boolean functions where the connected

Figure 2.12. Symmetry graph of the function of Example 2.6

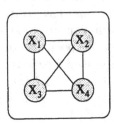

components of the graph form cliques, for incompletely specified functions there is not any structural property. On the contrary we can prove (see proof of Theorem 2.3 on page 55) that for every graph G with n nodes, there is an (incompletely specified) Boolean function $f : D \rightarrow \{0, 1\}$ such that the symmetry graph of f coincides with G. Thus, any undirected graph may occur as the symmetry graph of an incompletely specified function.

Even if f is symmetric in *all* pairs of variables x_i and x_j of a subset λ of the variable set of f, f is *not* necessarily symmetric in λ. This is illustrated by the following example:

EXAMPLE 2.6 Consider $f : D \rightarrow \{0, 1\}, D \subseteq \{0, 1\}^4$.

$$f(\epsilon) = \begin{cases} 1 & \text{for } \epsilon = (0, 0, 1, 1) \\ dc & \text{for } \epsilon = (0, 1, 0, 1), \epsilon = (0, 1, 1, 0), \\ & \quad \epsilon = (1, 0, 0, 1), \epsilon = (1, 0, 1, 0) \\ 0 & \text{for } \epsilon = (1, 1, 0, 0) \\ 0 & \text{otherwise} \end{cases}$$

It is easy to see that f is symmetric in all pairs of variables x_i and x_j, $i, j \in \{1, 2, 3, 4\}$. The symmetry graph of f is shown in Figure 2.12. It is a complete graph. For each completely specified extension f' of f which is symmetric in (x_1, x_3), $f'(0, 1, 1, 0) = 0$ holds and for each completely specified extension f'' of f which is symmetric in (x_2, x_4), $f''(0, 1, 1, 0) = 1$ holds. Hence there is no completely specified extension of f which is symmetric in (x_1, x_3) *and* (x_2, x_4) and therefore no extension which is symmetric in $\{x_1, x_2, x_3, x_4\}$.

Example 2.6 also points out another fact: If an incompletely specified Boolean function f is symmetric in all variable sets λ_i of a partition $P = \{\lambda_1, \ldots, \lambda_k\}$, it is *not* necessarily symmetric in P (choose $P = \{\{x_1, x_3\}, \{x_2, x_4\}\}$).

Strong Symmetry. As already mentioned, the difficulties with the detection of large symmetry groups of incompletely specified functions result from the fact that symmetry in pairs of variables does not form an equivalence relation on the variable set X and thus transitivity cannot be used to successively enlarge symmetry sets. In the following we introduce a stronger notion of symmetry of incompletely specified functions which results in an equivalence relation as in the case of completely specified functions:

DEFINITION 2.13 (STRONG SYMMETRY)
An incompletely specified Boolean function $f : D \to \{0,1\}$ is called strongly symmetric *in a pair of input variables (x_i, x_j) iff $\forall (\epsilon_1, \ldots, \epsilon_n) \in \{0,1\}^n$ either (a) or (b) holds.*

(a) $(\epsilon_1, \ldots, \epsilon_i, \ldots, \epsilon_j, \ldots, \epsilon_n) \notin D$ *and* $(\epsilon_1, \ldots, \epsilon_j, \ldots, \epsilon_i, \ldots, \epsilon_n) \notin D$

(b) $(\epsilon_1, \ldots, \epsilon_i, \ldots, \epsilon_j, \ldots, \epsilon_n) \in D$ *and* $(\epsilon_1, \ldots, \epsilon_j, \ldots, \epsilon_i, \ldots, \epsilon_n) \in D$
and $f(\epsilon_1, \ldots, \epsilon_i, \ldots, \epsilon_j, \ldots, \epsilon_n) = f(\epsilon_1, \ldots, \epsilon_j, \ldots, \epsilon_i, \ldots, \epsilon_n)$.

In contrast to *strong* symmetry of incompletely specified functions the symmetry defined so far is called *weak* symmetry. (Notice that for completely specified Boolean functions *strong* symmetry and *weak* symmetry are identical.)

The following lemma for strong symmetry follows directly from the definition:

LEMMA 2.5 *Strong symmetry in pairs of variables of an incompletely specified Boolean function $f : D \to \{0,1\}$ forms an equivalence relation on the variable set X of f.*

Due to Lemma 2.5 there is a unique minimum partition P of the set X of input variables such that f is strongly symmetric in P. As in the case of completely specified Boolean functions, f is strongly symmetric in a subset λ of X iff $\forall x_i, x_j \in \lambda$ f is strongly symmetric in (x_i, x_j). f is strongly symmetric in a partition $P = \{\lambda_1, \ldots, \lambda_k\}$ of X iff $\forall 1 \le i \le k$ f is strongly symmetric in λ_i.

Of course, if a function f is weakly symmetric in a partition P, it need not be strongly symmetric in P, but it follows directly from Definition 2.12 that there is an extension of f which is strongly symmetric in P.

Before we deal with the computation of extensions of incompletely specified Boolean functions which are strongly symmetric in minimum sized variable partitions, we will characterize weak and strong symmetry in variable partitions in more detail. To do this, we need the term of the 'weight class' of a given partition.

DEFINITION 2.14 (WEIGHT CLASS OF A PARTITION P)
Let $P = \{\lambda_1, \ldots, \lambda_k\}$ be a partition of $\{x_1, \ldots, x_n\}$.
We call $w^1(\epsilon_1, \ldots, \epsilon_n) = \sum_{i=1}^{n} \epsilon_i$ the 1–weight of $(\epsilon_1, \ldots, \epsilon_n)$ and $w^0(\epsilon_1, \ldots, \epsilon_n) = n - w^1(\epsilon_1, \ldots, \epsilon_n)$ the 0–weight of $(\epsilon_1, \ldots, \epsilon_n) \in \{0,1\}^n$.
For $\lambda_i = \{x_{i_1}, \ldots, x_{i_l}\}$, $w^1_{\lambda_i}(\epsilon_1, \ldots, \epsilon_n) = \sum_{j \in \{i_1, \ldots, i_l\}} \epsilon_j$ is the 1–weight of the 'λ_i–part' of $(\epsilon_1, \ldots, \epsilon_n)$.
$C^P_{w_1, \ldots, w_k} = \{(\epsilon_1, \ldots, \epsilon_n) \in \{0,1\}^n \,|\, w^1_{\lambda_i}(\epsilon_1, \ldots, \epsilon_n) = w_i, 1 \leq i \leq k\}$ is called weight class of the partition P with weights $(w_1, \ldots, w_k)$.

EXAMPLE 2.7 Let $P = \{\{x_1, x_2\}, \{x_3, x_4, x_5\}\}$. $C^P_{1,2}$ is the subset of all vectors of $\{0,1\}^n$ with a 1–weight 1 of the $\{x_1, x_2\}$–part and a 1–weight 2 of the $\{x_3, x_4, x_5\}$–part, i.e., the subset of all vectors with exactly one 1 in the first two components and exactly two 1's in the remaining components:

$$C^P_{1,2} = \{(0,1,0,1,1),(0,1,1,0,1),(0,1,1,1,0),$$
$$(1,0,0,1,1),(1,0,1,0,1),(1,0,1,1,0)\}.$$

If we consider a completely specified function f that is symmetric in a partition P, it follows directly that the function value of f is identical for all elements in a fixed weight class of P: The number of 1's for each symmetry group already determines the value of f. For incompletely specified functions the situation is a little bit more complicated: If strong symmetry is considered, permutation of bit positions belonging to the same symmetry group either does not change the value of f (if it is defined) or all input vectors in a fixed weight class belong to the don't care set. In the case of weak symmetry, by definition we have the existence of a symmetric completely specified extension, which shows the following: Permutation of bit positions belonging to the same symmetry group can only lead to input vectors having the same function value of f or to input vectors in the don't care set, but we never obtain function values 0 *and* 1 in the same weight class.

The characterization of weak and strong symmetry is summarized in the following lemma:

LEMMA 2.6 *Let $P = \{\lambda_1, \ldots, \lambda_k\}$ be a partition of $\{x_1, \ldots, x_n\}$. $f : D \to \{0,1\}$ is*

(1) strongly symmetric *in P iff*

$$\forall \, 1 \leq i \leq k \quad \forall \, 0 \leq w_i \leq |\lambda_i|$$

$$f(C^P_{w_1,\ldots,w_k}) = \begin{cases} \{0\} & or \\ \{1\} & or \\ \{dc\} \end{cases}$$

(2) (weakly) symmetric *in P iff*

$$\forall 1 \leq i \leq k \quad \forall 0 \leq w_i \leq |\lambda_i| \quad \{0,1\} \not\subseteq f(C^P_{w_1,\ldots,w_k}).$$

PROOF: See Appendix C.   □

The Problem 'Minimum Symmetry Partition' (MSP). As mentioned before, our goal is to solve the following problem **MSP** (Minimum Symmetry Partition):

Problem MSP (Minimum Symmetry Partition)

Given: Incompletely specified function $f : D \to \{0,1\}$, represented by ROBDDs for f_{on} and f_{dc}.

Find: Partition P of the set $X = \{x_1,\ldots,x_n\}$ such that

- f is symmetric in P and

- for any partition P' of X in which f is symmetric, the inequation $|P| \leq |P'|$ holds.

We can prove the following theorem by a polynomial–time transformation from the NP–complete problem 'Partition into Cliques' (**PC**) (see (Garey and Johnson, 1979)) to MSP:

THEOREM 2.3 *MSP is NP-hard.*

PROOF: See Appendix D.   □

Solution to MSP. Now we turn our attention to heuristic methods for the solution of MSP. Thereby we use a heuristic solution of 'Partition into Cliques' for the symmetry graph G^f_{sym} of f. However, the examples in Section 2.2.2.1 show that f is *not* symmetric in *all* partitions into cliques of G^f_{sym}. So the heuristic algorithm has to be changed in order to guarantee that f is symmetric in the resulting partition P. We will make use of the concept of *strong symmetry*, which — due to transitivity — allows us to incrementally reduce the size of the partition.

The heuristic algorithm we employ to solve the problem PC makes use of the following well-known lemma:

Figure 2.13. Graph $\overline{G}$ can be colored with 2 colors (black and gray nodes). Thus G can be partitioned into 2 cliques.

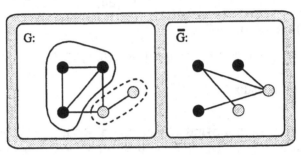

LEMMA 2.7 *A graph $G = (V, E)$ can be partitioned into k disjoint cliques iff $\overline{G} = (V, \overline{E})$ can be colored with k colors. ($\overline{G}$ is the inverse graph of G, which has the same node set V as G and an edge $\{v, w\}$ between two nodes v and w iff there is no edge $\{v, w\}$ in G, i.e., $\overline{E} = \{\{v, w\} \mid \{v, w\} \notin E\}$.)*

If $\{V_1, \ldots, V_k\}$ is a partition of V into k disjoint cliques, then in $\overline{G}$ the nodes of V_i can be given the same color, since they form an 'independent set' in $\overline{G}$. Inversely, nodes in $\overline{G}$ which have the same color form an independent set and thus a clique in G (see Figure 2.13).

Thus, heuristics for node coloring can be directly used for the solution of 'Partition into Cliques'. Our implementation is based on Brélaz' algorithm for node coloring (Brélaz, 1979) which has a run time of $O(N)$ in an implementation of Morgenstern (Morgenstern, 1992), where N denotes the number of nodes of the graph which have to be colored. It is a greedy algorithm which colors node by node and does not change the color of a node which is already colored. In the algorithm there are certain criteria for choosing the next node to be colored and the color to use for it in a clever way (see (Brélaz, 1979; Morgenstern, 1992)).

Figure 2.14 shows our heuristic algorithm for the problem MSP, which is derived from the Brélaz/Morgenstern heuristics for node coloring. The algorithm receives as input an incompletely specified function $f : D \to \{0, 1\}$, represented by ROBDDs for f_{on} and f_{dc} and computes as a result a partition P of $\{x_1, \ldots, x_n\}$, such that f is (weakly) symmetric in P. Thereby, first of all the symmetry graph G_{sym}^f of f (or the inverse graph $\overline{G_{sym}^f}$) is computed. The nodes of $\overline{G_{sym}^f}$ are the variables $x_1, \ldots, x_n$. These nodes are colored in the algorithm. Nodes of the same color form a clique in G_{sym}^f. Note that partition P (see line 3) has the property that it contains set $\{x_k\}$ for any uncolored node x_k and that nodes of the same color are in the same set of P, at any moment.

Figure 2.14. Algorithm to solve MSP.

Input: Incompletely specified function $f : D \to \{0,1\}$, $D \subseteq \{0,1\}^n$, represented by f_{on} and f_{dc}

Output: Partition P of $\{x_1, \ldots, x_n\}$, such that f is symmetric in P

Algorithm:

```
 1      Compute symmetry graph G^f_sym = (V, E) of f (or ~G^f_sym = (V, ~E)).
 2      ∀1 ≤ k ≤ n : color(x_k) := undef.
 3      P = {{x_1}, {x_2}, ..., {x_n}}
 4      node_candidate_set := {x_1, ..., x_n}
 5      while (node_candidate_set ≠ ∅) do
 6          /* f is strongly symmetric in P */
 7          Choose x_i ∈ node_candidate_set according to Brélaz/Morgenstern criterion
 8          color_candidate_set := {c | 1 ≤ c ≤ n, ∄x_j with {x_i, x_j} ∈ ~E and color(x_j) = c}
 9          while (color(x_i) = undef.) do
10              curr_color := min(color_candidate_set)
11              color(x_i) := curr_color
12              if (∃ colored node x_j with color(x_j) = color(x_i))
13                  then
14                      if (f symmetric in (x_i, x_j))
15                          then
16                              P := (P \ {[x_j], {x_i}}) ∪ {[x_j] ∪ {x_i}}
17                              /* f is symmetric in P */                      (*)
18                              Make f strongly symmetric in P.               (**)
19                          else
20                              color_candidate_set := color_candidate_set \ {curr_color}
21                              color(x_i) := undef.
22                      fi
23                  fi
24          od
25          node_candidate_set := node_candidate_set \ {x_i}
26      od
```

The crucial point of the algorithm is that the invariant 'f is strongly symmetric in P' of line 6 is always maintained.

Now let us take a look at the algorithm in more detail. At first glance, the set of all admissible colors for the next node x_i is the set of all colors between 1 and n except the colors of nodes which are adjacent to x_i in $\overline{G^f_{sym}}$. In the original Brélaz/Morgenstern algorithm the minimal color among these colors is chosen for x_i ($curr\_color$ in lines 10, 11). However, since we have to guarantee that f is symmetric in the partition P which results from coloring, it is possible that we are not allowed to assign x_i the color $curr\_color$. If there is already another node x_j which has been assigned $curr\_color$, then f has to be symmetric in the partition P' which results from union of $\{x_i\}$ and $[x_j]$ ($[x_j]$ denotes λ_q, if $x_j \in \lambda_q$ and $P = \{\lambda_1, \ldots, \lambda_k\}$). If there is such a node x_j, we have to test whether f is symmetric in (x_i, x_j) (line 14). Note that this test can have a negative result, since the don't care set of f is reduced during the algorithm. If f is not symmetric in (x_i, x_j), *curr_color* is removed from the set of color candidates for x_i (line 20) and the minimal color of the remaining set is chosen as the new color candidate (line 10). If the condition of line 14 is true, the new partition P results from the old partition P by union of $\{x_i\}$ and $[x_j]$ (line 16). Now f is symmetric in the new partition P (invariant (*) from line 17, see Lemma 2.8 below), and we can assign don't cares of f such that f becomes strongly symmetric in P (line 18).

The fact that the conditions given in the algorithm imply that f is now symmetric in the new partition P, will be shown in Lemma 2.8. In addition we have to point out how f can be made strongly symmetric in P (line 18). At the end we obtain an extension of the original incompletely specified Boolean function, which is strongly symmetric in the resulting partition P.

To prove invariant (*) in line 17, we need the following lemma:

LEMMA 2.8 *Let $f : D \to \{0, 1\}$ be strongly symmetric in P, $[x_i], [x_j] \in P$ two subsets with $\|[x_i]\| = 1$, and let f be symmetric in (x_i, x_j), then f is symmetric in $P' = (P \setminus \{[x_j], \{x_i\}\}) \bigcup \{[x_j] \cup \{x_i\}\}$.*

PROOF: Let $P = \{\lambda_1, \ldots, \lambda_k\}$ and w.l.o.g. $\lambda_1 = \{x_i\}$, $\lambda_2 = [x_j]$. Then we have $P' = \{\lambda_1 \cup \lambda_2, \lambda_3, \ldots, \lambda_k\}$.
Because of Lemma 2.6, we have to show that there is no weight class $C^{P'}_{w_2,\ldots,w_k}$ of P' with $\{0, 1\} \subseteq f(C^{P'}_{w_2,\ldots,w_k})$.

Case 1: $w_2 \geq 1$
    $C^{P'}_{w_2,\ldots,w_k}$ can be written as a disjoint union of two weight classes of P:

$$C^{P'}_{w_2,\ldots,w_k} = C^{P}_{0,w_2,\ldots,w_k} \cup C^{P}_{1,w_2-1,w_3,\ldots,w_k}.$$

Since f is strongly symmetric in P, $|f(C^{P}_{0,w_2,\ldots,w_k})| = |f(C^{P}_{1,w_2-1,w_3,\ldots,w_k})|$ $= 1$ holds according to Lemma 2.6. Suppose $\{0, 1\} \subseteq f(C^{P'}_{w_2,\ldots,w_k})$, then we have $f(C^{P}_{0,w_2,\ldots,w_k}) = c$ and $f(C^{P}_{1,w_2-1,\ldots,w_k}) = \bar{c}$ for $c \in \{0, 1\}$.
This leads to a contradiction to the condition that f is symmetric in x_i and x_j, since there are $\epsilon \in C^{P}_{0,w_2,\ldots,w_k}$ and $\delta \in C^{P}_{1,w_2-1,\ldots,w_k}$ such that ϵ results from δ only by exchange of the ith and jth component, but $f(\epsilon) = c$ and $f(\delta) = \bar{c}$.

Case 2: $w_2 = 0$
    $C^{P'}_{w_2,\ldots,w_k} = C^{P}_{0,w_2,\ldots,w_k}$ and $\{0, 1\} \not\subseteq f(C^{P'}_{w_2,\ldots,w_k})$ follows from the strong symmetry of f in P.

□

Figure 2.15. Procedure *make_strongly_symm*

Input: $f : D \to \{0,1\}$, represented by f_{on}, f_{off}, f_{dc}. f is (weakly) symmetric in (x_i, x_j).

Output: Extension f' of f (represented by f'_{on}, f'_{off}, f'_{dc}), which is strongly symmetric in (x_i, x_j) and has a maximum number of don't cares.

Algorithm:

1. $f'_{on} = \overline{x}_i\overline{x}_j f_{on\overline{x}_i\overline{x}_j} + x_i x_j f_{on x_i x_j} + (x_i\overline{x_j} + \overline{x}_i x_j)(f_{on x_i\overline{x_j}} + f_{on\overline{x}_i x_j})$

2. $f'_{off} = \overline{x}_i\overline{x}_j f_{off\overline{x}_i\overline{x}_j} + x_i x_j f_{off x_i x_j} + (x_i\overline{x_j} + \overline{x}_i x_j)(f_{off x_i\overline{x_j}} + f_{off\overline{x}_i x_j})$

3. $f'_{dc} = \overline{f'_{on} + f'_{off}}$

REMARK 2.5 *The statement of Lemma 2.8 is not correct if we substitute 'f strongly symmetric in P' with 'f (weakly) symmetric in P' or if we do not assume $\|[x_i]\| = 1$. But note that the given conditions coincide exactly with the conditions existing in the algorithm.*

To make the algorithm complete we must explain how f is made strongly symmetric in the partition P in line 18 of the algorithm. From the definition of symmetry of incompletely specified functions it is clear that it is possible to extend a function f, which is (weakly) symmetric in a partition P, to a function which is strongly symmetric in P. From the set of all extensions of f which are strongly symmetric in P, we choose the extension with a maximum number of don't cares. If f is (weakly) symmetric in a pair of variables (x_i, x_j), the extension f' of f, which is strongly symmetric in (x_i, x_j) and which has a maximum don't care set among all extensions of f with that property, can easily be computed from the ROBDD representations of f_{on}, f_{dc} and f_{off} by the procedure *make_strongly_symm* in Figure 2.15.

We can use a sequence of calls of the procedure *make_strongly_symm* to make f strongly symmetric in the partition P in line 18 of the algorithm. For this purpose we prove the following theorem:

THEOREM 2.4 *Let $f : D \to \{0,1\}$ be strongly symmetric in P, $\{x_i\}, [x_{j_1}] \in P$, $[x_{j_1}] = \{x_{j_1}, \ldots, x_{j_k}\}$, $f =: f^{(0)}$ symmetric in (x_i, x_{j_1}).*

$$f^{(1)} = make\_strongly\_symm(f^{(0)}, x_i, x_{j_1})$$

$$f^{(2)} \quad = \quad make\_strongly\_symm(f^{(1)}, x_i, x_{j_2})$$

$$\vdots$$

$$f^{(k)} \quad = \quad make\_strongly\_symm(f^{(k-1)}, x_i, x_{j_k}).$$

Then $f^{(k)}$ is strongly symmetric in

$$P' = (P \setminus \{[x_{j_1}], \{x_i\}\}) \bigcup \{[x_{j_1}] \cup \{x_i\}\}.$$

PROOF: See Appendix E.   □

There are examples where we need the complete sequence of calls given in the theorem. However, in many cases there is a $p < k$ such that $f^{(p)}$ does not differ from $f^{(p-1)}$. We can prove that the sequence of calls can be stopped in such cases with the result $f^{(k)} = f^{(p-1)}$ (Scholl, 1996).

2.2.2.2   Equivalence symmetry of incompletely specified functions

The approach given in the previous section, which deals with (nonequivalence) symmetry, can easily be extended to equivalence symmetry of incompletely specified Boolean functions. According to Section 2.1.4.4 the search for equivalence and nonequivalence symmetries of a function f can be reduced to the search for a polarity vector POL and a partition P, such that f_{POL} is (nonequivalence) symmetric in P.

Due to this fact the algorithm in Figure 2.14 can easily be changed, now considering also equivalence symmetry. Now the algorithm is not working based on the symmetry graph G^f_{sym} (or the inverse graph $\overline{G^f_{sym}}$), but on G^f_{esym} (cf. Section 2.1.4.4).

Figure 2.16 shows the resulting algorithm to solve the problem extended to equivalence symmetries. In addition to the current partition P we use a polarity vector $POL = (pol_1, \ldots, pol_n)$. The polarity vector is initialized by $(1, \ldots, 1)$ (line 4). Whenever a node x_i is colored, the polarity of x_i is set to pol_j, if there is a node x_j having the same color and if f is nonequivalence symmetric in (x_i, x_j) (line 18). On the other hand the polarity of x_i is set to $\overline{pol_j}$, if there is a node x_j having the same color and if f is equivalence symmetric in (x_i, x_j) (line 25). When the algorithm has finished its computation, we obtain a partition P, a polarity vector POL and an extension f of the original incompletely specified Boolean function with the property that f_{POL} is strongly symmetric in partition P.

Figure 2.16. Algorithm to handle both nonequivalence and equivalence symmetries.

Input: Incompletely specified function $f : D \rightarrow \{0,1\}$, $D \subseteq \{0,1\}^n$, represented by f_{on} and f_{dc}

Output: Partition P of $\{x_1,\ldots,x_n\}$ and polarity vector $POL \in \{0,1\}^n$, such that f_{POL} is (nonequivalence) symmetric in P

Algorithm:

```
1       Compute symmetry graph G^f_{esym} = (V, E) of f (or \overline{G^f_{esym}} = (V, \overline{E})).
2       ∀1 ≤ k ≤ n : color(x_k) := undef.
3       P = {{x_1}, {x_2},..., {x_n}}
4       POL = (pol_1,..., pol_n) = (1,...,1)
5       node_candidate_set := {x_1,...,x_n}
6       while (node_candidate_set ≠ ∅) do
7           /* f_POL is strongly symmetric in P */
8           Choose x_i ∈ node_candidate_set according to Brélaz/Morgenstern criterion
9           color_candidate_set := {c | 1 ≤ c ≤ n, ∄x_j with {x_i, x_j} ∈ \overline{E} and color(x_j) = c}
10          while (color(x_i) = undef.) do
11              curr_color := min(color_candidate_set)
12              color(x_i) := curr_color
13              if (∃ colored node x_j with color(x_j) = color(x_i))
14                  then
15                      if (f (nonequivalence) symmetric in (x_i, x_j))
16                          then
17                              P := (P \ {[x_j], {x_i}}) ⋃{[x_j] ∪ {x_i}}
18                              pol_i = pol_j
19                              /* f_POL is (nonequivalence) symmetric in P */
20                              Make f_POL strongly symmetric in P.
21                          else
22                              if (f equivalence symmetric in (x_i, x_j))
23                                  then
24                                      P := (P \ {[x_j], {x_i}}) ⋃{[x_j] ∪ {x_i}}
25                                      pol_i = \overline{pol_j}
26                                      /* f_POL is (nonequivalence) symmetric in P */
27                                      Make f_POL strongly symmetric in P.
28                                  else
29                                      color_candidate_set :=
30                                                  color_candidate_set \ {curr_color}
31                                      color(x_i) := undef.
32                              fi
33                      fi
34                  fi
35          od
36          node_candidate_set := node_candidate_set \ {x_i}
37      od
```

2.2.3 Combining communication minimization and symmetry

This section answers the question as to how the techniques from Sections 2.2.1 and 2.2.2 can be combined.

In the previous section we presented an algorithm to compute a minimum sized partition P of the input variables in which an incompletely specified function f is symmetric. In addition we assigned values to don't cares to make f *strongly* symmetric in P. Usually the result will still contain don't cares after this assignment.

Now we try to make use of these remaining don't cares by applying the technique of Chang (Chang et al., 1994) and Shiple (Shiple et al., 1994) to further reduce ROBDD sizes by 'communication minimization'. Since this method removes don't cares, we have to ask the question whether the method can destroy symmetries which were found earlier. An application of the two methods in this order makes sense only if the second method does not destroy the results of the first one.

The answer to this question, which is given in this section, is that we can preserve these symmetries using a slightly modified version of Chang's technique.

The algorithm proposed by Chang (Chang et al., 1994) reduces the number of nodes at every level of the ROBDD by minimizing the number of 'linking nodes' assigning as few don't cares as possible to either the on-set $ON(f)$ or the off-set $OFF(f)$. After the minimization of nodes at a certain level of the ROBDD they use the remaining don't cares to minimize the number of nodes at the next level. The cut line is moved from top to bottom in the ROBDD. We can prove that under certain conditions, this method does preserve strong symmetry:

LEMMA 2.9 *Let f be an incompletely specified Boolean function which is strongly symmetric in $P = \{\lambda_1, \ldots, \lambda_k\}$ and assume that the variable order of the ROBDD representing f is a symmetric order with the variables in λ_i before the variables in λ_{i+1} ($1 \leq i < k$). If we restrict the minimization of 'linking nodes' to cut lines between two symmetric groups λ_i and λ_{i+1}, then it preserves strong symmetry in P.*

PROOF: See Appendix F.   □

Since we will use such 'symmetry orders' to minimize ROBDD sizes (see Section 2.2.4), we merely have to restrict the minimization of linking nodes to cut lines between symmetric groups to guarantee that we will not lose any symmetries.

The following example shows that symmetries are not preserved if the conditions of Lemma 2.9 are not met:

EXAMPLE 2.8 Consider the incompletely specified function f with four input variables, which is defined by the decomposition matrix or the ROBDD for $ext(f)$ on the left hand side of Figure 2.17. It is easy to see that f is strongly symmetric in $\{x_1, x_2, x_3\}$. If the number of linking nodes below a cut line after

Figure 2.17. Symmetry in x_1 and x_3 is destroyed by minimizing the number of linking nodes with cut line after x_2.

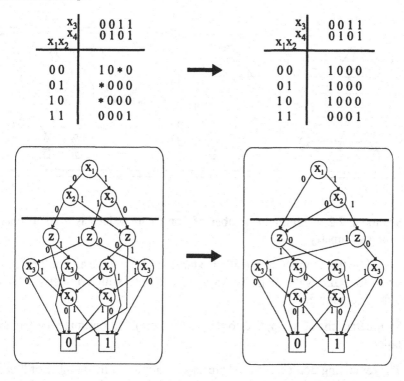

x_3	0 0 1 1
$x_1 x_2$ x_4	0 1 0 1
0 0	1 0 * 0
0 1	* 0 0 0
1 0	* 0 0 0
1 1	0 0 0 1

x_3	0 0 1 1
$x_1 x_2$ x_4	0 1 0 1
0 0	1 0 0 0
0 1	1 0 0 0
1 0	1 0 0 0
1 1	0 0 0 1

variable x_2 is minimized, the decomposition matrix or the ROBDD on the right hand side of Figure 2.17 results. The resulting function f' is *not* symmetric in x_1 and x_3 (since $f'(1000) \neq f'(0010)$).

2.2.4   Combining don't care minimization and variable reordering

Since locating symmetric variables side by side has proved to be a successful heuristic method to find good variable orders for completely specified functions, we use this method also during the search for extensions of incompletely specified functions which can be represented by small ROBDDs:

1. In a first step for an incompletely specified function f, a small partition $P = \{\lambda_1, \ldots, \lambda_k\}$ is computed, such that f is symmetric in P. An extension f' of f is computed, which is *strongly symmetric* in P. As described in

Figure 2.18. On the left hand side the method presented by Chang is illustrated (cut lines between all levels). On the right hand side our method is illustrated.

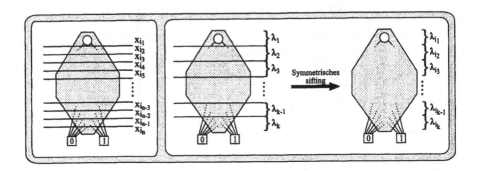

Section 2.2.2 a minimum number of don't care values is used to obtain strong symmetry.

2. Let f'' be the completely specified extension of f' where all don't cares are set to the same value, e.g. 0. An ROBDD for f'' with symmetry variable order is computed.

3. Symmetric sifting is applied to the ROBDD of f'' to optimize the variable order.

4. The remaining don't cares of f' are used to minimize the number of linking nodes. According to Lemma 2.9 we restrict the cut lines to cuts between groups of symmetric variables. Then the minimization of linking nodes does not destroy symmetries.

5. Since symmetric groups were not destroyed, we can apply symmetric sifting again to optimize the variable order after minimization of linking nodes.

Figure 2.18 illustrates our modification of Chang's technique to preserve symmetries.

Altogether we compute an extension of an incompletely specified Boolean function together with a variable order, which results in a small ROBDD representation.

2.2.5   Experimental results

We have carried out experiments to evaluate our method, which computes small ROBDD representations for extensions of incompletely specified Boolean functions.

Table 2.1. Experimental results. The table shows the number of nodes in the ROBDDs of each function. Numbers in parenthesis show the CPU times (measured on a SPARCstation 20 (96 MByte RAM)).

circuit	in	out	restrict nodes	restrict_s nodes	sym_s nodes	sym_group nodes	(time)	sym_cover nodes	(time)
5xp1	7	10	63	63	67	66	(0.2 s)	53	(0.5 s)
9symml	9	1	67	65	108	25	(0.3 s)	25	(0.4 s)
alu2	10	6	192	182	201	201	(0.7 s)	152	(2.6 s)
apex6	135	99	993	940	1033	983	(267.6 s)	612	(459.7 s)
apex7	49	37	730	716	814	728	(27.7 s)	340	(52.2 s)
b9	41	21	213	211	256	185	(8.6 s)	122	(11.5 s)
c8	28	18	110	98	156	95	(1.7 s)	70	(3.2 s)
example2	85	66	497	496	491	484	(69.2 s)	416	(119.4 s)
mux	21	1	32	32	34	29	(0.6 s)	29	(0.7 s)
pcler8	27	17	111	111	78	73	(1.9 s)	72	(3.3 s)
rd73	7	3	75	74	76	34	(0.3 s)	27	(0.4 s)
rd84	8	4	135	132	144	42	(0.7 s)	42	(0.7 s)
sao2	10	4	89	89	104	104	(0.4 s)	70	(0.8 s)
x4	94	71	814	812	829	633	(121.9 s)	485	(203.4 s)
z4ml	7	4	47	46	51	32	(0.2 s)	17	(0.3 s)
partmult3	9	6	70	65	152	35	(1.0 s)	29	(1.2 s)
partmult4	16	8	307	294	971	222	(49.5 s)	114	(50.6 s)
partmult5	25	10	857	843	4574	998	(1540.4 s)	365	(1548.4 s)
total			5402	5269	10139	4969		3040	

To generate incompletely specified functions from completely specified functions, we used a method proposed in (Chang et al., 1994): After collapsing each benchmark circuit to two level form, we randomly selected cubes in the on-set with a probability of 40% to be included in the don't care set[6]. The last three Boolean functions in Table 2.1 are partial multipliers $partmult_n$[7].

We performed three experiments: First of all, we applied symmetric sifting to the ROBDDs representing the on-set of each function. The results are shown in column 6 (*sym_s*) of Table 2.1. The entries are ROBDD sizes in terms of node counts.

---

[6]Because of this method of generating incompletely specified functions we had to confine ourselves to benchmark circuits which could be collapsed to two level form.

[7]The n^2 inputs are the bits of the n partial products and the $2n$ outputs are the product bits. The don't care set contains all input vectors which cannot occur for the reason that the input bits are not independent of each other, because they are conjunctions $a_i b_j$ of bits of the operands $(a_1,\ldots,a_n)$ and $(b_1,\ldots,b_n)$ of the multiplication.

Table 2.2. Experimental results. The table shows the number of nodes in the ROBDDs of each function with 10% don't cares.

circuit	in	out	sym_s	sym_group	sym_cover
5xp1	7	10	75	73	68
9symml	9	1	75	25	25
alu2	10	6	199	199	166
apex6	135	99	961	911	585
apex7	49	37	807	753	428
b9	41	21	203	195	141
c8	28	18	180	161	83
example2	85	66	547	540	464
mux	21	1	40	35	33
pcler8	27	17	83	83	81
rd73	7	3	65	35	31
rd84	8	4	126	42	42
sao2	10	4	106	106	79
x4	94	71	677	670	499
z4ml	7	4	50	30	17
total			4194	3858	2742

In a second experiment, we applied our algorithm to minimize the number of symmetric groups followed by symmetric sifting. Column *sym_group* of Table 2.1 shows the results. *sym_group* provides a partition $P = \{\lambda_1, \ldots, \lambda_k\}$ and an extension f' of the original function f, such that f' is strongly symmetric in P. On average, we were able to improve the ROBDD size by 51%.

In the last experiment we started with the results of *sym_group* and then went on with our modified version of the technique of Chang (Chang et al., 1994) and Shiple (Shiple et al., 1994) with minimization of linking nodes restricted to cut lines between symmetry groups, such that strong symmetries supplied by *sym_group* are not destroyed (see Lemma 2.9). Since symmetry groups are not destroyed, we can perform symmetric sifting after the node minimization with the same symmetric groups as before. Column *sym_cover* of Table 2.1 shows the resulting ROBDD sizes. On average, the new technique led to an improvement in the ROBDD sizes by 70%.

A comparison to the results of the *restrict* operator (Coudert et al., 1989a) (applied to ROBDDs whose variable order was optimized by regular *sifting*) in column *restrict* of Table 2.1 shows that our ROBDD sizes are on average 44% smaller. Even if sifting is called again after the *restrict* operator has been applied, the improvement is still more than 40% on average (see column *restrict_s*).

Finally, we carried out the same experiment once more, but this time the probability of a cube being included in the don't care set was reduced to 10% (instead of 40%)[8]. The numbers for *sym_s*, *sym_group* and *sym_cover* are given in Table 2.2 in columns 4, 5 and 6, respectively. It can easily be seen that the reduction ratio decreases when only a smaller number of don't cares is available, but with only 10% don't cares still more than 30% of the nodes can be saved on average.

---

[8]Note that the sizes of the don't care sets for the partial multipliers $partmult_n$ are fixed, since these don't care sets arise in a 'natural way' as described above.

Chapter 3

FUNCTIONAL DECOMPOSITION FOR COMPLETELY SPECIFIED SINGLE-OUTPUT FUNCTIONS

Functional decomposition as a technique to find realizations for Boolean functions was already introduced in the late fifties and early sixties by Ashenhurst (Ashenhurst, 1959), Curtis (Curtis, 1961), Roth and Karp (Roth and Karp, 1962; Karp, 1963). In recent years functional decomposition has attracted a lot of renewed interest due to several reasons:

- The increased capacities of today's computers as well as the development of new methods have made the method applicable to larger–scale problems.

- Functional decomposition is especially well suited for the synthesis of lookup-table based FPGAs architectures. During the last few years FPGAs (see Section 1.4) have become increasingly important.

- ROBDDs provide a data structure which is a compact representation for most Boolean functions occurring in practical applications. Algorithms for functional decomposition have been developed, which work directly based on ROBDDs, so that the decomposition algorithm works based on compact representations and not on function tables as in previous approaches.

In this chapter we present the basic method for decomposition of completely specified single-output functions. The use of ROBDDs makes the decomposition efficient. Decompositions based on ROBDD representations were first used in (Lai et al., 1993a; Lai et al., 1993b; Lai et al., 1994b; Sasao, 1993; Scholl and Molitor, 1994; Scholl and Molitor, 1995b; Scholl and Molitor, 1995a).

The following chapters give extensions of the basic method, which make the method applicable to multi-output functions (Chapter 4) and incompletely specified functions (Chapter 5).

Figure 3.1.   One-sided decomposition.

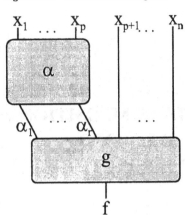

3.1.   Definitions

This section gives basic definitions for the decomposition of single-output functions. First, two different types of (disjoint) decompositions are defined: one-sided and two-sided decomposition.

DEFINITION 3.1 (ONE-SIDED DECOMPOSITION)
A one-sided (disjoint) decomposition of a Boolean function $f \in B_n$ with input variables $x_1, \ldots, x_p, x_{p+1}, \ldots, x_n$ $(0 < p < n)$ with respect to the variable set $\{x_1, \ldots, x_p\}$ is a representation of f of the form

$$f(x_1, \ldots, x_p, x_{p+1}, \ldots, x_n) =$$
$$g(\alpha_1(x_1, \ldots, x_p), \ldots, \alpha_r(x_1, \ldots, x_p), x_{p+1}, \ldots, x_n).$$

A one-sided decomposition is illustrated by Figure 3.1. Functions $\alpha_1, \ldots, \alpha_r$ compute an 'intermediate result' based on input variables $x_1, \ldots, x_p$ and function g computes the final function value of f based on this intermediate result and the remaining input variables $x_{p+1}, \ldots, x_n$.

Similarly, two-sided decompositions may also be defined:

DEFINITION 3.2 (TWO-SIDED DECOMPOSITION)
A two-sided (disjoint) decomposition of a Boolean function $f \in B_n$ with input variables $x_1, \ldots, x_p, x_{p+1}, \ldots, x_n$ $(0 < p < n)$ with respect to the variable partition $\{\{x_1, \ldots, x_p\}, \{x_{p+1}, \ldots, x_n\}\}$ is a representation of f of the form

$$f(x_1, \ldots, x_p, x_{p+1}, \ldots, x_n) =$$
$$g(\alpha_1(x_1, \ldots, x_p), \ldots, \alpha_r(x_1, \ldots, x_p),$$
$$\beta_1(x_{p+1}, \ldots, x_n), \ldots, \beta_s(x_{p+1}, \ldots, x_n)).$$

Figure 3.2. Two-sided decomposition.

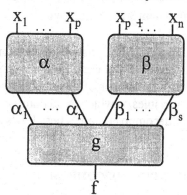

A two-sided decomposition is illustrated by Figure 3.2. Here functions $\alpha_1, \ldots,$ α_r compute an intermediate result based on input variables $x_1, \ldots, x_p$ and functions $\beta_1, \ldots, \beta_s$ an intermediate result based on input variables $x_{p+1}, \ldots, x_n$. Function g computes the final function value of f based on these two intermediate results.

NOTATION 3.1 *Functions α_i ($1 \leq i \leq r$) and β_j ($1 \leq j \leq s$) are called* decomposition functions, *function g is called* composition function. *In the case of one-sided decompositions the variables in set $\{x_1, \ldots, x_p\}$ of the definition above are denoted as 'bound variables', $\{x_1, \ldots, x_p\}$ as 'bound set', variables in $\{x_{p+1}, \ldots, x_n\}$ as 'free variables', $\{x_{p+1}, \ldots, x_n\}$ as 'free set'.*

If decompositions are used for logic synthesis, it is expedient to look for decompositions where the number of decomposition functions is as small as possible. This has the following advantages:

- If the number of decomposition functions is smaller, then we have to realize fewer (decomposition) functions, when the method is applied recursively. Thus we hope to minimize the total cost of the overall circuit.

- If the number of decomposition functions is smaller, the number of inputs of the composition function g is likewise smaller. The complexity of g is hopefully smaller, if g has a smaller number of inputs.

- Based on the netlist (or Ω–circuit), which is produced by logic synthesis, a layout of the circuit realizing the given Boolean function is computed. In many cases we can observe circuits in which the layout area is not affected primarily by the number of logic gates or cells, but rather the number of global wires which have to be implemented to connect logic gates. A decomposition with a small number of decomposition functions suggests a realization with a small number of global wires.

Of interest, then, are those decompositions where the number of decomposition functions is smaller than the number of input variables (for the case of two-sided decompositions) or where the number of decomposition functions is smaller than the number of input variables on which the decomposition functions depend (for the case of one-sided decompositions). These kinds of decompositions are called *nontrivial*.

DEFINITION 3.3 (NONTRIVIAL DECOMPOSITION)

■ *A one-sided decomposition*

$$f(x_1, \ldots, x_p, x_{p+1}, \ldots, x_n) =$$
$$g(\alpha_1(x_1, \ldots, x_p), \ldots, \alpha_r(x_1, \ldots, x_p), x_{p+1}, \ldots, x_n)$$

is nontrivial, *if* $r < p$.

■ *A two-sided decomposition*

$$f(x_1, \ldots, x_p, x_{p+1}, \ldots, x_n) =$$
$$g(\alpha_1(x_1, \ldots, x_p), \ldots, \alpha_r(x_1, \ldots, x_p),$$
$$\beta_1(x_{p+1}, \ldots, x_n), \ldots, \beta_s(x_{p+1}, \ldots, x_n))$$

is nontrivial, *if* $r + s < n$.

If only nontrivial decompositions of f are used and if the decomposition method is applied recursively to find realizations for the decomposition functions and the composition function, then all functions which are processed in the next recursive step have a smaller number of inputs than f.

Decompositions which use a *minimum* number of decomposition functions are called 'communication minimal':

DEFINITION 3.4 (COMMUNICATION MINIMAL DECOMPOSITION)

■ *A one-sided decomposition*

$$f(x_1, \ldots, x_p, x_{p+1}, \ldots, x_n) =$$
$$g(\alpha_1(x_1, \ldots, x_p), \ldots, \alpha_r(x_1, \ldots, x_p), x_{p+1}, \ldots, x_n)$$

is communication minimal *with respect to* $\{x_1, \ldots, x_p\}$, *if for all one-sided decompositions*

$$f(x_1, \ldots, x_p, x_{p+1}, \ldots, x_n) =$$
$$g'(\alpha_1'(x_1, \ldots, x_p), \ldots, \alpha_{r'}'(x_1, \ldots, x_p), x_{p+1}, \ldots, x_n)$$

$r \leq r'$ *holds.*

- *A two-sided decomposition*

$$f(x_1, \ldots, x_p, x_{p+1}, \ldots, x_n) =$$
$$g(\alpha_1(x_1, \ldots, x_p), \ldots, \alpha_r(x_1, \ldots, x_p),$$
$$\beta_1(x_{p+1}, \ldots, x_n), \ldots, \beta_s(x_{p+1}, \ldots, x_n))$$

is communication minimal *with respect to* $\{\{x_1, \ldots, x_p\}, \{x_{p+1}, \ldots, x_n\}\}$, *if for all two-sided decompositions*

$$f(x_1, \ldots, x_p, x_{p+1}, \ldots, x_n) =$$
$$g'(\alpha_1'(x_1, \ldots, x_p), \ldots, \alpha_{r'}'(x_1, \ldots, x_p),$$
$$\beta_1'(x_{p+1}, \ldots, x_n), \ldots, \beta_{s'}'(x_{p+1}, \ldots, x_n))$$

$r \leq r'$ *and* $s \leq s'$ *holds.*

In the following we will deal only with nontrivial *and* communication minimal decompositions.

3.2. Decomposition and Field Programmable Gate Arrays

Functional decomposition can be applied recursively to synthesize realizations of Boolean functions by circuits over a fixed library like B_2, STD etc. (see Section 1.1). Recursion stops when the problem is small enough to allow a trivial mapping to cell library elements.

However, functional decomposition is most widely used for the synthesis of FPGAs, especially for look-up table based FPGAs. Although functional decomposition using communication minimal decompositions is not restricted to FPGA synthesis, it is especially suited for the synthesis of look-up table based FPGAs. As already described in Section 1.4, look-up table based FPGAs such as the *Xilinx XC3000* device or the *Altera APEX 20K* device, are able to realize arbitrary functions $lut : \{0,1\}^b \rightarrow \{0,1\}$ up to a certain number b of inputs using look-up tables with b inputs[1].

---

[1]For the *Xilinx XC3000* device $b = 5$, for *Altera APEX 20K* device $b = 4$.

If in Figure 3.1, e.g., the number p of inputs of the decomposition functions $\alpha_1, \ldots, \alpha_r$ is not larger than this limit b, then $\alpha_1, \ldots, \alpha_r$ can be realized simply by r look-up tables. If p is larger than b, then functional decomposition has to be applied recursively to realize $\alpha_1, \ldots, \alpha_r$. In every case it seems to be reasonable to use communication minimal decompositions, such that the number of outputs of block α and the number of inputs of block g in Figure 3.1 is minimized.

3.3.   Minimizing the number of decomposition functions

For the reasons mentioned above we are interested in considering decompositions where the number of decomposition functions is as small as possible.

First, we are looking for a possibility to find a communication minimal decomposition provided that the variable set $X^{(1)}$ is already given.

Here the decomposition matrix as defined in Definition 2.2 on page 25 is a useful means.

The minimum number of decomposition functions to decompose a single-output function f with respect to a variable set $X^{(1)}$ (or with respect to a partition $\{X^{(1)}, X^{(2)}\}$ of the input variables) can be determined based on the decomposition matrix $Z(X^{(1)}, f)$ of f. For one-sided decompositions the number of different row patterns $nrp(X^{(1)}, f)$ is of interest, for two-sided decompositions also the number of different column patterns, which is denoted by $ncp(X^{(1)}, f)$:

NOTATION 3.2 *Let $f \in B_n$ with input variables $x_1, \ldots, x_p, x_{p+1}, \ldots, x_n$. Let $Z(X^{(1)}, f)$ be the decomposition matrix of f with respect to variable set $X^{(1)} := \{x_1, \ldots, x_p\}$. Then the number of different column patterns of $Z(X^{(1)}, f)$ is denoted by $ncp(X^{(1)}, f)$. Similarly to the case of equal row patterns, equality of column patterns of $Z(X^{(1)}, f)$ provides an equivalence relation on $\{0, 1\}^{n-|X^{(1)}|}$ and the number of equivalence classes is $ncp(X^{(1)}, f)$.*

The minimum number of decomposition functions needed to decompose f with respect to $X^{(1)}$ (or with respect to $\{X^{(1)}, X^{(2)}\}$) results easily from $nrp(X^{(1)}, f)$ (and $ncp(X^{(1)}, f)$). This is demonstrated by the following example:

EXAMPLE 3.1 Function vgl_4 is defined as

$$vgl_4(x_1, \ldots, x_4, y_1, \ldots, y_4) = \begin{cases} 1, & \text{if } (x_1, x_2, x_3, x_4) = (y_1, y_2, y_3, y_4) \\ 0, & \text{if } (x_1, x_2, x_3, x_4) \neq (y_1, y_2, y_3, y_4). \end{cases}$$

Thus vgl_4 compares the binary number formed by (x_1, x_2, x_3, x_4) and the binary number formed by (y_1, y_2, y_3, y_4) and produces 1, when the two binary numbers are equal.

A possible two-sided decomposition with respect to $\{\{x_1, x_2, y_1, y_2\}, \{x_3, x_4, y_3, y_4\}\}$ is

$$vgl_4(x_1, x_2, x_3, x_4, y_1, y_2, y_3, y_4) =$$
$$and_2(vgl_2(x_1, x_2, y_1, y_2), vgl_2(x_3, x_4, y_3, y_4))$$

$$\text{with } vgl_2(x_1, x_2, y_1, y_2) = \left\{ \begin{array}{ll} 1, & \text{if } (x_1, x_2) = (y_1, y_2) \\ 0, & \text{if } (x_1, x_2) \neq (y_1, y_2) \end{array} \right.$$

Based on variables $\{x_1, x_2, y_1, y_2\}$, this realization computes whether the first two bits of the binary numbers, which have to be compared, are equal or not (two items of information, encoded by one bit of the decomposition function) and accordingly based on variables $\{x_3, x_4, y_3, y_4\}$, it computes whether the last two bits of these binary number are equal or not. The composition function (and_2) 'composes' the two intermediate results to the final function value of vgl_4.

The number of items of information which have to be computed by the decomposition functions based on the variables of these two sets can be derived from the decomposition matrix in Figure 3.3.

There are two different row patterns in the decomposition matrix. The 0–rows indicate that $(x_1, x_2) \neq (y_1, y_2)$, and the other rows indicate that $(x_1, x_2) = (y_1, y_2)$. (In the same way 0–columns indicate that $(x_3, x_4) \neq (y_3, y_4)$ and the other columns that $(x_3, x_4) = (y_3, y_4)$.)

In the general case we have $nrp(X^{(1)}, f)$ different items of information given by the different row patterns, which can be encoded by $\lceil \log(nrp(X^{(1)}, f)) \rceil$ different bits or decomposition functions and in the case of two-sided decompositions we have also $ncp(X^{(1)}, f)$ different items of information given by the different column patterns, which can be encoded by $\lceil \log(ncp(X^{(1)}, f)) \rceil$ different decomposition functions. These observations lead to the following Theorems 3.1 and 3.2, which give necessary and sufficient conditions for the existence of decompositions with a given number of decomposition functions.

THEOREM 3.1 (CURTIS '61) *Let $f \in B_n$ be a function with input variables $x_1, \ldots, x_p, x_{p+1}, \ldots, x_n$. Then there is a one-sided decomposition of f with respect to $X^{(1)} = \{x_1, \ldots, x_p\}$ of form*

$$f(x_1, \ldots, x_p, x_{p+1}, \ldots, x_n) =$$
$$g(\alpha_1(x_1, \ldots, x_p), \ldots, \alpha_r(x_1, \ldots, x_p), x_{p+1}, \ldots, x_n),$$

Figure 3.3.   Decomposition matrix for vgl_4

$x_1x_2y_1y_2$ \	x_3 0	0	0	0	0	0	0	1	1	1	1	1	1	1	1
x_4 0	0	0	0	1	1	1	1	0	0	0	0	1	1	1	1
y_3 0	0	1	1	0	0	1	1	0	0	1	1	0	0	1	1
y_4 0	1	0	1	0	1	0	1	0	1	0	1	0	1	0	1

$x_1x_2y_1y_2$																
0 0 0 0	1	0	0	0	0	1	0	0	0	0	1	0	0	0	0	1
0 0 0 1	0	0	0	0	0	0	0	0	0	0	0	0	0	0	0	0
0 0 1 0	0	0	0	0	0	0	0	0	0	0	0	0	0	0	0	0
0 0 1 1	0	0	0	0	0	0	0	0	0	0	0	0	0	0	0	0
0 1 0 0	0	0	0	0	0	0	0	0	0	0	0	0	0	0	0	0
0 1 0 1	1	0	0	0	0	1	0	0	0	0	1	0	0	0	0	1
0 1 1 0	0	0	0	0	0	0	0	0	0	0	0	0	0	0	0	0
0 1 1 1	0	0	0	0	0	0	0	0	0	0	0	0	0	0	0	0
1 0 0 0	0	0	0	0	0	0	0	0	0	0	0	0	0	0	0	0
1 0 0 1	0	0	0	0	0	0	0	0	0	0	0	0	0	0	0	0
1 0 1 0	1	0	0	0	0	1	0	0	0	0	1	0	0	0	0	1
1 0 1 1	0	0	0	0	0	0	0	0	0	0	0	0	0	0	0	0
1 1 0 0	0	0	0	0	0	0	0	0	0	0	0	0	0	0	0	0
1 1 0 1	0	0	0	0	0	0	0	0	0	0	0	0	0	0	0	0
1 1 1 0	0	0	0	0	0	0	0	0	0	0	0	0	0	0	0	0
1 1 1 1	1	0	0	0	0	1	0	0	0	0	1	0	0	0	0	1

if and only if

$$r \geq \log(nrp(X^{(1)}, f)).$$

PROOF:

"‘$\Longrightarrow$'": Suppose that f can be decomposed by

$$f(x_1, \ldots, x_p, x_{p+1}, \ldots, x_n) =$$
$$g(\alpha_1(x_1, \ldots, x_p), \ldots, \alpha_r(x_1, \ldots, x_p), x_{p+1}, \ldots, x_n).$$

Then function

$$\alpha = (\alpha_1, \ldots, \alpha_r) \in B_{p,r}$$

can have at most 2^r different function values.

Now suppose that

$$r < \log(nrp(X^{(1)}, f), \text{ i.e., } 2^r < nrp(X^{(1)}, f).$$

Then there are $\delta^{(1)} = (\delta_1^{(1)}, \ldots, \delta_p^{(1)})$ and $\delta^{(2)} = (\delta_1^{(2)}, \ldots, \delta_p^{(2)})$, such that

$$\alpha(\delta^{(1)}) = \alpha(\delta^{(2)}),$$

but the corresponding row patterns in the decomposition matrix for $\delta^{(1)}$ and $\delta^{(2)}$ are different.

When the corresponding row patterns are different, there is $\epsilon^{(0)} = (\epsilon_{p+1}^{(0)}, \ldots, \epsilon_n^{(0)})$ with

$$f(\delta_1^{(1)}, \ldots, \delta_p^{(1)}, \epsilon_{p+1}^{(0)}, \ldots, \epsilon_n^{(0)}) \neq f(\delta_1^{(2)}, \ldots, \delta_p^{(2)}, \epsilon_{p+1}^{(0)}, \ldots, \epsilon_n^{(0)}).$$

Since g has to be a well–defined Boolean function and since $\alpha(\delta^{(1)}) = \alpha(\delta^{(2)})$, it holds

$$g(\alpha_1(\delta_1^{(1)}, \ldots, \delta_p^{(1)}), \ldots, \alpha_r(\delta_1^{(1)}, \ldots, \delta_p^{(1)}), \epsilon_{p+1}^{(0)}, \ldots, \epsilon_n^{(0)}) =$$
$$g(\alpha_1(\delta_1^{(2)}, \ldots, \delta_p^{(2)}), \ldots, \alpha_r(\delta_1^{(2)}, \ldots, \delta_p^{(2)}), \epsilon_{p+1}^{(0)}, \ldots, \epsilon_n^{(0)}).$$

This contradicts the assumption that g and α constitute a decomposition of f. Thus $r \geq \log(nrp(X, f)$ must hold.

"$\Longleftarrow$": Let

$$r \geq \log(nrp(X, f)), \text{ i.e., } 2^r \geq nrp(X, f).$$

Then it is possible to define $\alpha \in B_{p,r}$, such that

$$\alpha(\delta^{(1)}) \neq \alpha(\delta^{(2)})$$

for all $\delta^{(1)}, \delta^{(2)} \in \{0,1\}^p$ with different row patterns in the decomposition matrix.

The composition function g can be defined according to the following rule: For all $(\epsilon_1, \ldots, \epsilon_n) \in \{0,1\}^n$

$$g(\alpha_1(\epsilon_1, \ldots, \epsilon_p), \ldots, \alpha_r(\epsilon_1, \ldots, \epsilon_p), \epsilon_{p+1}, \ldots, \epsilon_n) =$$
$$f(\epsilon_1, \ldots, \epsilon_p, \epsilon_{p+1}, \ldots, \epsilon_n). \qquad (\star)$$

If $(a_1, \ldots, a_r)$ does not occur in the image of α, then $g(a_1, \ldots, a_r, x_{p+1}, \ldots, x_n)$ is not defined or can be chosen arbitrarily.

It is easy to see that g is well–defined according to rule $(\star)$:

If g were not well–defined, then there would be $\delta^{(1)}$ and $\delta^{(2)}$ in $\{0,1\}^p$ with $\alpha(\delta^{(1)}) = \alpha(\delta^{(2)})$, and $\epsilon^{(0)} \in \{0,1\}^{n-p}$ with $f(\delta^{(1)}, \epsilon^{(0)}) \neq f(\delta^{(2)}, \epsilon^{(0)})$. But this would mean that the row patterns belonging to $\delta^{(1)}$ and $\delta^{(2)}$ were different and thus $\alpha(\delta^{(1)}) = \alpha(\delta^{(2)})$ contradicts the definition of α.

$\square$

The following corollary follows from the proof of Theorem 3.1:

COROLLARY 3.1 *Let $f \in B_n$ be a function with input variables $x_1, \ldots, x_p$, $x_{p+1}, \ldots, x_n$, for $1 \le i \le r$ let $\alpha_i \in B_p$ and let $Z(X^{(1)}, f)$ be the decomposition matrix of f with respect to $X^{(1)} = \{x_1, \ldots, x_p\}$. Then there is a one-sided decomposition of f with respect to $X^{(1)}$ of form*

$$f(x_1, \ldots, x_p, x_{p+1}, \ldots, x_n) =$$
$$g(\alpha_1(x_1, \ldots, x_p), \ldots, \alpha_r(x_1, \ldots, x_p), x_{p+1}, \ldots, x_n),$$

if function $\alpha = (\alpha_1, \ldots, \alpha_r)$ has the following property:

If there is an $\epsilon^{(0)} \in \{0,1\}^{n-p}$ for some pair $\delta^{(1)}, \delta^{(2)} \in \{0,1\}^p$ with

$$f(\delta^{(1)}, \epsilon^{(0)}) \ne f(\delta^{(2)}, \epsilon^{(0)})$$

(i.e., if row patterns of rows $int(\delta^{(1)})$ and $int(\delta^{(2)})$ of $Z(X^{(1)}, f)$ are different), then

$$\alpha(\delta^{(1)}) \ne \alpha(\delta^{(2)}).$$

Similarly we can prove the following theorem for two-sided decompositions:

THEOREM 3.2 *Let $f \in B_n$ be a function with input variables $x_1, \ldots, x_p$, $x_{p+1}, \ldots, x_n$. Let $X^{(1)} = \{x_1, \ldots, x_p\}$ and $X^{(2)} = \{x_{p+1}, \ldots, x_n\}$. Then there is a two-sided decomposition of f with respect to variable partition $\{X^{(1)}, X^{(2)}\}$ of form*

$$f(x_1, \ldots, x_p, x_{p+1}, \ldots, x_n) =$$
$$g(\alpha_1(x_1, \ldots, x_p), \ldots, \alpha_r(x_1, \ldots, x_p),$$
$$\beta_1(x_{p+1}, \ldots, x_n), \ldots, \beta_s(x_{p+1}, \ldots, x_n))$$

if and only if

$$r \ge \log(nrp(X^{(1)}, f)) \ and \ s \ge \log(ncp(X^{(1)}, f)).$$

PROOF:

"$\Longrightarrow$": If we assume

$$r < \log(nrp(X^{(1)}, f)) \ or \ s < \log(ncp(X^{(1)}, f)),$$

then we can derive a contradiction in the same way as in the proof of the previous Theorem 3.1.

"$\Longleftarrow$": Let

$$r \geq \log(nrp(X^{(1)}, f)), \text{ i.e., } 2^r \geq nrp(X^{(1)}, f) \text{ and}$$

$$s \geq \log(ncp(X^{(1)}, f)), \text{ i.e., } 2^s \geq ncp(X^{(1)}, f).$$

Then we can define $\alpha \in B_{p,r}$, such that

$$\alpha(\delta^{(1)}) \neq \alpha(\delta^{(2)})$$

for all $\delta^{(1)}, \delta^{(2)} \in \{0, 1\}^p$ with different row patterns in the decomposition matrix.

In the same way we can define $\beta \in B_{q,s}$, such that

$$\beta(\epsilon^{(1)}) \neq \beta(\epsilon^{(2)})$$

for all $\epsilon^{(1)}, \epsilon^{(2)} \in \{0, 1\}^{n-p}$ with different column patterns in the decomposition matrix.

Then we can define the composition function g according to the following rule

$$\begin{aligned}
g(\alpha_1(\epsilon_1, \ldots, \epsilon_p), &\ldots, \alpha_r(\epsilon_1, \ldots, \epsilon_p), \\
\beta_1(\epsilon_{p+1}, \ldots, \epsilon_n), &\ldots, \beta_s(\epsilon_{p+1}, \ldots, \epsilon_n)) = \\
&= f(\epsilon_1, \ldots, \epsilon_p, \epsilon_{p+1}, \ldots, \epsilon_n) \qquad (\star)
\end{aligned}$$

for all $(\epsilon_1, \ldots, \epsilon_p, \epsilon_{p+1}, \ldots, \epsilon_n) \in \{0, 1\}^n$.
If $(a_1, \ldots, a_r)$ does not occur in the image of α or $(b_1, \ldots, b_s)$ does not occur in the image of β, then $g(a_1, \ldots, a_r, b_1, \ldots, b_s)$ is undefined or can be chosen arbitrarily.

It is easy to see that g is well–defined according to rule $(\star)$:

Suppose g were not well–defined, then there would be $(\delta^{(1)}, \epsilon^{(1)})$ and $(\delta^{(2)}, \epsilon^{(2)})$ in $\{0, 1\}^n$ with

$$\alpha(\delta^{(1)}) = \alpha(\delta^{(2)}) \text{ and } \beta(\epsilon^{(1)}) = \beta(\epsilon^{(2)}) \qquad (\star\star)$$

and

$$f(\delta^{(1)}, \epsilon^{(1)}) \neq f(\delta^{(2)}, \epsilon^{(2)}).$$

That is, we would have

$$f(\delta^{(1)}, \epsilon^{(1)}) = \epsilon \in \{0, 1\} \text{ and } f(\delta^{(2)}, \epsilon^{(2)}) = \bar{\epsilon}.$$

Now it cannot be the case that *both* the row patterns of rows belonging to $\delta^{(1)}$ and $\delta^{(2)}$ are equal *and* the column patterns of columns belonging to $\epsilon^{(1)}$ and $\epsilon^{(2)}$ are equal:

From the fact that the rows for $\delta^{(1)}$ and $\delta^{(2)}$ are equal, we can conclude

$$f(\delta^{(2)}, \epsilon^{(1)}) = f(\delta^{(1)}, \epsilon^{(1)}) = \epsilon.$$

Because of

$$f(\delta^{(2)}, \epsilon^{(2)}) = \bar{\epsilon} \text{ we have } f(\delta^{(2)}, \epsilon^{(1)}) \neq f(\delta^{(2)}, \epsilon^{(2)}),$$

and therefore the column patterns belonging to $\epsilon^{(1)}$ and $\epsilon^{(2)}$ are not equal.

Analogously we can conclude from the fact that the columns belonging to $\epsilon^{(1)}$ and $\epsilon^{(2)}$ are equal:

$$f(\delta^{(1)}, \epsilon^{(2)}) = f(\delta^{(1)}, \epsilon^{(1)}) = \epsilon.$$

and thus

$$f(\delta^{(1)}, \epsilon^{(2)}) \neq f(\delta^{(2)}, \epsilon^{(2)}) = \bar{\epsilon}.$$

This means that the row patterns belonging to $\delta^{(1)}$ and $\delta^{(2)}$ are different.

Altogether the row patterns belonging to $\delta^{(1)}$ and $\delta^{(2)}$ are different or the column patterns belonging to $\epsilon^{(1)}$ and $\epsilon^{(2)}$ are different.

In the first case we have, according to the definition of α,

$$\alpha(\delta^{(1)}) \neq \alpha(\delta^{(2)}) \quad \Longrightarrow \text{ Contradiction to } (\star\star),$$

in the second case we have, according to the definition of β,

$$\beta(\epsilon^{(1)}) \neq \beta(\epsilon^{(2)}) \quad \Longrightarrow \text{ Contradiction to } (\star\star).$$

Thus the assumption that g is not well–defined by rule $(\star)$ cannot be true.

$$\square$$

Theorems 3.1 and 3.2 give the minimum numbers of decomposition functions for one-sided decompositions and two-sided decompositions, when the variable set $X^{(1)}$ or the variable partition $\{X^{(1)}, X^{(2)}\}$ is already given. In general the number of decomposition functions in a communication minimal decomposition certainly depends strongly on the choice of this variable set or variable partition.

EXAMPLE 3.1 (CONTINUED):

If we choose for function vgl_4 of Example 3.1 the partition $\{\{x_1, x_2, x_3, x_4\}, \{y_1, y_2, y_3, y_4\}\}$ instead of $\{\{x_1, x_2, y_1, y_2\}, \{x_3, x_4, y_3, y_4\}\}$, then we need 4 decomposition functions depending on $\{x_1, \ldots, x_4\}$ and 4 decomposition functions depending on $\{y_1, \ldots, y_4\}$, respectively. The two 4-bit-numbers which we have to compare are located in separate subsets of the input variables, so

Figure 3.4. Decomposition matrix for vgl_4 decomposed with respect to $\{\{x_1, x_2, x_3, x_4\},$ $\{y_1, y_2, y_3, y_4\}\}$.

	y_1	0	0	0	0	0	0	0	0	1	1	1	1	1	1	1	1
	y_2	0	0	0	0	1	1	1	1	0	0	0	0	1	1	1	1
	y_3	0	0	1	1	0	0	1	1	0	0	1	1	0	0	1	1
	y_4	0	1	0	1	0	1	0	1	0	1	0	1	0	1	0	1
$x_1 x_2 x_3 x_4$																	
0 0 0 0		1	0	0	0	0	0	0	0	0	0	0	0	0	0	0	0
0 0 0 1		0	1	0	0	0	0	0	0	0	0	0	0	0	0	0	0
0 0 1 0		0	0	1	0	0	0	0	0	0	0	0	0	0	0	0	0
0 0 1 1		0	0	0	1	0	0	0	0	0	0	0	0	0	0	0	0
0 1 0 0		0	0	0	0	1	0	0	0	0	0	0	0	0	0	0	0
0 1 0 1		0	0	0	0	0	1	0	0	0	0	0	0	0	0	0	0
0 1 1 0		0	0	0	0	0	0	1	0	0	0	0	0	0	0	0	0
0 1 1 1		0	0	0	0	0	0	0	1	0	0	0	0	0	0	0	0
1 0 0 0		0	0	0	0	0	0	0	0	1	0	0	0	0	0	0	0
1 0 0 1		0	0	0	0	0	0	0	0	0	1	0	0	0	0	0	0
1 0 1 0		0	0	0	0	0	0	0	0	0	0	1	0	0	0	0	0
1 0 1 1		0	0	0	0	0	0	0	0	0	0	0	1	0	0	0	0
1 1 0 0		0	0	0	0	0	0	0	0	0	0	0	0	1	0	0	0
1 1 0 1		0	0	0	0	0	0	0	0	0	0	0	0	0	1	0	0
1 1 1 0		0	0	0	0	0	0	0	0	0	0	0	0	0	0	1	0
1 1 1 1		0	0	0	0	0	0	0	0	0	0	0	0	0	0	0	1

that the decomposition functions cannot perform any meaningful preprocessing based on these subsets. A possible choice of the decomposition functions would be

$$\alpha_i(x_1, \ldots, x_4) = x_i \text{ and } \beta_i(y_1, \ldots, y_4) = y_i$$

for $1 \leq i \leq 4$, such that the composition function would be vgl_4 itself. This fact is also obvious from the decomposition matrix for this decomposition, which is shown in Figure 3.4. The decomposition matrix is the identity matrix with 16 row patterns and 16 column patterns. Thus we need 4 decomposition functions depending on $\{x_1, \ldots, x_4\}$ and 4 decomposition functions depending on $\{y_1, \ldots, y_4\}$.

3.4. Nontrivial decomposability — theoretical considerations

In this section we look into the general problem of nontrivial decomposability. Nontrivial decompositions are important for an application of decomposition in a recursive synthesis procedure, since for nontrivial decompositions decomposition and composition functions will have fewer variables than the original function. Here we can confine ourselves mainly to one-sided decompositions, since we can directly conclude the following from the definition of nontrivial decomposability:

If there is a nontrivial one-sided decomposition of a function $f \in B_n$ with input variables $X^{(1)} \cup X^{(2)}$ ($X^{(1)} \cap X^{(2)} = \emptyset$) with respect to $X^{(1)}$, then there is also a nontrivial two-sided decomposition with respect to partition $\{X^{(1)}, X^{(2)}\}$. If there is a nontrivial two-sided decomposition with respect to $\{X^{(1)}, X^{(2)}\}$, then there is a nontrivial one-sided decomposition with respect to $X^{(1)}$ *or a* nontrivial one-sided decomposition with respect to $X^{(2)}$.

The following corollary is a trivial conclusion from Theorems 3.1 and 3.2:

COROLLARY 3.2 *$f \in B_n$ is nontrivially decomposable with respect to $X^{(1)}$* $= \{x_1, \ldots, x_p\}$, *iff*

$$nrp(X^{(1)}, f) \leq 2^{p-1}.$$

$f \in B_n$ is nontrivially decomposable with respect to $\{X^{(1)}, X^{(2)}\} = \{\{x_1, \ldots, x_p\}, \{x_{p+1}, \ldots, x_n\}\}$, iff

$$nrp(X^{(1)}, f) \leq 2^{p-1} \ \ \text{or} \ \ ncp(X^{(1)}, f) \leq 2^{(n-p)-1}.$$

The next lemma says that it is easier to find nontrivial one-sided decompositions with respect to large variable sets than nontrivial one-sided decompositions with respect to small variable sets.

LEMMA 3.1 *If there is no nontrivial one-sided decomposition of $f \in B_n$ with respect to $X^{(1)} = \{x_1, \ldots, x_p\}$, then there is also no nontrivial one-sided decomposition with respect to subsets of $X^{(1)}$. Or in other words: If f is nontrivially decomposable with respect to $X^{(1)} = \{x_1, \ldots, x_p\}$, then f is nontrivially decomposable with respect to each superset of variables containing $X^{(1)}$.*

PROOF: Let $f \in B_n$ with input variables $x_1, \ldots, x_n$ be nontrivially decomposable with respect to $X^{(1)} = \{x_1, \ldots, x_p\}$, i.e., $nrp(X^{(1)}, f) \leq 2^{p-1}$. It suffices to show that f is nontrivially decomposable with respect to $X^{(1)'} = \{x_1, \ldots, x_p, x_{p+1}\}$, i.e., that $nrp(X^{(1)'}, f) \leq 2^p$.

Let $Z(X^{(1)}, f)$ be the decomposition matrix with respect to $X^{(1)}$, $Z(X^{(1)'}, f)$ the decomposition matrix with respect to $X^{(1)'}$.

If in $Z(X^{(1)}, f)$ the row patterns of rows for $\epsilon^{(1)} = (\epsilon_1^{(1)}, \ldots, \epsilon_p^{(1)})$ and $\epsilon^{(2)} = (\epsilon_1^{(2)}, \ldots, \epsilon_p^{(2)})$ are equal, then

$$f(\epsilon^{(1)}, \delta) = f(\epsilon^{(2)}, \delta)$$

for all $\delta = (\delta_{p+1}, \ldots, \delta_n) \in \{0, 1\}^{n-p}$ and thus

$$f(\epsilon^{(1)}, \delta_{p+1}, \delta') = f(\epsilon^{(2)}, \delta_{p+1}, \delta')$$

for all $\delta_{p+1} \in \{0, 1\}$, $\delta' = (\delta_{p+2}, \ldots, \delta_n) \in \{0, 1\}^{n-p-1}$.
Thus in $Z(X^{(1)'}, f)$ the pairs of rows for

$$(\epsilon_1^{(1)}, \ldots, \epsilon_p^{(1)}, 0) \text{ and } (\epsilon_1^{(2)}, \ldots, \epsilon_p^{(2)}, 0)$$

and for

$$(\epsilon_1^{(1)}, \ldots, \epsilon_p^{(1)}, 1) \text{ and } (\epsilon_1^{(2)}, \ldots, \epsilon_p^{(2)}, 1)$$

are equal.

Consequently $Z(X^{(1)'}, f)$ can have at most twice as many different row patterns as $Z(X^{(1)}, f)$, i.e.,

$$nrp(X^{(1)'}, f) \leq 2 \cdot nrp(X^{(1)}, f) \leq 2^p.$$

This relation between the decomposition matrices $Z(X^{(1)}, f)$ and $Z(X^{(1)'}, f)$ is illustrated in the following figure:

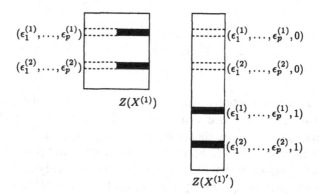

It is even true that each function $f \in B_n$ ($n \geq 4$) is nontrivially decomposable, if the variable set for decomposition is large enough:

LEMMA 3.2 *For each function $f \in B_n$ there exists a nontrivial one-sided decomposition $f(x_1, \ldots, x_p, x_{p+1}, \ldots, x_n) = g(\alpha_1(x_1, \ldots, x_p), \ldots, \alpha_r(x_1, \ldots, x_p), x_{p+1}, \ldots, x_n)$, if $p \geq 2^{n-p} + 1$.*

PROOF:
The rows of the decomposition matrix with respect to $X^{(1)} = \{x_1, \ldots, x_p\}$

have length 2^{n-p}. There are exactly $2^{2^{n-p}}$ different elements of $\{0,1\}^{2^{n-p}}$.

$\implies$ There are at most $2^{2^{n-p}}$ pairwise different row patterns in $Z(X^{(1)}, f)$.

$\implies$ $nrp(X^{(1)}, f) \leq 2^{2^{n-p}}$

$\implies$ $nrp(X^{(1)}, f) \leq 2^{2^{n-p}} \leq 2^{p-1}$ because of $p \geq 2^{n-p} + 1$

$\implies$ f is nontrivially decomposable with respect to $X^{(1)}$. $\square$

COROLLARY 3.3 *If $f \in B_n$ with $n \geq 4$, then there always exists a nontrivial one-sided decomposition.*

REMARK 3.1 *For functions $f \in B_n$ with $n \geq 4$ and input variables $\{x_1, \ldots, x_n\}$ the so-called* Shannon decomposition *represents a special nontrivial (but not necessarily communication minimal) decomposition with respect to variable set $\{x_1, \ldots, x_{n-1}\}$ (since obviously $n - 1 \geq 2^1 + 1$).*

The Shannon decomposition of f with respect to x_n is defined as

$$f(x_1, \ldots, x_n) = g(\alpha_1(x_1, \ldots, x_{n-1}), \alpha_2(x_1, \ldots, x_{n-1}), x_n),$$

where

$$\begin{aligned}
\alpha_1(x_1, \ldots, x_{n-1}) &= f(x_1, \ldots, x_{n-1}, 0) \\
\alpha_2(x_1, \ldots, x_{n-1}) &= f(x_1, \ldots, x_{n-1}, 1) \\
g(a_1, a_2, x_n) &= a_1 \cdot \overline{x_n} + a_2 \cdot x_n.
\end{aligned}$$

All Boolean functions with $n \geq 4$ input variables can be nontrivially decomposed (e.g. with respect to a variable set of size $p = n - 1$). However, if smaller variable sets $X^{(1)}$ are chosen for the decomposition, nontrivial decomposability of a function $f \in B_n$ becomes a special structural property of the functions, which guarantees that the function has a substantially smaller complexity than the complexity $\frac{2^n}{n}$, which is exceeded by almost all randomly chosen functions according to Shannon (see Section 1.5). This follows from the following lemma:

LEMMA 3.3 *If a function $f \in B_n$ has a nontrivial one-sided decomposition with respect to variable set $\{x_1, \ldots, x_p\}$, then every circuit realization which makes use of this decomposition has a B_2–complexity smaller than $\frac{2^n}{n}$, if n and p are sufficiently large and*

$$p \leq (n-1) - \log \frac{n(n-1)}{n-2} - \epsilon \text{ for } \epsilon > 0.$$

PROOF: Let $p \leq (n-1) - \log \frac{n(n-1)}{n-2} - \epsilon$.

Then, for $c > 1$ small enough, we have: $p \leq (n-1) - \log \frac{cn(n-1)}{(2-c)n-2}$.

This is equivalent to:

$$\Longleftrightarrow \quad n - p \geq \left(\log \frac{cn(n-1)}{(2-c)n-2}\right) + 1$$
$$\Longleftrightarrow \quad 2^{n-p-1} \geq \frac{cn(n-1)}{(2-c)n-2}$$
$$\Longleftrightarrow \quad cn(n-1) \leq 2^{n-p-1}(2n - 2 - cn)$$
$$\Longleftrightarrow \quad c \leq \frac{2^{n-p}(n-1) - 2^{n-p-1}cn}{n(n-1)}$$
$$\Longleftrightarrow \quad c(1 + \frac{2^{n-p-1}}{n-1}) \leq \frac{2^{n-p}}{n} \quad (\star)$$

The condition $p \leq (n-1) - \log \frac{n(n-1)}{n-2} - \epsilon$ is used later on in the proof in the form of inequality $(\star)$.

Assume that there is a nontrivial decomposition of form

$$f(x_1, \ldots, x_p, x_{p+1}, \ldots, x_n) =$$
$$g(\alpha_1(x_1, \ldots, x_p), \ldots, \alpha_r(x_1, \ldots, x_p), x_{p+1}, \ldots, x_n).$$

According to Theorem 1.3 given by Lupanov we can estimate the complexity of decomposition functions and the complexity of the composition function by

$$C_{B_2}(\alpha_i) \leq \frac{2^p}{p} + o\left(\frac{2^p}{p}\right)$$

and

$$C_{B_2}(g) \leq \frac{2^{r+n-p}}{r+n-p} + o\left(\frac{2^{r+n-p}}{r+n-p}\right) \leq \frac{2^{n-1}}{n-1} + o\left(\frac{2^{n-1}}{n-1}\right),$$

respectively.

If p and n are sufficiently large, this implies

$$C_{B_2}(\alpha_i) \leq \frac{2^p}{p} + o\left(\frac{2^p}{p}\right) \leq c\frac{2^p}{p}$$

and

$$C_{B_2}(g) \leq \frac{2^{n-1}}{n-1} + o\left(\frac{2^{n-1}}{n-1}\right) \leq c\frac{2^{n-1}}{n-1}.$$

The overall cost of a circuit S, which uses the given decomposition, can be estimated by

$$C_{B_2}(S) \leq rc\frac{2^p}{p} + c\frac{2^{n-1}}{n-1}$$

$$
\begin{aligned}
&\leq\ c\left((p-1)\frac{2^p}{p}+\frac{2^{n-1}}{n-1}\right) \\
&=\ c2^p\left(\frac{p-1}{p}+\frac{2^{n-p-1}}{n-1}\right) \\
&<\ 2^p c\left(1+\frac{2^{n-p-1}}{n-1}\right) \\
&\leq\ 2^p\frac{2^{n-p}}{n}\qquad \text{because of } (\star) \\
&=\ \frac{2^n}{n}
\end{aligned}
$$

$\square$

According to Lemma 3.3, a function which has a nontrivial one–sided decomposition with respect to a small variable set has a B_2–complexity smaller than $\frac{2^n}{n}$. Together with Shannon's counting arguments (Theorem 1.2, p. 20) we can conclude that it is not likely that a randomly chosen function has a nontrivial decomposition with respect to a small variable set. However, as already mentioned in Section 1.5, we do not have to deal with randomly chosen functions in practical examples. This observation is also validated by experimental results for the decomposition of benchmark functions given in the following chapters. According to Lemma 3.2, nontrivial one–sided decompositions always exist with respect to *large* variable sets; according to Lemma 3.3, one–sided decomposability with respect to *small* variable sets is a *special property* of certain Boolean functions.

A Boolean function has a nontrivial two–sided decomposition with respect to $\{X^{(1)}, X^{(2)}\}$ if and only if it has a nontrivial one–sided decomposition with respect to $X^{(1)}$ *or* a nontrivial one–sided decomposition with respect to $X^{(2)}$. Thus nontrivial two–sided decomposability is certainly a special functional property for those variable partitions where both sets of the partition are of about the same size. Such decompositions are called 'balanced':

DEFINITION 3.5 (BALANCED DECOMPOSITIONS)
A two-sided decomposition of $f \in B_n$ with respect to variable partition $\{\{x_1, \ldots, x_p\}, \{x_{p+1}, \ldots, x_n\}\}$ is called balanced, *if*

$$
p = \left\lfloor \frac{n}{2} \right\rfloor \ or \ p = \left\lceil \frac{n}{2} \right\rceil .
$$

Balanced decompositions are of interest, because the recursive utilization of communication minimal balanced decompositions leads in many cases to realizations with a small depth.

Of course it is not always possible to find nontrivial balanced decompositions. In this case it makes sense to use one decomposition step with non-balanced decomposition, where a partitioning into sets of different size takes place (cf. Lemma 3.2). In the next recursive step nontrivial balanced decompositions can exist again. The following example shows such a case:

EXAMPLE 3.2 Let $f \in B_5$ be defined by

$$f(x_0, \ldots, x_4) = \overline{x_0} \cdot (x_1 \oplus x_2) \cdot (x_3 \oplus x_4) + x_0 \cdot (x_1 + x_3) \cdot (x_2 + x_4)$$

for all $(x_0, \ldots, x_4) \in \{0, 1\}^5$.

It is easy to see (e.g. by considering all decomposition matrices for all $\binom{5}{3}$ balanced variable partitions) that there is no nontrivial balanced decomposition for f.

However, after one Shannon decomposition has been performed, the decomposition functions can be decomposed by a nontrivial balanced decomposition:

$$f(x_0, \ldots, x_4) = \overline{x_0} \cdot \alpha_1(x_1, \ldots, x_4) + x_0 \cdot \alpha_2(x_1, \ldots, x_4)$$

with decomposition functions α_1 and α_2 with

$$\alpha_1(x_1, \ldots, x_4) = (x_1 \oplus x_2) \cdot (x_3 \oplus x_4) \text{ and}$$

$$\alpha_2(x_1, \ldots, x_4) = (x_1 + x_3) \cdot (x_2 + x_4).$$

For α_1 there is a nontrivial balanced decomposition with respect to $\{\{x_1, x_2\}, \{x_3, x_4\}\}$ and for α_2 there is a nontrivial balanced decomposition with respect to $\{\{x_1, x_3\}, \{x_2, x_4\}\}$. The resulting circuit is given in Figure 3.5.

In the following we will show that the use of decompositions can be very profitable in logic synthesis. However, we cannot guarantee, even for Boolean functions with *excellent* decomposability properties, that the use of decompositions will necessarily lead to optimal realizations. The following lemma, which can be concluded from a theorem by W. J. Paul (Paul, 1976), gives evidence of this fact (for the proof see (Scholl and Molitor, 1993)):

LEMMA 3.4 *For every $\epsilon > 0$ there is an $n \in I\!N$ and a Boolean function $G \in B_{2n}$ with the following property: There is a communication minimal balanced decomposition of G with respect to the variable partition $A = \{\{x_1, \ldots, x_n\}, \{x_{n+1}, \ldots, x_{2n}\}\}$ with exactly 2 decomposition functions. However, for the cost of every realization S for G which makes use of a communication minimal balanced decomposition with respect to A,*

$$C_{B_2}(S) \geq \frac{2}{1 + \epsilon} C_{B_2}(G) + 1$$

holds.

Figure 3.5. Circuit for example function f (Example 3.2).

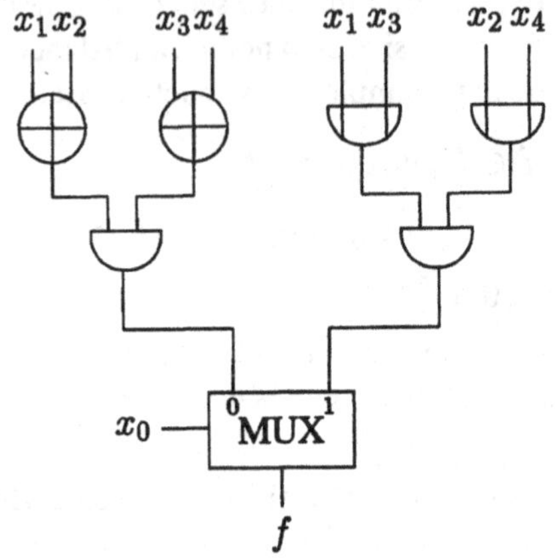

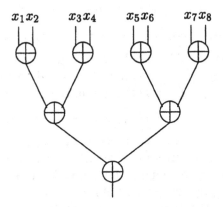

Figure 3.6. Realization of $exor_8$

3.5.   Examples of circuits derived by decomposition

In this section we illustrate the effect of recursive decompositions with a detailed discussion of two simple example functions. The circuits were computed automatically based on two-sided decompositions.

EXAMPLE 3.3 Our first example is an *exor* function with 8 inputs. As shown in Figure 3.6, our algorithm computes a balanced tree of 7 $exor_2$ gates. In the first recursive step the variables were partitioned into sets $\{x_1, \ldots, x_4\}$ and $\{x_5, \ldots, x_8\}$. Both the decomposition function depending on $\{x_1, \ldots, x_4\}$ and

Figure 3.7. Realization of s_4^6.

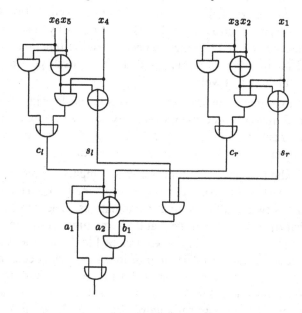

the decomposition function depending on $\{x_5, \ldots, x_8\}$ are $exor_4$ functions; the composition function is an $exor_2$ function. Then the decomposition functions are decomposed recursively resulting in realizations using 3 $exor_2$ gates. It is easy to see that the computed realization uses a minimum number of cells of B_2.

EXAMPLE 3.4 Our second example is a threshold function s_4^6, which is defined as follows:

$$s_4^6(x_1, \ldots, x_6) = 1 \iff \sum_{i=1}^{6} x_i \geq 4.$$

The resulting circuit is given in Figure 3.7.

In the first step our algorithm partitions the input variables into sets $\{x_1, x_2, x_3\}$ and $\{x_4, x_5, x_6\}$. (Methods of finding good partitions will be given in Section 3.7.) There are two decomposition functions c_r and s_r depending on $\{x_1, x_2, x_3\}$. These functions are just carry and sum bits of the binary addition of x_1, x_2, and x_3. In the same way, decomposition functions c_l and s_l depending on variables $\{x_4, x_5, x_6\}$ are carry and sum bits of the binary addition of x_4, x_5, and x_6. Considering the realization for (c_r, s_r), we can see that (c_r, s_r)

is realized by a full adder circuit. The full adder results from a decomposition of c_r and s_r with respect to $\{\{x_1\}, \{x_2, x_3\}\}$. Here $x_2 \oplus x_3$ is used both in the decomposition of c_r and s_r. (The use of this decomposition function for two different output functions c_r and s_r did not occur by chance: In Chapter 4 we will present a method to compute decomposition functions which can be used in the decomposition of as many output functions as possible.) (c_l, s_l) is realized by a full adder as well.

The composition function in the first recursive step has four inputs c_l, s_l, c_r and s_r. When the function is processed recursively, the variable partition $\{\{c_l, c_r\}, \{s_l, s_r\}\}$ is computed. For $\{c_l, c_r\}$ two decomposition functions a_1 and a_2 are computed and for $\{s_l, s_r\}$ one decomposition function b_1.

The corresponding composition function with 3 inputs a_1, a_2 and b_1 is recursively decomposed again. In this decomposition our algorithm uses the fact that this function has two don't care input combinations. The don't cares are due to the fact that a_1 and a_2 cannot be 1 at the same time. If the don't care input combinations $(1, 1, 0)$ and $(1, 1, 1)$ were mapped to 0, then the resulting function would not have a nontrivial decomposition. Nontrivial decompositions are possible only if the function value for $(1, 1, 0)$ is set to 1. Our implementation defines the function values for $(1, 1, 0)$ and $(1, 1, 1)$ to be 1 and thus obtains a nontrivial decomposition with variable partition $\{\{a_1\}, \{a_2, b_1\}\}$. This function is then realized by the following subcircuit:

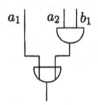

Methods for a systematic exploitation of don't cares during decomposition are given in Chapter 5.

If costs are computed using R_2–complexity with $R_2 = B_2 \setminus \{exor, equiv\}$ (see Section 1.1), the computed realization S for s_4^6 has

$$C_{R_2}(S) = 7C_{R_2}(and) + 3C_{R_2}(or) + 5C_{R_2}(exor) = 7 + 3 + 5 \cdot 3 = 25.$$

For comparison we consider the minimal sum-of-products realization for s_4^6. It is given by

$$P = \bigvee_{1 \leq i_1 \leq ... \leq i_4 \leq 6} x_{i_1} ... x_{i_4}.$$

The cost of this realization is

$$C_{R_2}(P) = 15C_{R_2}(and_4) + C_{R_2}(or_{15}) = 15 \cdot 3 + 14 = 59.$$

Table 3.1. Run times for the synthesis of n–bit–adders $adder_n$.

circuit	in	out	run time of mulop	run time of mulopII
$adder_4$	8	4	1.30 s	0.20 s
$adder_8$	16	8	18 h 38 min	1.15 s
$adder_{16}$	32	16	—	11.55 s
$adder_{32}$	64	32	—	3 min 8 s
$adder_{64}$	128	64	—	31 min 16 s

3.6. Decomposition and ROBDD representations

As already mentioned, the use of ROBDDs in functional decomposition was a crucial step in making decomposition efficient.

Whereas function tables and decomposition matrices have an exponential size in the number of inputs of the Boolean function, ROBDDs provide a much more compact representation. Although this is not true for Boolean functions of B_n *on average*, it is true for many Boolean functions occurring in practical applications.

Decompositions can be performed based on ROBDD representations rather than using function tables or decomposition matrices. Table 3.1 shows that it is profitable to deal with decompositions based on ROBDD representations. As an example we give run times to synthesize adders of different bit widths by decomposition. Column 4 (*mulop*) gives the run times of our former tool *mulop* (Molitor and Scholl, 1994), which was based on decomposition matrices, and column 5 gives the run times of our ROBDD based tool *mulopII* (Scholl and Molitor, 1994; Scholl and Molitor, 1995b; Scholl and Molitor, 1995a) (run times measured in CPU seconds of a SUN Sparc2). The much higher run times of the version based on function tables and decomposition matrices can easily be explained by the larger amounts of data which have to be processed. A simple computation shows that the size of a function table for a 16–bit–adder (32 inputs, 16 outputs, 1 bit per table entry) would be 8 gigabytes, for a 32–bit–adder already about 69 billion gigabytes. It is precisely because of the representation sizes needed that problems of this magnitude cannot be processed by a table based tool.

3.6.1   Minimum number of decomposition functions

The first step into the intended direction was already made in Section 2.1.2. In this section a relationship was established between the number $nrp(X^{(1)}, f)$ of different row patterns in a decomposition matrix and the number of linking nodes immediately below a cut line in an ROBDD with an appropriate variable order. To compute the minimum number of decomposition functions in a one-sided decomposition of a function f with respect to $X^{(1)} = \{x_1, \ldots, x_p\}$ (according to Theorem 3.1) we need to determine $nrp(X^{(1)}, f)$. If f is represented by an ROBDD with $x_1, \ldots, x_p$ as the first p variables in the variable order, then $nrp(X^{(1)}, f)$ is equal to the number of linking nodes immediately below a cut line after the first p variables in the ROBDD. This means that the number of decomposition functions in a communication minimal decomposition can easily be computed based on an ROBDD with an appropriate variable order: We simply have to perform a depth-first traversal of the ROBDD to determine the number of different linking nodes.

The computation of good variable partitions (and good ROBDD orders) will be the subject of Section 3.7.

3.6.2   Computing decomposition and composition functions

To compute decompositions based on ROBDD representations it is certainly not enough to determine the minimum number of decomposition functions based on ROBDDs. Additionally we have to compute decomposition and composition functions efficiently based on the ROBDD representation. The proof of Theorem 3.1 shows how decomposition and composition functions in a communication minimal decomposition can be computed in principle, but the question remains as to how to compute them based on ROBDD representations without using function tables.

According to the proof of Theorem 3.1, decomposition and composition functions of one-sided decompositions can be found based on the decomposition matrix $Z(X^{(1)}, f)$: The decomposition functions have to be defined in such a way that they assign different function values or codes to the indices of rows with different row patterns. (In this process we have some degree of freedom in the exact choice of the codes and thus the decomposition functions. Section 3.8 and Chapter 4 deal with the exploitation of this degree of freedom.) Then the composition function g in the decomposition $f(x_1, \ldots, x_p, x_{p+1}, \ldots, x_n)$ $= g(\alpha_1(x_1, \ldots, x_p), \ldots, \alpha_r(x_1, \ldots, x_p), x_{p+1}, \ldots, x_n)$ results automatically from the condition that for all $\epsilon \in \{0,1\}^n$ $g(\alpha_1(\epsilon_1, \ldots, \epsilon_p), \ldots, \alpha_r(\epsilon_1, \ldots, \epsilon_p), \epsilon_{p+1}, \ldots, \epsilon_n) = f(\epsilon_1, \ldots, \epsilon_p, \epsilon_{p+1}, \ldots, \epsilon_n)$. Note that g may be an incompletely specified function, since some vectors $(a_1, \ldots, a_r) \in \{0,1\}^r$ may not

Figure 3.8.   Decomposition of function $f(x_1, \ldots, x_4) = \overline{x_4}x_2x_3 + x_4(x_1 \oplus x_2)$ with respect to $\{x_1, x_2, x_3\}$. 2 decomposition functions α_1 and α_2 and the composition function g are given. The resulting circuit is shown below.

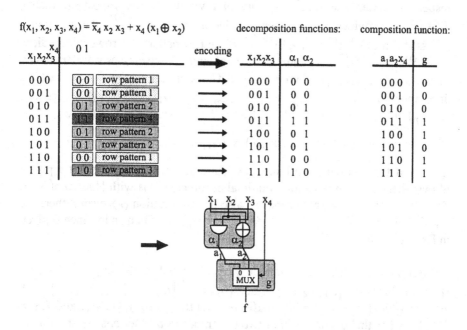

be in the image of $(\alpha_1, \ldots, \alpha_r)$. (These vectors again represent some degree of freedom, which will be exploited by the methods in Chapter 5.)

With the following example we show how decomposition and composition functions are constructed based on the decomposition matrix. Then we will show how these functions can be determined based on ROBDDs.

EXAMPLE 3.5   Figure 3.8 shows the decomposition of a function $f(x_1, \ldots, x_4) = \overline{x_4}x_2x_3 + x_4(x_1 \oplus x_2)$ with respect to $\{x_1, x_2, x_3\}$. The corresponding decomposition matrix has exactly 4 different row patterns, such that a communication minimal decomposition needs 2 decomposition functions α_1 and α_2. The function value $(0, 0)$ of (α_1, α_2) appears only for indices of the decomposition matrix with row pattern 1, $(0, 1)$ only for rows with pattern 2, etc. The composition function results from the choice of decomposition functions α_1 and α_2.

Before we demonstrate the computation of decomposition and composition functions based on ROBDDs, we will introduce a restriction to the choice of decomposition functions, i.e., a restriction to the so-called 'strict' decomposition functions.

As mentioned in Notation 2.2, during decomposition of a function f with respect to variable set $X^{(1)}$ equality of rows in the decomposition matrix $Z(X^{(1)}, f)$ provides an equivalence relation on $\{0,1\}^{|X^{(1)}|}$, if one defines $\epsilon^{(1)} \equiv \epsilon^{(2)}$ for $\epsilon^{(1)}, \epsilon^{(2)} \in \{0,1\}^{|X^{(1)}|}$ iff the row patterns of rows with indices $int(\epsilon^{(1)})$ and $int(\epsilon^{(2)})$ are equal. The set of equivalence classes was denoted by $\{K_1, \ldots, K_{nrp(X^{(1)}, f)}\}$. The condition of Corollary 3.1 on page 77 says that in a decomposition $f(x_1, \ldots, x_p, x_{p+1}, \ldots, x_n) = g(\alpha_1(x_1, \ldots, x_p), \ldots, \alpha_r(x_1, \ldots, x_p), x_{p+1}, \ldots, x_n)$ for all $\epsilon \in K_i$ and $\delta \in K_j$ with $i \neq j$ condition

$$(\alpha_1, \ldots, \alpha_r)(\epsilon) \neq (\alpha_1, \ldots, \alpha_r)(\delta) \qquad (\star)$$

holds. If the number of different row patterns $nrp(X^{(1)}, f)$ is not a power of two, then for communication minimal decompositions with $\lceil \log(nrp(X^{(1)}, f)) \rceil$ decomposition functions we can also fulfill condition $(\star)$, even if there are $\epsilon \neq \delta \in K_i$ with $(\alpha_1, \ldots, \alpha_r)(\epsilon) \neq (\alpha_1, \ldots, \alpha_r)(\delta)$. Such an instance is given in Example 3.6.

EXAMPLE 3.6 Consider a function $f \in B_n$ which is decomposed with respect to $X^{(1)} = \{x_1, x_2, x_3\}$. Let $\{0,1\}^{|X^{(1)}|}/\equiv \; = \{K_1, K_2, K_3\}$ with $K_1 = \{(000)\}$, $K_2 = \{(001), (010), (011), (100), (101), (110)\}$, and $K_3 = \{(111)\}$. For this decomposition two decomposition functions α_1 and α_2 are needed. The choice of $\alpha = (\alpha_1, \alpha_2)$ with $\alpha(K_1) = (0,0)$, $\alpha(\{(001), (010), (011)\}) = (01)$, $\alpha(\{(100), (101), (110)\}) = (10)$ and $\alpha(K_3) = (11)$ obviously fulfills the condition of Corollary 3.1, such that α_1 and α_2 can be used as decomposition functions in a decomposition of f with respect to $X^{(1)}$. α assigns *different* codes to elements of *different* classes K_i, but not the same code to all elements of K_2.

In Example 3.5 the decomposition functions assign the same code to all elements of the same equivalence class:

EXAMPLE 3.5 (CONTINUED):
We have $K_1 = \{(000), (001), (110)\}$, $K_2 = \{(010), (100), (101)\}$ $K_3 = \{(111)\}$ and $K_4 = \{(011)\}$. Moreover we have $\alpha(K_1) = (0,0)$, $\alpha(K_2) = (0,1)$, $\alpha(K_3) = (1,0)$, and $\alpha(K_4) = (1,1)$.

Decompositions with the property of Example 3.5 are called *strict*. Strict decompositions have the advantage that the computation of decomposition and composition functions based on ROBDDs can be performed more easily. Other advantages of strict decomposition functions are given in Section 3.8.1.

DEFINITION 3.6 (STRICT DECOMPOSITION)
Let $f \in B_n$ be a Boolean function, let $f(x_1, \ldots, x_p, x_{p+1}, \ldots, x_n) = g(\alpha_1(x_1, \ldots, x_p), \ldots, \alpha_r(x_1, \ldots, x_p), x_{p+1}, \ldots, x_n)$ be a decomposition of f with respect to $X^{(1)} = \{x_1, \ldots, x_p\}$, and let $\{0,1\}^p/_\equiv = \{K_1, \ldots, K_{nrp(X^{(1)}, f)}\}$ be the partition of $\{0,1\}^p$ into equivalence classes induced by equality of row patterns in $Z(X^{(1)}, f)$. A decomposition function α_i is strict, if for all $\epsilon, \delta \in K_j$ $(1 \leq j \leq nrp(X^{(1)}, f))$ $\alpha_i(\epsilon) = \alpha_i(\delta)$. A decomposition is strict, if all its decomposition functions are strict.

A strict decomposition can be viewed as an encoding of equivalence classes $K_i \in \{0,1\}^p/_\equiv$ by α. Each equivalence class is assigned a fixed code.

LEMMA 3.5
Let $f \in B_n$ and let $f(x_1, \ldots, x_p, x_{p+1}, \ldots, x_n) = g(\alpha_1(x_1, \ldots, x_p), \ldots, \alpha_r(x_1, \ldots, x_p), x_{p+1}, \ldots, x_n)$ be a decomposition of f with respect to $X^{(1)} = \{x_1, \ldots, x_p\}$. If $nrp(X^{(1)}, f)$ is a power of two, then each communication minimal decomposition of f with respect to $X^{(1)}$ is a strict decomposition.

PROOF: Let $nrp(X^{(1)}, f) = 2^r$, $\{0,1\}^p/_\equiv = \{K_1, \ldots, K_{2^r}\}$.
According to Theorem 3.1 the number of decomposition functions in a communication minimal decomposition is equal to r.
Corollary 3.1 says that there must not be any $\epsilon \in K_i$ and $\delta \in K_j$, $i \neq j$, with $(\alpha_1, \ldots, \alpha_r)(\epsilon) = (\alpha_1, \ldots, \alpha_r)(\delta)$. If there is K_i $(1 \leq i \leq 2^r)$ with $\epsilon^{(1)} \neq \epsilon^{(2)} \in K_i$, $\alpha(\epsilon^{(1)}) \neq \alpha(\epsilon^{(2)})$, then the number of different function values of α would be larger than 2^r. But obviously there are at most 2^r different values in the image of $\alpha = (\alpha_1, \ldots, \alpha_r)$. $\qquad\square$

If strict decompositions are used, it is possible to determine decomposition and composition functions easily, based on ROBDDs without using larger representations like function tables or decomposition matrices. We will demonstrate this construction in the following and illustrate it in Example 3.7.

As in Section 2.1.2, we make use of the fact that for each row pattern m_i in the decomposition matrix $Z(X^{(1)}, f)$ of a function f there exists a unique 'linking node' n_i in the ROBDD of f, such that the subgraph with root n_i represents the function whose function table is represented by row pattern m_i. (Assume w.l.o.g. $X^{(1)} = \{x_1, \ldots, x_p\}$ and variable order $x_1 <_{index} \cdots <_{index} x_n$ in the ROBDD.) The set of all row indices of $Z(X^{(1)}, f)$ with row pattern m_i is given by the equivalence class K_i of $\{0,1\}^p/_\equiv$. Then the set of all

$(\epsilon_1, \ldots, \epsilon_p) \in \{0, 1\}^p$ with the property that linking node n_i is reached by $(\epsilon_1, \ldots, \epsilon_p)$ is exactly the set K_i. If node n_i is replaced by the constant 1 and all other linking nodes are replaced by the constant 0, then we obtain a (possibly non-reduced) OBDD for the characteristic function of K_i. [2]

A strict decomposition function α_j is defined by its function values $w_j^i \in \{0, 1\}$ for equivalence classes K_i ($\alpha_j(K_i) = \{w_j^i\}$). To obtain an OBDD for α_j, it suffices to replace linking nodes in the original ROBDD by the constants 0 and 1: Linking node n_i is replaced by the constant 0, if $w_j^i = 0$, and by the constant 1, if $w_j^i = 1$. The replacement results in a (possibly non-reduced) OBDD $A_j = (G' = (V, E), m)$ for α_j, which can be reduced to an ROBDD in time $O(|V| \cdot \log(|V|))$. Thus ROBDDs for decomposition functions $\alpha_1, \ldots, \alpha_r$ can be computed in a simple manner from encodings $(w_1^i, \ldots, w_r^i)$ for linking nodes n_i (or for equivalence classes K_i).

The ROBDD for composition function g can also be computed easily: g depends on new variables $a_1, \ldots, a_r$ and variables $x_{p+1}, \ldots, x_n$. We construct the ROBDD for g with variable order $a_1 <_{index} \cdots <_{index} a_r <_{index} x_{p+1} <_{index} \ldots <_{index} x_n$. Suppose that the node reached by $(\epsilon_1, \ldots, \epsilon_p)$ in the original ROBDD for f is linking node n_i and $\alpha = (\alpha_1, \ldots, \alpha_r)$ assigns $(w_1^i, \ldots, w_r^i)$ to input $(\epsilon_1, \ldots, \epsilon_p) \in K_i$. Since for all $(\delta_{p+1}, \ldots, \delta_n) \in \{0, 1\}^{n-p}$

$$
\begin{aligned}
f(\epsilon_1, \ldots, \epsilon_p, \delta_{p+1}, \ldots, \delta_n) &= \\
&= g(\alpha_1(\epsilon_1, \ldots, \epsilon_p), \ldots, \alpha_r(\epsilon_1, \ldots, \epsilon_p), \delta_{p+1}, \ldots, \delta_n) \\
&= g(w_1^i, \ldots, w_r^i, \delta_{p+1}, \ldots, \delta_n),
\end{aligned}
$$

the cofactors $f_{x_1^{\epsilon_1} \ldots x_p^{\epsilon_p}}$ and $g_{a_1^{w_1^i} \ldots a_r^{w_r^i}}$ of f and g have to be identical. Thus the node reached by $(w_1^i, \ldots, w_r^i)$ in the ROBDD for g has to be the subgraph with root n_i in the ROBDD for f, which consequently represents cofactor $f_{x_1^{\epsilon_1} \ldots x_p^{\epsilon_p}}$ $= g_{a_1^{w_1^i} \ldots a_r^{w_r^i}}$. That means that we obtain an OBDD for g by copying the 'lower part' of ROBDD for f containing the subgraphs, whose roots are the linking nodes, and by constructing the 'upper part' with variables $a_1, \ldots a_r$, such that linking nodes n_i are reached by $(w_1^i, \ldots, w_r^i)$ ($1 \leq i \leq nrp(\{x_1, \ldots, x_p\}, f)$). This is done in the following way:

1. Construct a balanced binary tree of depth r.

2. Label nodes with depth 0 by variable name a_1, nodes with depth 1 by variable name a_2 etc., nodes with depth $r - 1$ by variable name a_r.

---

[2] The characteristic function of K_i is a function $c \in B_p$ with $c(\epsilon_1, \ldots, \epsilon_p) = 1$, if $(\epsilon_1, \ldots, \epsilon_p) \in K_i$, and $c(\epsilon_1, \ldots, \epsilon_p) = 0$ otherwise.

3. For each node, label one outgoing edge 0 and the other 1.

4. For $1 \leq i \leq nrp(\{x_1, \ldots, x_p\}, f)$: Replace the node with depth r which is reached by $(w_1^i, \ldots, w_r^i)$, by the linking node n_i.

5. For a communication minimal decomposition, where the number of different linking nodes (or row patterns in the decomposition matrix) is a power of two (i.e., $nrp(\{x_1, \ldots, x_p\}, f) = 2^r$), this leads to a reduced OBDD, since the number of leaves in the binary tree is exactly equal to the number of linking nodes and since the ROBDD of f was already reduced.

If the number of linking nodes is not a power of two, there are still nodes with depth r, which have not been replaced by linking nodes. These nodes are reached by vectors $(\delta_1, \ldots, \delta_r)$, which do not occur in the image of $\alpha = (\alpha_1, \ldots, \alpha_r)$. They can be replaced by arbitrary nodes (e.g. by the constant 0), since these vectors $(\delta_1, \ldots, \delta_r)$ represent don't cares of the composition function. After that a reduction step may be necessary to reduce the OBDD to an ROBDD.

Example 3.7 illustrates the computation of decomposition and composition functions based on ROBDDs:

EXAMPLE 3.7
Figure 3.9 shows the decomposition of function $f(x_1, \ldots, x_4) = \overline{x_4}x_2x_3 + x_4(x_1 \oplus x_2)$ from Example 3.5 with respect to $\{x_1, x_2, x_3\}$. There are 4 different linking nodes $n_1, \ldots, n_4$ (from left to right), such that a communication minimal decomposition needs 2 decomposition functions α_1 and α_2. The corresponding equivalence classes K_1, K_2, K_3 and K_4 are encoded by $(0, 0)$, $(0, 1)$, $(1, 0)$ and $(1, 1)$, respectively.

Thus the OBDD for α_1 is obtained by replacing n_1 by the first code bit of $(0, 0)$, n_2 by the first code bit of $(0, 1)$, n_3 by the first code bit of $(1, 0)$ and n_4 by the first code bit of $(1, 1)$. In an analogous manner the OBDD for α_2 results from replacements by the second code bits. To obtain an ROBDD for α_1 and α_2 the resulting OBDDs still have to be reduced.

We obtain the ROBDD for composition function g by removing the upper part above the cut line followed by the construction of a binary tree, such that node n_1 is reached by $(0, 0)$, node n_2 is reached by $(0, 1)$ etc.

The computation of decomposition and composition functions for two–sided decompositions can be reduced to one-sided decompositions (more precisely to a series of two one-sided decompositions). Details on the computation of more general k–sided decompositions with $k > 1$ can be found in (Scholl, 1996).

Figure 3.9. Decomposition of function $f(x_1, \ldots, x_4) = \overline{x_4}x_2x_3 + x_4(x_1 \oplus x_2)$ with respect to $\{x_1, x_2, x_3\}$ based on ROBDDs.

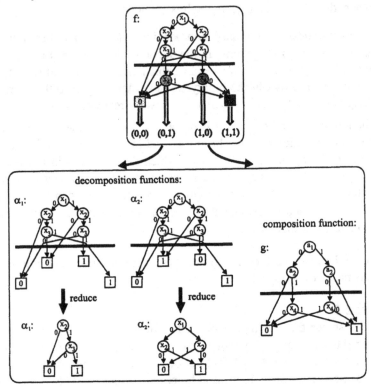

3.6.3   ROBDD size and decomposability

In Section 3.6.1 we studied the relationship between the number of linking nodes in the ROBDD below a cut line after variables $x_1, \ldots, x_p$ and the minimum number of decomposition functions of a decomposition with respect to variable set $\{x_1, \ldots, x_p\}$. We can conclude from this relationship that (assuming a large number n of input variables) a function $f \in B_n$ has good decomposition properties, when f has a compact ROBDD representation.

Let us consider sequences $(f_n)_{n \in \mathbb{N}}$ of Boolean functions $f_n \in B_n$ and their ROBDD representations. For large n it will only be possible to construct ROBDDs for functions f_n, if the ROBDD sizes for the sequence are limited by a polynomial of small degree k. If the number of ROBDD nodes for $f_n \in B_n$ under a given variable order is limited by $P(n) = \sum_{i=0}^{k} a_i n^i$, then the number of linking nodes is limited by $\frac{1}{2}(P(n) + 1)$. [3]

Then the minimum number of decomposition functions is not larger than

$$\lceil \log(\tfrac{1}{2}(P(n) + 1)) \rceil \;\; \leq \;\; \log(P(n))$$

$$\leq \;\; \log\!\left(a_{max} \cdot \frac{n^{k+1} - 1}{n - 1}\right) \text{ with } a_{max} = \max_{0 \leq i \leq k} a_i$$

$$< \;\; \log(a_{max}) + (k+1)\log(n) - \log(n - 1)$$

If we decompose the function f with respect to variable set $\{x_1, \ldots, x_p\}$ with $p = \lceil \frac{n}{2} \rceil$, e.g., then (assuming sufficiently large n, sufficiently small degree k of the polynomial and sufficiently small constants a_i) the upper limit $\log(a_{max}) + (k+1)\log(n) - \log(n-1)$ for the number of decomposition functions is much smaller than $p = \lceil \frac{n}{2} \rceil$, such that we certainly have a nontrivial decomposition.

Thus it will definitely be a good idea to look for small ROBDD representations to find good decompositions. In the case of completely specified functions an optimization of the variable order comes into consideration here; in the case of incompletely specified functions also an additional exploitation of don't cares.

However, the inverse is not generally true: It is possible to construct functions with good decomposition properties, which do not show a compact ROBDD representation. To construct such functions we use Lemma 3.6, which follows from Shannon's counting arguments (see Section 1.5):

LEMMA 3.6 *For sufficiently large n, the ROBDD size (i.e., the number of nodes in the ROBDD) is larger than $\frac{2^n}{3n}$ for almost all functions $f \in B_n$ — regardless of the variable order of the ROBDD.*

PROOF: (Sketch)
According to Shannon's Theorem 1.2, for sufficiently large n almost all functions $f \in B_n$ have a B_2 complexity larger than $\frac{2^n}{n}$. By replacing non-terminal nodes by multiplexers, an ROBDD for a function f can be interpreted as a multiplexer circuit which realizes f. The select inputs of the multiplexers are driven by the input variables which form the labels of the respective non-terminal nodes (cf. Figure 3.10). Each multiplexer can be realized by three 2–input–cells from B_2.   □

---

[3]This results from the fact that each non-terminal node in the ROBDD has outdegree 2. If there are k linking nodes below the cut line, there have to be (at least) $k - 1$ nodes above the cut line. Thus, a ROBDD of size s has at most $\frac{1}{2}(s + 1)$ linking nodes.

Figure 3.10.   ROBDD for $f(x_1, \ldots, x_4) = x_2 + x_1 x_3 x_4$ and corresponding multiplexer circuit.

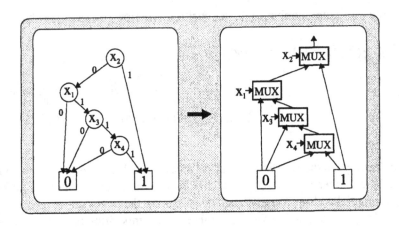

Now the following lemma shows that there are functions f in B_n which have good decomposition properties, but do not have any compact ROBDD representation:

LEMMA 3.7 *For sufficiently large n there are functions in B_{2n} whose ROBDD sizes are larger than $\frac{2^n}{3n}$ (regardless of the variable order of the ROBDD), and which have a decomposition of the form*

$$f(x_1, \ldots, x_n, y_1, \ldots, y_n) = g(\alpha_1(x_1, \ldots, x_n), y_1, \ldots, y_n)$$

(with only one decomposition function).[4]

PROOF: Let $f_1 \in B_n$ be a function whose ROBDD size is larger than $\frac{2^n}{3n}$ (regardless of the variable order). Let f_2 and f_3 be two different functions from B_n, such that w.l.o.g. there is an $(\epsilon_1, \ldots, \epsilon_n) \in \{0, 1\}^n$, such that $f_2(\epsilon_1, \ldots, \epsilon_n) = 0$ and $f_3(\epsilon_1, \ldots, \epsilon_n) = 1$.
The function $f \in B_{2n}$, which is defined by

$$f(x_1, \ldots, x_n, y_1, \ldots, y_n) =$$
$$\overline{f_1(x_1, \ldots, x_n)} \cdot f_2(y_1, \ldots, y_n) + f_1(x_1, \ldots, x_n) \cdot f_3(y_1, \ldots, y_n)$$

fulfills the properties required by the lemma:

---

[4]Here f has good decomposition properties, but a large ROBDD size. However good decomposition properties are not guaranteed for decomposition and composition functions (with regard to a recursive application of the decomposition method).

1. The size of an ROBDD for f is larger than $\frac{2^n}{3n}$ (regardless of the variable order):

   It holds that $f_{y_1^{\epsilon_1} \ldots y_n^{\epsilon_n}} = f_1$ and the ROBDD of f_1 has more than $\frac{2^n}{3n}$ different nodes (regardless of the variable order). Thus the ROBDD for f also has more than $\frac{2^n}{3n}$ nodes, because the ROBDD of a cofactor of f (respecting the same variable order) cannot be larger than the ROBDD for f. ¶

2. It is clear that f has the decomposition given in the lemma. For example, choose

   - $\alpha_1 = f_1$ and
   - $g \in B_{n+1}$ with $g(a, y_1, \ldots, y_n) = mux(a, f_2(y_1, \ldots, y_n), f_3(y_1, \ldots, y_n))$, and $mux \in B_3$ defined by

   $$mux(s, x_1, x_2) = \begin{cases} x_1, & \text{if } s = 0 \\ x_2, & \text{if } s = 1. \end{cases}$$

   $\square$

3.7. Variable partitioning

As already shown in Example 3.1, the number of decomposition functions in a communication minimal decomposition strongly depends on the choice of the underlying variable set or variable partition. This section deals with the search for good variable sets or partitions which lead to decompositions, where the number of decomposition functions is as small as possible.

3.7.1 Partitioning for bound set of fixed size

First we suppose that we are looking for one-sided decompositions of a function $f \in B_n$ with a fixed size p of the bound set, i.e., the decomposition is done with respect to a variable set of size p.

If the number $\binom{n}{p}$ is not too large, we can determine for all $\binom{n}{p}$ subsets $X^{(1)}$ of the variable set X the number $\lceil \log(nrp(X^{(1)}, f)) \rceil$ of decomposition functions, and then we select the subset $X^{(1)}$ with the minimum number of decomposition functions.

If f is given by an ROBDD F, the number $nrp(X^{(1)}, f)$ for the set $X^{(1)}$ consisting of the first p variables in the variable order is computed (as shown in

---

¶ We obtain an OBDD of a cofactor $f_{x^{\epsilon_i}}$ from the ROBDD of f by removing nodes with label x_i and by connecting all incoming edges of a removed node n with the ϵ_i-son of n. After that the OBDD possibly has to be reduced to an ROBDD. In every case we obtain an ROBDD which is not larger than the ROBDD for f.

Section 3.6.1) by counting linking nodes immediately below a cut line in the ROBDD after the first p variables. (The linking nodes can be determined during a depth-first traversal of the ROBDD in linear time in the size of the ROBDD.) To determine $nrp(Y^{(1)}, f)$ for other subsets $Y^{(1)}$ of X with p elements, the variable order has to be changed, such that the variables in $Y^{(1)}$ are located before the other variables in the variable order. To compute $nrp(Y^{(1)}, f)$ for all subsets $Y^{(1)}$ of X with p elements, starting from $F_1 = F$, a sequence of $\binom{n}{p}$ different ROBDDs $F_1, \ldots, F_{\binom{n}{p}}$ is generated by exchanging pairs of variables in the order of the current ROBDD F_i, such that the subsets $X_i^{(1)}$ of the first p variables in the order of F_i are pairwise different. We select the variable set $X_i^{(1)}$ where $nrp(X_i^{(1)}, f)$ is minimal. To exchange a pair of variables $x_{index(k)}$ and $x_{index(j)}$ in the order of F_i either ROBDD operations are used[6] or the variable exchange is reduced to a repeated exchange of adjacent variables.[7]

Enumeration of all partitions of fixed size was already used in the functional decomposition method (Murgai et al., 1991), which is integrated in *sis* (Sentovich et al., 1992). [8] In contrast to methods described here this decomposition method is not based on ROBDD representations, but on sum-of-products representations. The enumerative approach was adapted for ROBDD based functional decomposition also by Lai et al. in (Lai et al., 1993b) and by Sasao (Sasao, 1993).

However if n and p are large, we cannot expect to be able to generate all $\binom{n}{p}$ ROBDDs to determine an appropriate variable set. For large sizes of n and p we have to be content with determining a good bound set heuristically.

A heuristic solution to the bound set selection problem was also given in (Shen et al., 1995). However this solution works only based on a sum-of-products representation of the function which has to be decomposed. Here we focus on ROBDD representations.

Although there are functions with good decomposition properties, which do not have a compact ROBDD representation (see Section 3.6.3), it would be a good heuristic method to start the search for good decompositions of a Boolean function F with small ROBDD representations of f. To obtain small ROBDD representations of f, variable reordering heuristics such as *sifting* and *symmetric sifting* (see also Sections 2.1.3 and 2.1.4) can be used. After variable reordering we obtain an ROBDD F' for f and a variable set $X_1^{(1)}$ with p variables, which

---

[6]One possibility is to compute the ROBDD F_i' for $f' = \overline{x_{index(k)}} \, \overline{x_{index(j)}} \cdot f_{\overline{x_{index(k)}} \, \overline{x_{index(j)}}} + \overline{x_{index(k)}} x_{index(j)} \cdot f_{x_{index(k)} \overline{x_{index(j)}}} + x_{index(k)} \overline{x_{index(j)}} \cdot f_{\overline{x_{index(k)}} x_{index(j)}} + x_{index(k)} x_{index(j)} \cdot f_{x_{index(k)} x_{index(j)}}$ and then to exchange labels $x_{index(k)}$ and $x_{index(j)}$. A second possibility is to use the exchange operation as defined in (Wang et al., 1993).

[7]Exchanging variables $x_{index(k)}$ and $x_{index(k+1)}$ is a local operation, i.e., affects only nodes labeled $x_{index(k)}$ and $x_{index(k+1)}$ and can be performed easily (Rudell, 1993).

[8]If there exists no nontrivial decomposition of this fixed size, Shannon decomposition is used instead.

consists of the first p variables in the order of F'. In a decomposition with respect to $X_1^{(1)}$ $\lceil \log(nrp(X_1^{(1)}, f)) \rceil$ different decomposition functions are needed.

Now we use a greedy search to find a subset of p variables needing fewer decomposition functions. In this greedy approach we try to minimize the number of decomposition functions by variable exchanges. We determine the pair of variables $x_k \in X_1^{(1)}$ and $x_j \in X \setminus X_1^{(1)}$, such that the number of decomposition functions in a decomposition with respect to $X_2^{(1)} = (X_1^{(1)} \setminus \{x_k\}) \cup \{x_j\}$ is locally minimal and exchange x_k and x_j in the variable order. The process is continued until $\log(nrp(X_i^{(1)}, f))$ no longer improves or until a given number of increases of $\log(nrp(X_i^{(1)}, f))$ compared to $\log(nrp(X_{i-1}^{(1)}, f))$ has been observed. Then we select the bound set $X_j^{(1)}$, where $\log(nrp(X_j^{(1)}, f))$ was the smallest.

If n is too large, the determination of the locally optimal variable pair for the exchange mentioned above can be too expensive. If we cannot examine all pairs of variables $x_k \in X_i^{(1)}$ and $x_j \in X \setminus X_i^{(1)}$, then an alternative method is to proceed in two steps to approximate the optimal variable pair: In a first step we choose $x_k \in X_i^{(1)}$, such that $nrp(X_i^{(1)} \setminus \{x_k\}, f)$ is minimal and we set $X_{i+1}^{(1)'} := X_i^{(1)} \setminus \{x_k\}$. In a second step we choose $x_j \in X \setminus X_{i+1}^{(1)'}$ such that $nrp(X_{i+1}^{(1)'} \cup \{x_j\})$ is minimal and use $X_{i+1}^{(1)} := X_{i+1}^{(1)'} \cup \{x_j\}$ as the new bound set. A similar method was used by Schlichtmann in (Schlichtmann, 1993).

Up to now we have assumed that the number p of variables in the bound set was given. This can be the case if we confine our search to balanced decompositions or if we target a special technology which suggests a certain choice of p.[9] However it can be the case that decompositions with other sizes p of the bound set will lead to much better realizations or that there is no variable set of size p which leads to a nontrivial decomposition. Since we restrict our search to nontrivial decompositions[10], p has to be changed in this case. According to Corollary 3.3 (for $n \geq 4$) $p = n - 1$ would certainly be suitable.

3.7.2 Partitioning for bound set of variable size

Another possibility is to determine p dynamically instead of using a fixed p:

If f is given by an ROBDD F with variable order $<_{index}$, then $nrp(f, \{x_{index(1)}, \ldots, x_{index(p)}\})$ can be determined by a depth-first traversal of the ROBDD. It is

---

[9]E.g. if lookup table based FPGAs (Field Programmable Gate Arrays) are used, Boolean functions up to a given constant b of inputs can be realized by one functional unit (LUT = lookup table). In that case the choice $p = b$ will be preferred.

[10]Otherwise a termination of the decomposition method, which is applied recursively to decomposition and composition functions, is not guaranteed.

clear that it is also possible to count the number of linking nodes below *different* cut lines in parallel (after variable $x_{index(1)}$, after $x_{index(2)}$ etc.). Thus it is possible to compute $nrp(f, \{x_{index(1)}, \ldots, x_{index(p)}\})$ for all $p = 1, \ldots, n-1$ during one traversal of the ROBDD.

To arrive at a good choice for p, it is no longer enough to choose the realization with the minimum number $\lceil \log(nrp(f, \{x_{index(1)}, \ldots, x_{index(p)}\})) \rceil$ of decomposition functions. Instead, the cost of a realization using a decomposition with respect to $\{x_{index(1)}, \ldots, x_{index(p)}\}$ now has to be roughly estimated to obtain a decision for a certain size p of the bound set. If $r_p = \lceil \log(nrp(f, \{x_{index(1)}, \ldots, x_{index(p)}\})) \rceil$ is the minimum number of decomposition functions in a decomposition with respect to $\{x_{index(1)}, \ldots, x_{index(p)}\}$, then r_p different decomposition functions from B_p and one composition function from B_{n-p+r_p} have to be realized. When $cost(in)$ denotes the estimated cost of a Boolean function with in inputs and one output, then the estimated cost of a decomposition with respect to $\{x_{index(1)}, \ldots, x_{index(p)}\}$ is

$$est\_cost_p(<_{index}) = \begin{cases} sharing(r_p) \cdot r_p \cdot cost(p) \\ \qquad + cost(n - p + r_p), & \text{if } r_p < p, \\ \infty, & \text{if } r_p = p, \end{cases}$$

where $sharing(r_p)$ is a factor which takes into account that identical subcircuits can be reused in the realization of r_p different Boolean functions (cf. Chapter 4). Thus *sharing* has to be a monotonic decreasing function with $sharing(r_p) \leq 1$ (e.g. $sharing(r_p) = (r_p)^c$ with $-1 < c \leq 0$).[11] We can refine our cost estimation for nontrivial decompositions by additionally taking into consideration the fact that a strict decomposition function of f cannot depend on variables which f does not depend on, now resulting in

$$est\_cost_p(<_{index}) = sharing(r_p) \cdot r_p \cdot$$
$$cost(ess\_var(\{x_{index(1)}, \ldots, x_{index(p)}\}))$$
$$+ cost(ess\_var(\{x_{index(p+1)}, \ldots, x_{index(n)}\}) + r_p) \qquad (3.5)$$

where for each subset Y of the variables of f $ess\_var(Y)$ gives the number of variables from Y which f essentially depends on.

For a pessimistic cost estimation we can choose $cost(n) = \frac{2^n}{n}$ (see Shannon effect, Chapter 1); in any case $cost$ should be a monotonic increasing function (e.g. $cost(n) = n$, $cost(n) = n^2$, or $cost(n) = n^3$).[12] Moreover, the choice of $cost$ should reflect the special needs of the technology in which the circuits

---

[11]The experimental results in Chapters 4 and 5 were obtained with $sharing(r_p) = (r_p)^{-0.1}$.
[12]The experimental results in Chapters 4 and 5 were obtained with $cost(n) = 0.3n^2$ for $n \geq 3$.

are realized: If lookup table based FPGAs are used which can realize functions up to b inputs by one LUT, then *cost* will estimate the number of LUT's and $cost(n)$ will be 1 for $n \leq b$.

For a fixed variable order $<_{index}$ the size p of the bound set is chosen, where $est\_cost_p(<_{index})$ is minimal. In the following we will refer to this 'optimal choice' of p for variable order $<_{index}$ as $p_{opt}(<_{index})$ and to the corresponding cost estimation as $min\_cost(<_{index}) = min_{2 \leq p \leq n-1} est\_cost_p(<_{index})$. To determine a good bound set for decomposition we use a greedy method similar to the method for fixed p:

1. Starting with an ROBDD F for $f \in B_n$, use a variable reordering algorithm to minimize the ROBDD size. The result is an ROBDD F_1 with variable order $<_{index_1}$. Determine $p_{opt}(<_{index_1})$ and $min\_cost(<_{index_1})$. Set $i = 1$ and *failed* $= 0$.

2. Determine the pair of variables $x_{index_i(k)}$, $x_{index_i(j)}$ $(k \neq j)$, such that after exchanging $x_{index_i(k)}$ and $x_{index_i(j)}$ for the resulting variable order $<_{index_{i+1}}$ $min\_cost(<_{index_{i+1}})$ is minimal[13]. The resulting ROBDD after variable exchange is denoted as F_{i+1}.

3. If $min\_cost(<_{index_i}) \leq min\_cost(<_{index_{i+1}})$: Set *failed* $=$ *failed* $+ 1$.

4. If *failed* exceeds a given limit: Stop, otherwise: Set $i = i + 1$ and continue with 2..

When the algorithm above stops, the variable order $<_{index_i}$ where $min\_cost(<_{index_i})$ was minimal is chosen. The computed bound set is set to $\{x_{index_i(1)}, \ldots, x_{index_i(p_{opt}(<_{index_i}))}\}$.

3.7.3 Other cost functions

Of course, cost functions other than the function of formula (3.5) can also be used. Formula (3.5) is used to estimate the area of the realization obtained by decomposition. With this formula for look-up table based FPGAs, the number of look-up tables can be estimated and for realizations by Ω–circuits the number of cells can be estimated.

Another possibility is to estimate the delay or depth of the resulting circuit as well, or some combination of area and delay as an alternative optimization goal.

---

[13]Consider only variable pairs, such that the resulting variable order did not occur in $<_{index_1}, \ldots, <_{index_i}$.

A bound set selection, which also takes delay into consideration, was presented by Legl et al. in (Legl et al., 1996a).

Here we use a unit delay model for look-up table based FPGAs, i.e., it is assumed that one look-up table (LUT) has a delay of 1. The delay of a LUT based FPGA in the unit delay model is the same as the depth of the corresponding Ω–circuit according to Definition 1.11 on page 6, when Ω is the set of all functions which can be realized by one LUT. Now each signal s in the FPGA circuit is associated with an 'arrival time' $at(s)$. Usually primary inputs are associated with arrival time 0. For each signal s we consider all paths from some primary input to this signal. Each path has a delay which is equal to the arrival time of the input plus the number of LUTs on the path. The arrival time $at(s)$ for s is defined as the maximum of all delays for paths from a primary input to signal s.

If the arrival times of the primary inputs are known and if the recursive procedure, which realizes decomposition and composition functions recursively, does process decomposition functions first and composition functions after that, then we know the arrival times for the inputs in each application of the recursive procedure. Now let us assume that a function $f \in B_n$ with input variables $x_1, \ldots, x_n$ has to be decomposed and that the variables x_i (which correspond to signals in the circuit) are associated with arrival times $at(x_i)$ $(1 \le i \le n)$. f is given by an ROBDD F with variable order $<_{index}$. Moreover, assume a bound set $X^{(1)} = \{x_{index(1)}, \ldots, x_{index(p)}\}$ and a free set $X^{(2)} = \{x_{index(p+1)}, \ldots, x_{index(n)}\}$ and assume $r_p = \lceil \log(nrp(f, \{x_{index(1)}, \ldots, x_{index(p)}\})) \rceil$.

To estimate the delay cost of a decomposition with bound set $X^{(1)}$, we need an estimate $delay(k)$ for the delay of realizations for functions with k input variables. Since functions with $k \le b$ input variables can be realized by one LUT, we set $delay(k) = 1$ for $k \le b$. For the general case the following estimate is used in (Legl et al., 1996a):

$$delay(k) = \begin{cases} 1, & \text{if } k \le b \\ k - b + 1, & \text{if } k > b. \end{cases}$$

This delay estimate $delay(k)$ is equal to the delay of a LUT circuit obtained by an iterated Shannon decomposition and thus is a worst-case estimate for the propagation delay in the unit delay model (at least if $b \ge 3$ and thus a multiplexer can be realized by one LUT).

Using this delay estimate we can estimate the delay cost of a decomposition with bound set $X^{(1)}$:

$$delay\_cost_p(<_{index}) =$$
$$= \max \left[\max_{x_i \in X^{(1)}} at(x_i) + delay(ess\_var(X^{(1)})), \right.$$

$$\left. \max_{x_j \in X^{(2)}} at(x_j) \right] + delay(r_p + ess\_var(X^{(2)})) \qquad (3.6)$$

In this formula $delay(ess\_var(X^{(1)}))$ estimates the delay of the decomposition functions and $delay(r_p + ess\_var(X^{(2)}))$ estimates the delay of the composition function.

Of course, to consider both area and delay, we can also use some combination of the area estimation of formula (3.5) and the delay estimation of formula (3.6) (e.g. $\sigma \cdot est\_cost_p(<_{index}) + (1 - \sigma) \cdot delay\_cost_p(<_{index})$ for some $\sigma \in (0, 1)$) as an optimization criterion for the choice of the variable order and the size of the bound set.

3.8. The encoding problem

In *strict* decompositions of a single-output function f with respect to $X^{(1)} = \{x_1, \ldots, x_p\}$, the computation of decomposition functions $\alpha_1, \ldots, \alpha_r$ and a composition function g can be viewed as an encoding of equivalence classes $K_i \in \{0, 1\}^p/\equiv$ by $\alpha = (\alpha_1, \ldots \alpha_r)$. When a decomposition is not strict, then *subsets* of the equivalence classes K_i are encoded by α.

In the decompositions, which were first considered by Ashenhurst (Ashenhurst, 1959), the encoding process did not play an important role: The so–called *simple decompositions* or *Ashenhurst decompositions* are communication minimal decompositions of the form

$$f(x_1, \ldots, x_p, x_{p+1}, \ldots, x_n) = g(\alpha_1(x_1, \ldots, x_p), x_{p+1}, \ldots, x_n),$$

i.e., there is only one decomposition function and $\{0, 1\}^p/\equiv$ has exactly two equivalence classes K_1 and K_2. Thus there are only two possibilities for the encoding:

$$\alpha_1(K_1) = 0 \text{ and } \alpha_1(K_2) = 1$$

or

$$\alpha_1'(K_1) = 1 \text{ and } \alpha_1'(K_2) = 0.$$

Since $\alpha_1' = \overline{\alpha_1}$, there are only two different decompositions due to the different encodings in Ashenhurst decompositions:

$$f(x_1, \ldots, x_p, x_{p+1}, \ldots, x_n) = g(\alpha_1(x_1, \ldots, x_p), x_{p+1}, \ldots, x_n)$$

and

$$f(x_1, \ldots, x_p, x_{p+1}, \ldots, x_n) = g'(\overline{\alpha_1}(x_1, \ldots, x_p), x_{p+1}, \ldots, x_n)$$

Figure 3.11.   Decompositions with different encodings.

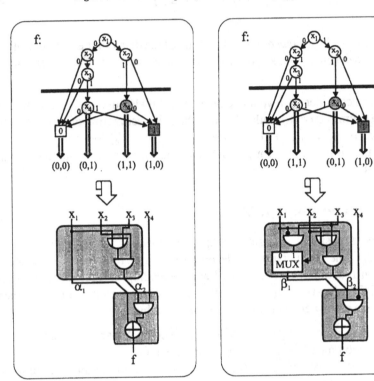

with $g'(a_1, x_{p+1}, \ldots, x_n) = g(\overline{a_1}, x_{p+1}, \ldots, x_n)$ for all $(a_1, x_{p+1}, \ldots, x_n) \in \{0,1\}^{n-p+1}$. When, for example, a lookup table based FPGA is used to realize f, the two possible solutions lead to a circuit with the same cost.

However, if the number of equivalence classes in $\{0,1\}^p/\equiv$ is greater than 2, encoding becomes an important problem. The next example shows two different strict decompositions of the same functions, where different encodings lead to realizations with different costs:

EXAMPLE 3.8 Figure 3.11 shows two decompositions of a function f with respect to variables $\{x_1, x_2, x_3\}$. There are $nrp(\{x_1, x_2, x_3\}, f) = 4$ different equivalence classes of the equivalence relation $\equiv$: $K_1 = \{(0,0,0), (0,0,1), (0, 1,0)\}$, $K_2 = \{(0,1,1)\}$, $K_3 = \{(1,1,0),(1,1,1)\}$ and $K_4 = \{(1,0,0), (1,0,1)\}$. The decomposition on the left hand side is performed with $\alpha(K_1) = (0,0)$, $\alpha(K_2) = (0,1)$, $\alpha(K_3) = (1,1)$ and $\alpha(K_4) = (1,0)$. The decomposition on the right hand side is performed with $\beta(K_1) = (0,0)$, $\beta(K_2) = (1,1)$, $\beta(K_3) = (0,1)$ and $\beta(K_4) = (1,0)$. Figure 3.11 shows that the different encodings of the equivalence classes lead to realizations with different costs: In the first decomposition $\alpha_1(x_1, x_2, x_3) = x_1$, such that no gates are needed to realize α_1, but in the second decomposition this is not true for β_1. The different costs are only due to a different encoding of classes K_2 and K_3 in the two decompositions.

This small example shows already that there is a degree of freedom in the choice of decomposition functions to decompose a Boolean function and that it can be utilized to arrive at better solutions.

There are several possible approaches for operating with the freedom mentioned above. In the following sections we give an overview of various methods.

3.8.1   Strict decompositions and symmetries

When $nrp(X^{(1)}, f)$ is not a power of two, we are free to use strict or non-strict decomposition functions in communication minimal decompositions. However, we generally prefer the selection of *strict* decompositions (Scholl, 1996; Scholl, 1997; Scholl, 1998), since strict decompositions have the property that they preserve symmetry properties of the decomposed functions. Symmetries are important structural properties of Boolean functions occurring in practical applications and they are in many cases responsible for the fact that Boolean functions can be realized at far below the worst case cost (cf. Section 1.5). In the following we will first look into the relevance of symmetries, especially for decompositions, and then we will prove that strict decompositions preserve symmetry properties.

3.8.1.1   Relevance of symmetries for decomposition

Here we will show how symmetries of completely specified Boolean functions can be used in the decomposition method. Information about utilization of symmetries of incompletely specified Boolean functions can be found in Chapter 5 (in combination with results of Section 2.2.2).

If we know that a Boolean function $f \in B_n$ is symmetric in a subset $X^{(1)} = \{x_1, \ldots, x_p\}$ of the inputs variables ($p > 2$), i.e., f does not change when variables from $X^{(1)}$ are exchanged, then we can make use of this knowledge to perform a nontrivial decomposition of f with respect to $X^{(1)}$. It is clear that the decomposition matrix $Z(X^{(1)}, f)$ with respect to $X^{(1)}$ can have at most $(p + 1)$ different row patterns, since for all $(\epsilon_1, \ldots, \epsilon_p, \epsilon_{p+1}, \ldots, \epsilon_n)$ and $(\delta_1, \ldots, \delta_p, \epsilon_{p+1}, \ldots, \epsilon_n) \in \{0, 1\}^n$ with the same 1–weight $w^1_{X^{(1)}}$ of the $X^{(1)}$–part, the function value of f is the same. Thus rows of the decomposition matrix with indices having the same 1–weight will have the same row pattern. So f (with $p > 2$) can be nontrivially decomposed in the form

$$f(x_1, \ldots, x_p, x_{p+1}, \ldots, x_n) =$$
$$g(\alpha_1(x_1, \ldots, x_p), \ldots, \alpha_r(x_1, \ldots, x_p), x_{p+1}, \ldots, x_n)$$

with $r \leq \lceil \log(p+1) \rceil$.

Analogous results hold when f is symmetric not in *all* pairs of variables of $X^{(1)}$ and when we also consider equivalence symmetries. We obtain the following theorem:

THEOREM 3.3 *Let $f \in B_n$ be a completely specified Boolean function with variable set $X = \{x_1, \ldots, x_n\}$ and let $X^{(1)} = \{x_1, \ldots, x_p\}$ be a subset of X. Let $\sim_{esym}$ be an equivalence relation defined on $X^{(1)}$ with*

$$x_i \sim_{esym} x_j \quad \Longleftrightarrow \quad f \text{ is extended symmetric}^{14} \text{in } (x_i, x_j).$$

Let $P_{\sim_{esym}} = \{\mu_1, \ldots, \mu_l\}$ be the partition of $X^{(1)}$, which is formed by the equivalence classes of the relation $\sim_{esym}$. Moreover w.l.o.g. let $\mu_1, \ldots, \mu_t$ ($0 \leq t \leq l$) be the equivalence classes which contain multiply symmetric variable pairs (which implies that f is multiply symmetric in all pairs of variables in μ_i ($1 \leq i \leq t$)).[15]
Then we have the following upper bound on the number of different row patterns of the decomposition matrix $Z(X^{(1)}, f)$ of f with respect to $X^{(1)}$:

$$nrp(X^{(1)}, f) \leq 2^t \cdot (|\mu_{t+1}| + 1) \cdot (|\mu_{t+2}| + 1) \cdot \ldots \cdot (|\mu_l| + 1).$$

Thus there is a decomposition of f with respect to $X^{(1)}$ having at most

$$\lceil \log(2^t \cdot (|\mu_{t+1}|+1) \cdot (|\mu_{t+2}|+1) \cdot \ldots \cdot (|\mu_l|+1)) \rceil = \left\lceil \sum_{i=t+1}^{l} \log(|\mu_i| + 1) \right\rceil + t$$

decomposition functions.

PROOF: See Appendix G.   □

If large symmetry sets of a function f are known (whereby symmetry here means nonequivalence *or* equivalence symmetry)), then Theorem 3.3 suggests putting variables of these symmetry sets together into the bound set $X^{(1)}$. Although it is

---

[14]According to Section 2.1.4.4, f is extended symmetric in a pair of variables if f is (nonequivalence) symmetric *or* equivalence symmetric in this pair of variables.
[15]see Lemma 2.4, page 38

possible to construct examples of functions, where putting pairs of symmetric variables together in the bound set does not lead to a decomposition with a minimum number of decomposition functions (Scholl and Molitor, 1993), it seems to be a good heuristic method to combine symmetric variables into the bound set, because — according to Theorem 3.3 — large symmetry sets lead to a small number of decomposition functions a-priori.

3.8.1.2 Advantages of strict decompositions

Boolean functions occurring in practical applications show certain structural properties which can be exploited to obtain good realizations. Examples of such structural properties are, among other things, independency of some input variables or symmetries. The importance of symmetry properties for decomposition was already discussed in the previous section. *Strict* decompositions are able to preserve such structural properties. For example, all strict decomposition functions of a Boolean function which is decomposed with respect to subset $\{x_1, \ldots, x_p\}$ of the variables and which does not depend on a variable $x_i \in \{x_1, \ldots, x_p\}$ do not depend on x_i either. If the function is symmetric in a pair (x_i, x_j) of variables in $\{x_1, \ldots, x_p\}$, then all strict decomposition functions are symmetric in (x_i, x_j).

This property of strict decomposition functions can easily be proved for more general symmetries as well. Before we prove this property, we will define a more general notion of symmetry, the so–called *invariance under permutations of* $\{0,1\}^n$ (Hotz, 1974).

DEFINITION 3.7 *Let* $S_{\{0,1\}^n}$ *be the set of all permutations of* $\{0,1\}^n$. *A Boolean function* $f \in B_n$ *is invariant under* $\tau \in S_{\{0,1\}^n}$ *if and only if for all* $(\epsilon_1, \ldots, \epsilon_n) \in \{0,1\}^n$ *the equation* $f(\epsilon_1, \ldots, \epsilon_n) = f(\tau(\epsilon_1, \ldots, \epsilon_n))$ *holds.* f *is invariant under a subset* $H \subset S_{\{0,1\}^n}$ *if and only if* f *is invariant under all* $\tau \in H$.

REMARK 3.2 *G–symmetry as defined in Definition 2.5 on page 30 is a special case of invariance under a subset of* $S_{\{0,1\}^n}$.

For our result on strict decomposition functions we need permutations of $S_{\{0,1\}^p}$ (p is the number of bound set variables), which, however, can easily be embedded into $S_{\{0,1\}^n}$ (n is the total number of variables of a function f):

NOTATION 3.3 *Each permutation* τ *of* $S_{\{0,1\}^p}$ *can be embedded into* $S_{\{0,1\}^n}$ ($n > p$) *using the embedding* $eb_{p,n}(\tau)$ *with*

$$eb_{p,n}(\tau)(\epsilon_1, \ldots, \epsilon_n) = (\delta_1, \ldots, \delta_p, \epsilon_{p+1}, \ldots, \epsilon_n)$$

$$\text{with } (\delta_1, \ldots, \delta_p) = \tau(\epsilon_1, \ldots, \epsilon_p) \; \forall (\epsilon_1, \ldots, \epsilon_n) \in \{0,1\}^n.$$

Subsets G of $\mathbf{S}_{\{0,1\}^p}$ can be embedded into $\mathbf{S}_{\{0,1\}^n}$ by

$$eb_{p,n}(G) = \{eb_{p,n}(\tau) \mid \tau \in G\}.$$

Now we can prove the following Theorem:

THEOREM 3.4 *Let $f \in B_n$ be invariant under $eb_{p,n}(G)$, $G \subseteq \mathbf{S}_{\{0,1\}^p}$. Then every strict decomposition function of a decomposition of f with respect to $\{x_1, \ldots, x_p\}$ is invariant under G.*

PROOF: Choose an arbitrary $\tau \in G$. We have to prove that every strict decomposition function α_i of a decomposition of f with respect to $\{x_1, \ldots, x_p\}$ is invariant under τ.

Let $(\epsilon_1, \ldots, \epsilon_p) \in \{0,1\}^p$ be arbitrarily chosen and let $(\delta_1, \ldots, \delta_p) = \tau(\epsilon_1, \ldots, \epsilon_p)$. Then for all $(\gamma_{p+1}, \ldots, \gamma_n) \in \{0,1\}^{n-p}$

$$
\begin{aligned}
f(\epsilon_1, \ldots, \epsilon_p, \gamma_{p+1}, \ldots, \gamma_n) &= f(eb_{p,n}(\tau)(\epsilon_1, \ldots, \epsilon_p, \gamma_{p+1}, \ldots, \gamma_n)) \\
&= f(\delta_1, \ldots, \delta_p, \gamma_{p+1}, \ldots, \gamma_n),
\end{aligned}
$$

since f is invariant under $eb_{p,n}(\tau)$.

Consequently $(\epsilon_1, \ldots, \epsilon_p) \equiv (\delta_1, \ldots, \delta_p)$, where $\equiv$ is the equivalence relation, which is induced on $\{0,1\}^p$ by equality of row patterns in $Z(\{x_1, \ldots, x_p\}, f)$.

According to the definition of strict decomposition functions, for every strict decomposition function α_i of a decomposition of f with respect to $\{x_1, \ldots, x_p\}$

$$\alpha_i(\epsilon_1, \ldots, \epsilon_p) = \alpha_i(\delta_1, \ldots, \delta_p) = \alpha_i(\tau(\epsilon_1, \ldots, \epsilon_p)).$$

Since $(\epsilon_1, \ldots, \epsilon_p) \in \{0,1\}^p$ was arbitrarily chosen, α_i is invariant under τ. $\square$

Theorem 3.4 is proved for arbitrary subsets $G \subseteq \mathbf{S}_{\{0,1\}^p}$. Of course it is also true for more restricted subsets $G \subseteq \mathbf{P}_p \subseteq \mathbf{S}_{\{0,1\}^p}$, where (as in Section 2.1.4.1) $\mathbf{P}_p$ is a subset (or subgroup) of $\mathbf{S}_{\{0,1\}^p}$ which is generated by permutations $\sigma_{ik}(1 \le i < k \le p)$ and $\nu_i(1 \le i \le p)$ with $\forall \alpha \in \{0,1\}^p$ $\sigma_{ik}(\alpha_1, \ldots, \alpha_i, \ldots, \alpha_k, \ldots, \alpha_p) = (\alpha_1, \ldots, \alpha_k, \ldots, \alpha_i, \ldots, \alpha_p)$ and $\nu_i(\alpha_1, \ldots, \alpha_i, \ldots, \alpha_p) = (\alpha_1, \ldots, \overline{\alpha_i}, \ldots, \alpha_p)$. Thus our theorem implies that strict decompositions also preserve G–symmetries and — since equivalence and nonequivalence symmetries are special cases of G–symmetries — they also preserve equivalence and nonequivalence symmetries. This leads to the following corollary:

COROLLARY 3.4 *Let $f \in B_n$ be $eb_{p,n}(G)$–symmetric for $G \subseteq \mathbf{P}_p$. Then every strict decomposition function in a decomposition of f with respect to*

$\{x_1, \ldots, x_p\}$ *is G–symmetric for* $G \subseteq \mathbf{P}_p$. *In particular strict decomposition functions inherit properties such as independency of variables* $x_i \in \{x_1, \ldots, x_p\}$ *or symmetries in pairs of variables* $x_i, x_j \in \{x_1, \ldots, x_p\}$ *from the function f.*

3.8.2   Optimizing composition functions by input encoding

As shown above the restriction to strict decomposition functions helps to reduce the complexity of *decomposition* functions, assuming that the function which is being decomposed has structural properties which can be preserved by strict decomposition functions.

In (Murgai et al., 1994) Murgai et al. used another approach to reduce the complexity of the *composition* function. In their work they do not consider a minimization of the complexity of the decomposition function, since they assume a realization by lookup-table FPGAs, where the lookup tables have 5 inputs, and they use only decompositions with the number of bound set variables p equal to 5. In this case each decomposition function can be implemented by exactly one lookup table and the complexity of the decomposition functions does not have to be taken into account. However, if larger functions are decomposed, the probability of finding nontrivial decompositions with 5 inputs decreases (not only for random functions, but also for functions occurring in practical applications, see also Lemma 3.1 on page 82). Murgai et al. circumvent the problem by using Shannon decompositions, when there is no decomposition with fixed bound set size $p = 5$. However, in the general case, when we decompose functions with large numbers of inputs and do not restrict ourselves to decompositions with 5 inputs or Shannon decompositions, a consideration of the complexity of the decomposition functions is also needed.

Murgai et al. make use of existing algorithms for the symbolic input encoding problem (Saldanha et al., 1992) to minimize the complexity of the composition function. When a single–output function f is decomposed with bound variables $x_1, \ldots, x_p$ and free variables $x_{p+1}, \ldots, x_n$, then the composition function $g_{symbolic}$ *before encoding* is viewed as a function with variables $x_{p+1}, \ldots, x_n$ and an additional symbolic input variable X.

When a strict encoding is used, then the symbolic input variable X can take $nrp(\{x_1, \ldots, x_p\}, f)$ different symbolic values, such that there is exactly one symbolic value S_i for each equivalence class K_i. The function $g_{symbolic}$ is defined in such a way that

$$g_{symbolic}(S_i, \epsilon_{p+1}, \ldots, \epsilon_n) = f(\epsilon_1, \ldots, \epsilon_p, \epsilon_{p+1}, \ldots, \epsilon_n)$$

for arbitrary $(\epsilon_1, \ldots, \epsilon_p) \in K_i$.

The encoding problem consists of finding a good binary encoding for the equivalence classes, i.e., a good binary encoding for the symbolic values $S_1, \ldots,$ $S_{nrp(\{x_1,\ldots,x_p\},f)}$, such that each symbolic value is assigned a *distinct* binary code. The decision which binary code to choose for the symbolic value S_i, is equivalent to the question, which binary code should be produced by the decomposition function α for inputs $\epsilon \in K_i$.

The binary encoding of the symbolic values is done by a constraint satisfier (Saldanha et al., 1992), which can solve the symbolic input encoding problem and was originally presented for application to the state encoding problem for Finite State Machines. After encoding, a composition function g with binary variables results.

Murgai et al. also propose a version for non–strict encodings: Then the symbolic input variable X of $g_{symbolic}$ can have 2^p different symbolic values, one symbolic value for each vector in $\{0,1\}^p$. [16] Here not all symbolic values need to be encoded by *distinct* binary codes. Two symbolic values S_δ and S_ϵ for δ and $\epsilon \in \{0,1\}^p$ need to be encoded by different binary values only if they belong to different equivalence classes (cf. Corollary 3.1).

Of course, it is necessary to have some cost estimation for the composition function g to minimize the cost of g by symbolic input encoding. In (Murgai et al., 1994) the 'literal count' [17] of g is used, since the encoding methods for the minimization of Finite State Machines, which were used to compute g, make use of this cost function. However, this is not a realistic cost function, since g is not realized by a sum-of-products, but by applying decomposition recursively. As already noticed in (Murgai et al., 1994), for this reason minimizing the literal count of g may lead to FPGA realizations with an increased number of look-up tables. So a more realistic cost function for symbolic input encoding would be desirable.

3.8.3   Optimizing composition functions by look-ahead

Jiang et al. (Jiang et al., 1998) also present a method to encode equivalence classes to minimize the complexity of the composition function g. However, they do not use the literal count of g as a cost function like (Murgai et al., 1994); rather their goal is to find a (strict) encoding of the equivalence classes, which leads to a good decomposability of g in the next recursive decomposition step.

---

[16] The use of 2^p different symbolic values is possible, only since in (Murgai et al., 1994) the number of bound variables is limited to a small number $p = 5$.

[17] The literal count is defined for a sum-of-products representation of a Boolean function (not for the Boolean function itself). It is equal to the number of literals (positive or negative variables) in a sum-of-products representation p. That is, it is equal to $C(p) + 1$ as defined in Definition 1.15 on page 9 (except for the constant functions 0 and 1).

When a Boolean function $f \in B_n$ is decomposed, they perform a random encoding of the equivalence classes first, leading to a composition function g'. Based on this random encoding they compute for the decomposition of the composition function g' a good partition of its input variables into a bound set $X_{g'}^{(1)}$ and a free set $X_{g'}^{(2)}$. After that, a new encoding is determined for f, which (heuristically) minimizes the number of decomposition functions $\lceil \log(nrp(X_{g'}^{(1)}, g)) \rceil$ for the *new* composition function g using the bound set $X_{g'}^{(1)}$ *found before*. The random encoding for f has to be performed first, since the computation of a good encoding with respect to the later decomposition of the composition function g only works when a bound set for g is already known.

Suppose that f is decomposed with bound set $X^{(1)}$, free set $X^{(2)}$ and the number of equivalence classes is $c = nrp(X^{(1)}, f)$, i.e., there are c different cofactors $f_{K_1}, \ldots, f_{K_c}$ of f with respect to the variables in $X^{(1)}$, where each cofactor f_{K_j} is obtained by assigning an (arbitrary) constant $\epsilon \in K_j$ to the variables in $X^{(1)}$.

We need $\lceil \log c \rceil$ different decomposition functions and for the composition function g we have (apart from the variables in $X^{(2)}$) $\lceil \log c \rceil$ additional input variables $a_1, \ldots, a_{\lceil \log c \rceil}$, which we will call the 'a–variables' in the following. We assume that a partition of these variables into a bound set $X_{g'}^{(1)} = \{a_1, \ldots, a_r, x_{b_1}, \ldots, x_{b_{s_1}}\}$ and a free set $X_{g'}^{(2)} = \{a_{r+1}, \ldots, a_{\lceil \log c \rceil}, x_{h_1}, \ldots, x_{h_{s_2}}\}$ has already been computed during the random encoding step. [18]

As illustrated by Figure 3.12, the decomposition matrix for g with respect to $X_{g'}^{(1)} = \{a_1, \ldots, a_r, x_{b_1}, \ldots, x_{b_{s_1}}\}$ contains 2^r different blocks of 2^{s_1} rows, where each row block contains the rows with a fixed assignment to variables $a_1, \ldots, a_r$. Similarly the matrix contains $2^{\lceil \log c \rceil - r}$ different blocks of 2^{s_2} columns, where each column block contains the columns with a fixed assignment to variables $a_{r+1}, \ldots, a_{\lceil \log c \rceil}$. The submatrices belonging to a fixed row block and a fixed column block represent cofactors of g with respect to all a–variables.

In the following we will see that the submatrices of $Z(X_{g'}^{(1)}, g)$ belonging to a fixed row block and a fixed column block are decomposition matrices of the cofactors f_{K_j}: Suppose that the code for equivalence class K_j of f is $(\delta_1, \ldots, \delta_r, \delta_{r+1}, \ldots, \delta_{\lceil \log c \rceil})$ (i.e., $\alpha(K_j) = \{(\delta_1, \ldots, \delta_r, \delta_{r+1}, \ldots, \delta_{\lceil \log c \rceil})\}$). Then the cofactor of g where $a_1, \ldots, a_r, a_{r+1}, \ldots, a_{\lceil \log c \rceil}$ are set to $(\delta_1, \ldots, \delta_r,$

---

[18] $\{x_{b_1}, \ldots, x_{b_{s_1}}\} \cup \{x_{h_1}, \ldots, x_{h_{s_2}}\} = X^{(2)}$. To simplify notations in the following, we choose the *first r* a–variables to be included in $X_{g'}^{(1)}$. It will be clear that for the method given in (Jiang et al., 1998) only the *number* of a–variables included in $X_{g'}^{(1)}$ is of interest.

Figure 3.12.   Matrix $Z(X_{g'}^{(1)}, g)$.

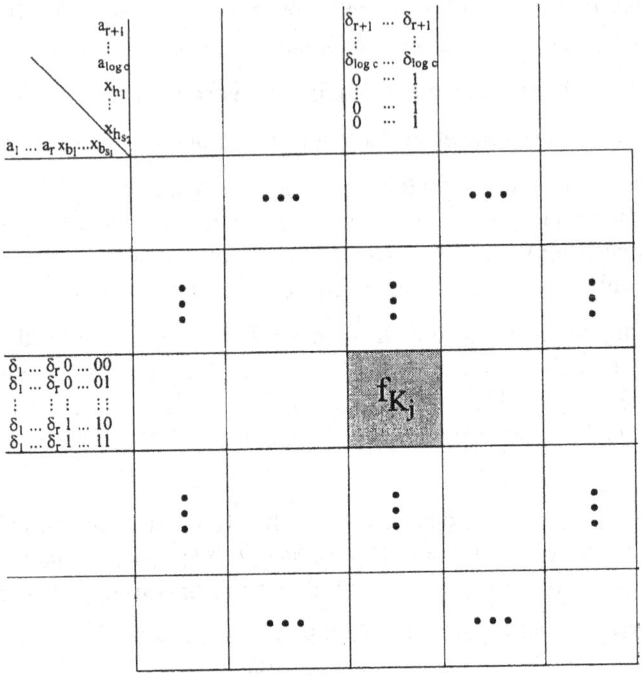

$\delta_{r+1}, \ldots, \delta_{\lceil \log c \rceil})$ is equal to f_{K_j}. The submatrix of $Z(X_{g'}^{(1)}, g)$ formed by the row block with $a_1, \ldots, a_r$ fixed to $(\delta_1, \ldots, \delta_r)$ and the column block with $a_{r+1}, \ldots, a_{\lceil \log c \rceil}$ fixed to $(\delta_{r+1}, \ldots, \delta_{\lceil \log c \rceil})$ is a representation of this cofactor of g and thus a representation of f_{K_j}. More precisely, this submatrix is equal to the decomposition matrix $Z(\{x_{b_1}, \ldots, x_{b_{s_1}}\}, f_{K_j})$ of f_{K_j} with respect to $\{x_{b_1}, \ldots, x_{b_{s_1}}\}$.

Thus encoding equivalence class K_j by $(\delta_1, \ldots, \delta_r, \delta_{r+1}, \ldots, \delta_{\lceil \log c \rceil})$ means placing the decomposition matrix $Z(\{x_{b_1}, \ldots, x_{b_{s_1}}\}, f_{K_j})$ of f_{K_j} into row block $(\delta_1, \ldots, \delta_r)$ and column block $(\delta_{r+1}, \ldots, \delta_{\lceil \log c \rceil})$ of the decomposition matrix $Z(X_{g'}^{(1)}, g)$ of g. In this sense finding an encoding for the equivalence classes means finding an arrangement of the decomposition matrices $Z(\{x_{b_1}, \ldots, x_{b_{s_1}}\}, f_{K_j})$ for $1 \le j \le c$ in the matrix $Z(X_{g'}^{(1)}, g)$ of g. [19]

---

[19]Here we assume that c is a power of 2. If c is not a power of 2, then there are submatrices which represent don't cares of g. These don't cares are assigned to minimize the number of different row patterns in $Z(X_{g'}^{(1)}, g)$ (cf. Section 2.2.1 and Chapter 5).

The task is now to find such an arrangement that the number $nrp(X_{g'}^{(1)}, g)$ of different row patterns in $Z(X_{g'}^{(1)}, g)$ is minimized. In this way, for the decomposition of f, an encoding of equivalence classes K_j is chosen, such that the composition function g can be decomposed using a small number of decomposition functions (assuming a bound set $X_{g'}^{(1)}$). To achieve this goal the following observations are of interest:

- If the sets of row patterns of $Z(\{x_{b_1}, \ldots, x_{b_{s_1}}\}, f_{K_i})$ and of $Z(\{x_{b_1}, \ldots, x_{b_{s_1}}\}, f_{K_j})$ are equal or similar, then these submatrices can be stacked into one *column* block, since this tends to keep the *total* number of different row patterns in $Z(X_{g'}^{(1)}, g)$ small — at least if only this column block is considered.

- If two submatrices $Z(\{x_{b_1}, \ldots, x_{b_{s_1}}\}, f_{K_i})$ and $Z(\{x_{b_1}, \ldots, x_{b_{s_1}}\}, f_{K_j})$ are put into the same *row* block, then in $Z(X_{g'}^{(1)}, g)$ two row patterns for rows k and l in this row block are different, if the patterns for rows k and l are different in $Z(\{x_{b_1}, \ldots, x_{b_{s_1}}\}, f_{K_i})$ or in $Z(\{x_{b_1}, \ldots, x_{b_{s_1}}\}, f_{K_j})$. If rows k and l of $Z(\{x_{b_1}, \ldots, x_{b_{s_1}}\}, f_{K_i})$ have the same row patterns and other submatrices have to be placed in the same row block, then those submatrices $Z(\{x_{b_1}, \ldots, x_{b_{s_1}}\}, f_{K_j})$ should be preferred for which the row patterns of rows k and l are equal as well.

Jiang et al. give heuristics which try to minimize $nrp(X_{g'}^{(1)}, g)$ based on the observations above by translating them into graph problems.

3.8.4   Support minimization for decomposition functions

Huang et al. (Huang et al., 1995) noticed that the choice of decomposition functions can be important even in the restricted case that lookup-table-based FPGAs with b-input LUTs are used and the decomposition is performed with respect to b variables. As described in Section 1.4, such architectures often provide Configurable Logic Blocks (CLBs) that can realize either one b-input function or two $(b-1)$–input functions, which depend on at most b different variables (e.g. Xilinx XC3000). In this case it makes sense to compute decomposition functions which depend only on $b-1$ inputs such that as many decomposition functions as possible can be pairwise combined into one CLB.

Legl et al. (Legl et al., 1996b) used a more general approach to compute decomposition functions with minimal support. [20] There is no restriction to the

---

[20] The support of a function f is the set of variables f essentially depends on, see Definition 2.1 on page 24.

number p of variables in the bound set of the decomposition and the optimization goal is not only to use many decomposition functions with $p-1$ variables, but to use decomposition functions with minimal support. It is assumed that a decomposition function with a smaller support can be realized at lower cost in further decompositions. In the extreme case of decomposition functions which depend only on one variable, no LUT is needed at all for the realization.

3.8.4.1 Support minimization for two-output CLBs

In the following we will describe the method by Huang (Huang et al., 1995) of computing decomposition functions which are independent of at least one variable. In (Huang et al., 1995) *strict* decomposition functions are used and the size of the bound set is fixed to 5, such that the method looks for decomposition functions with at most 4 variables.

Suppose that we decompose a function $f \in B_n$ and that a bound set $X^{(1)}$ of $p = 5$ variables is already selected. Identical row patterns of the decomposition matrix partition the elements of $\{0,1\}^p$ into equivalence classes $K_1, \ldots, K_{nrp(X^{(1)}, f)}$.

In the first step Huang et al. compute the set of all candidate functions of B_p with the following properties:

- They are strict in the sense that they assign the same value to all elements of an equivalence class K_j ($\forall 1 \le j \le nrp(X^{(1)}, f)$).

- There is at least one variable x_i in $X^{(1)}$ the candidate function does not depend on.

For a fixed variable $x_i \in X^{(1)}$ a strict candidate function α *not* depending on x_i has the following property:

If $(\epsilon_1, \ldots, \epsilon_i, \ldots, \epsilon_p) \in K_j$ and $(\epsilon_1, \ldots, \overline{\epsilon_i}, \ldots, \epsilon_p) \in K_l$ then $\alpha(K_j) = \alpha(K_l)$.

Now consider a graph $G = (V, E)$ where the nodes are equivalence classes K_j (i.e., $V = \{0,1\}^p/_{\equiv}$) and there is an edge between K_j and K_l if and only if there is a $(\epsilon_1, \ldots, \epsilon_i, \ldots, \epsilon_p) \in K_j$, such that $(\epsilon_1, \ldots, \overline{\epsilon_i}, \ldots, \epsilon_p) \in K_l$. Let CC_m be a connected component of G and let $U(CC_m) = \bigcup_{K_j \in CC_m} K_j$. Then it is clear that α has to assign the same function value to all elements of $U(CC_m)$, if α should be strict and independent from variable x_i. Different strict candidate functions not depending on x_i only differ in the values which they assign to the sets $U(CC_m)$. If there are cc connected components of G, then there are exactly 2^{cc} different strict candidate functions not depending on x_i.

This computation is repeated for all variables $x_i \in X^{(1)}$ and after the first step the set of all candidate functions which are independent of at least one variable has been computed. Let the number of all candidate functions be ncf.

In the second step a decomposition is computed which contains as many strict candidate functions as possible. In a first attempt the algorithm tries to compute a decomposition where all $r := \lceil \log(nrp(X^{(1)}, f)) \rceil$ decomposition functions are from this candidate set, i.e., all decomposition functions are independent of at least one variable. If this is not possible, it tries in a second attempt to compute decompositions with exactly $r - 1$ decomposition functions from the candidate set and so on. During an attempt to compute decompositions with t decomposition functions from the candidate set, $\binom{ncf}{t}$ different combinations of candidate functions are enumerated to check whether these combinations of decomposition functions can be used in communication minimal decompositions. The search space of this enumeration can be pruned when, during the search for t appropriate decomposition functions from the candidate set, it turns out that a subset of $s < t$ candidate functions cannot be completed to a communication minimal decomposition using $r - s$ additional decomposition functions* (even if these $r - s$ functions do not necessarily come from the candidate set). In this case all subsets of functions containing this set of s candidate functions do not need to be considered in the future. A necessary and sufficient criterion for deciding whether a subset of decomposition functions can be completed to a communication minimal decomposition will be given in Chapter 4, Lemma 4.1 (page 135).

The approach of Huang et al. of explicitly representing the set of all candidate functions profits from the fact that the size p of the bound set is fixed to a small number of 5 variables. If the size p of the bound set increases, then the sizes of the sets of candidate functions will also increase, such that even a pruned enumeration of all subsets of such a set will be too time-consuming. When p is fixed to 5, alternative strategies are needed to decompose functions which cannot be nontrivially decomposed with p bound variables.

3.8.4.2   Support minimization for the general case

Legl, Wurth and Eckl (Legl et al., 1996b) presented a decomposition method which is more general: It is also used for bound sets of sizes larger than 5 and it *minimizes* the number of variables which the decomposition functions depend on. This approach extends the search for decomposition functions to non–strict decompositions, such that the search space for support minimal decomposition functions is increased, but on the other hand, in doing so, the symmetry preserving properties of strict decomposition functions are not exploited (see Section 3.8.1).

Their method of decomposing a single-output function $f \in B_n$ with respect to a bound set $X^{(1)}$ has the following properties:

- Decomposition functions are computed step by step, always selecting single-output decomposition functions with minimal support.

- In each step the set of all functions is computed which can be used as decomposition functions in a communication minimal decomposition, assuming that the decomposition functions already selected remain fixed. These functions are called *assignable* functions. The set of these functions is represented *implicitly* by a single ROBDD.

- In addition, for each bound set variable x_i, the set of all functions depending only on $X^{(1)} \setminus \{x_i\}$ is also represented implicitly. To obtain decomposition functions which are independent of x_i, this set is intersected with the set of assignable functions mentioned above.

- If there are several possibilities of decomposition functions not depending on all variables of $X^{(1)}$ to choose from, then that function which depends on a minimal number of variables is selected.

The method of representing assignable functions implicitly by an ROBDD was introduced in (Wurth et al., 1995) and will be described in Section 4.5.2.2. To represent a set $S \subseteq B_p$ of functions, a 'characteristic function' $\chi_S \in B_{2^p}$ is used. $\chi_S(\epsilon_0, \ldots, \epsilon_{2^p-1}) = 1$ if and only if the function h_ϵ with function table $(\epsilon_0, \ldots, \epsilon_{2^p-1})$ is in S.[21] χ_S is represented by an ROBDD with 2^p variables $e_0, \ldots, e_{2^p-1}$.

Assume that we have a representation $\chi_{assignable}$ of the assignable candidate functions for the current decomposition step. Then for each variable x_i, the set of all functions in B_p which do not depend on x_i is computed based on the following consideration: Let $(\delta_1, \ldots, \delta_i, \ldots, \delta_p)$ and $(\delta_1, \ldots, \overline{\delta_i}, \ldots, \delta_p)$ be two elements of $\{0, 1\}^p$ which differ only in the value for variable x_i. Then for a function h, which does not depend on x_i, the function values $h(\delta_1, \ldots, \delta_i, \ldots, \delta_p)$ and $h(\delta_1, \ldots, \overline{\delta_i}, \ldots, \delta_p)$ are equal. The set of all functions with equal function values for $(\delta_1, \ldots, \delta_i, \ldots, \delta_p)$ and $(\delta_1, \ldots, \overline{\delta_i}, \ldots, \delta_p)$ is represented by the formula

$$e_{int(\delta_1, \ldots, \delta_i, \ldots, \delta_p)} \equiv e_{int(\delta_1, \ldots, \overline{\delta_i}, \ldots, \delta_p)}.$$

If changing the value of variable x_i does not change the function value of h *for all* $\delta \in \{0, 1\}^p$, then h is independent of x_i. Thus the set I_{x_i} of functions not

---

[21]That is, $h_\epsilon(\delta_1, \ldots, \delta_p) = \epsilon_{int(\delta_1, \ldots, \delta_p)}$ with $int(\delta_1, \ldots, \delta_p) = \sum_{j=1}^{p} \delta_j \cdot 2^{p-j}$.

depending on x_i is represented by

$$\chi_{I_{x_i}}(e_0,\ldots,e_{2^p-1}) = \bigwedge_{(\delta_1,\ldots,\delta_i,\ldots,\delta_p)} e_{int(\delta_1,\ldots,\delta_i,\ldots,\delta_p)} \equiv e_{int(\delta_1,\ldots,\overline{\delta_i},\ldots,\delta_p)}.$$

To obtain a representation $\chi_{assignable,x_i}$ of all assignable functions which do not depend on x_i we simply have to intersect the two characteristic functions:

$$\chi_{assignable,x_i} = \chi_{assignable} \cdot \chi_{I_{x_i}}.$$

Any element of the ON–set of $\chi_{assignable,x_i}$ represents a function table of an assignable function which is independent of x_i. To find a decomposition function which depends on a minimal number of variables, we are looking for a vector $(\epsilon_0,\ldots,\epsilon_{2^p-1})$ which is in the ON–set of as many characteristic functions $\chi_{assignable,x_i}$ as possible $(1 \le i \le p)$. This problem can be solved implicitly using Kam's *Lmax* algorithm (Kam et al., 1994).

Of course the efficiency of the proposed algorithm is limited by the large number of 2^p variables, which are needed for an implicit representation of the characteristic functions given above. To reduce the number of variables, Legl et al. propose a method which 'trades optimality for efficiency': A variable is used, not for each element in $\{0,1\}^p$, but only for certain subsets of $\{0,1\}^p$, such that the function values have to be equal for all elements of these subsets. This results in a restriction of the search space of candidate functions. More details can be found in (Legl et al., 1996b).

3.8.5 Common decomposition functions in multi-output decompositions

Another important strategy for the selection of decomposition functions comes into play when multi-output functions $f \in B_{n,m}$ $(m > 1)$ are decomposed: To reduce the overall cost of the implementation it is desirable to choose decomposition functions which can be shared among several output functions $f_i \in B_n$ $(1 \le i \le m)$. If it is possible to find decomposition functions which can be used in the communication minimal decomposition of *several* output functions f_i, then these decomposition functions have to be realized only once, but they can be used for several decompositions. Chapter 4 gives a detailed presentation of methods to achieve this goal.

Chapter 4

FUNCTIONAL DECOMPOSITION FOR COMPLETELY SPECIFIED MULTI-OUTPUT FUNCTIONS

The efficiency of good realizations of Boolean functions is often based on the fact that subcircuits are 'used more than once'. Identifying those subcircuits which several parts of the overall circuit may take advantage of, is a fundamental task in multi–level logic synthesis (see Figure 4.1).

Circuits for functions with several outputs do not realize the single–output functions separately. On the contrary, results which have been computed by subcircuits in the realization of *one* single–output function are reused in the realization of *other* single–output functions.

Nor for logic synthesis by decomposition is it advisable to realize the single outputs of a function $f \in B_{n,m}$ ($m > 1$) separately as single–output functions $f_1, \ldots, f_m$. To reuse *identical* subcircuits in the realization of *several* output functions, we *compute* decomposition functions which can be used in the communication minimal decompositions of several output functions (see Figure 4.2).

The identification of reusable subcircuits is essential for multi–level logic synthesis. This chapter gives a detailed presentation of our methods for solving this problem for decompositions of Boolean functions.[1] Alternative methods for the same problem are discussed in Section 4.5.[2]

For practical applicability it is important that the methods are not based on decomposition matrices and function tables, but rather on ROBDD representations.

---

[1] See also (Scholl and Molitor, 1993; Molitor and Scholl, 1994; Scholl and Molitor, 1994; Scholl and Molitor, 1995b; Scholl and Molitor, 1995a).

[2] See also (Lai et al., 1993b; Lai et al., 1994b; Lai et al., 1994a; Lai et al., 1996; Wurth et al., 1995; Sawada et al., 1995).

Figure 4.1. Identification of reusable subcircuits.

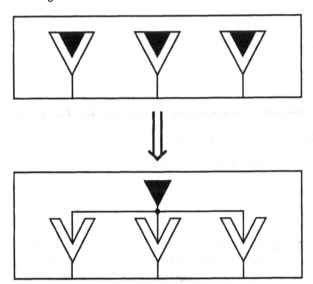

The chapter begins with the presentation of a small example as a motivation for the approach to compute common, reusable decomposition functions for multi–output functions (Section 4.1). The example demonstrates how the degree of freedom in the selection of decomposition functions can be used in the decomposition of multi–output functions to obtain reusable subcircuits (as common decomposition functions of several output functions).

Afterwards in Section 4.2, the method for partitioning input variables to find good bound sets for decomposition (see Section 3.7) is generalized for the decomposition of functions with more than one output.

After studying the complexity of the problem to compute common decomposition functions of several output functions, we finally derive an algorithm for solving the problem in Section 4.3. Here it turns out that the restriction to common *strict* decomposition functions not only serves to preserve structural properties of the original function (see Section 3.8.1), but it also makes a considerable contribution to speeding up the computation.

4.1.   A motivating example

This section illustrates the basic idea of the approach in Example 4.1:

EXAMPLE 4.1 Let $f \in B_{4,3}$ be a multi–output Boolean function with $f_1(x_1, \ldots, x_4) = \overline{x_4}x_2x_3 + x_4(x_1 \oplus x_2)$, $f_2(x_1, \ldots, x_4) = \overline{x_4}(x_1 \oplus x_2) + x_4(x_2$

Figure 4.2.   Decomposition of functions with several outputs.

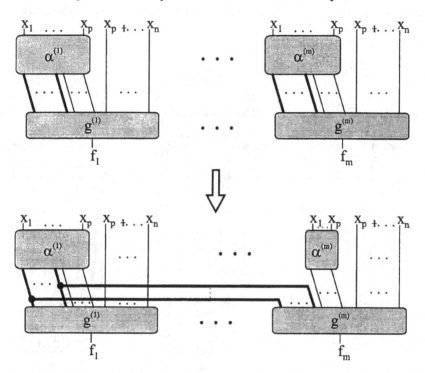

$+x_3$) and $f_3(x_1, \ldots, x_4) = \overline{x_4}(x_2 + x_3) + x_1 x_4$. f_1, f_2 and f_3 are decomposed with respect to the bound set $\{x_1, x_2, x_3\}$ (see Figure 4.3). There are 4 linking nodes, respectively, such that 2 decomposition functions are needed in each of the three decompositions. As already noticed in Section 3.8, there is some degree of freedom in the encoding of the linking nodes and the corresponding equivalence classes of $\{0, 1\}^3$: Figure 4.3 gives an encoding which leads to the realization also shown in the same figure. Apparently all decomposition functions $\alpha_1^{(1)}$, $\alpha_2^{(1)}$, $\alpha_1^{(2)}$, $\alpha_2^{(2)}$, $\alpha_1^{(3)}$ and $\alpha_2^{(3)}$ are pairwise different for this encoding.

Now we have the aim of choosing an encoding in such a way that the number of decomposition functions which can be used as common decomposition functions of f_1, f_2 and f_3 is maximized. Figure 4.4 shows an appropriate encoding. For this encoding both the decomposition functions $\alpha_2^{(1)}$ and $\alpha_1^{(2)}$ and the decomposition functions $\alpha_2^{(2)}$ and $\alpha_1^{(3)}$ are pairwise identical and thus they have to be realized only once.

Figure 4.3. Decomposition of the Boolean function from Example 4.1. $f_1(x_1, \ldots, x_4) = \overline{x_4} x_2 x_3 + x_4 (x_1 \oplus x_2)$, $f_2(x_1, \ldots, x_4) = \overline{x_4}(x_1 \oplus x_2) + x_4(x_2 + x_3)$ and $f_3(x_1, \ldots, x_4) = \overline{x_4}(x_2 + x_3) + x_1 x_4$.

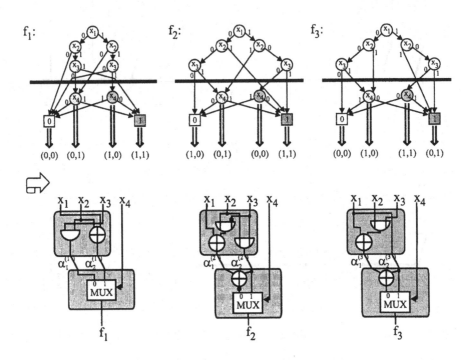

In contrast to other solutions such as (Hwang et al., 1990) the method of *computing* common decomposition functions, which is presented in Section 4.3, does not compare subfunctions in the hope of finding some subfunctions which are identical by chance, but we are really working towards finding identical subfunctions. Encodings like the one in the example above are *computed* with the aim of obtaining as many decomposition functions as possible which can be used as *common* decomposition functions in the decomposition of several output functions.

However, this approach is not only important for processing multi–output Boolean functions. Even if a Boolean function originally has only one output, we will usually obtain multi–output Boolean functions when we apply decomposition recursively to the decomposition functions.

As an alternative to our approach of computing communication minimal decompositions of all output functions of a multi–output Boolean function with the additional constraint to compute as many common decomposition functions as possible, we could also think of a decomposition of a multi–output function

Figure 4.4. Decomposition of the Boolean function from example 4.1. The encoding chosen here is different from that of Figure 4.3.

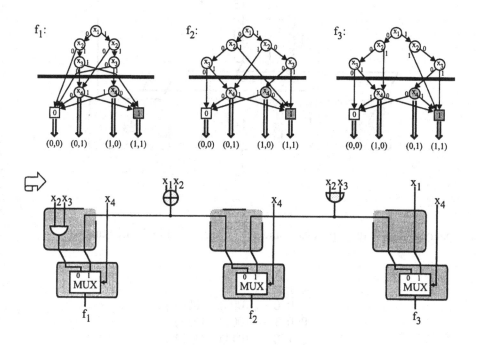

as a whole: In this case a multi–output function $f \in B_{n,m}$ is decomposed into a decomposition function $\alpha \in B_{p,r}$ and a composition function $g \in B_{n-p+r,m}$ such that

$$f(x_1, \ldots, x_p, x_{p+1}, \ldots, x_n) =$$
$$g(\alpha_1(x_1, \ldots, x_p), \ldots, \alpha_r(x_1, \ldots, x_p), x_{p+1}, \ldots, x_n). \qquad (\star)$$

This approach was chosen by Lai et al. in (Lai et al., 1994b). The number of decomposition functions results from the decomposition matrix of f in this case as well. Here the entries of the decomposition matrix are not elements of $\{0, 1\}$, but elements of $\{0, 1\}^m$. The decomposition matrix for the function of

Figure 4.5.   Decomposition of the function of Example 4.1 (all outputs together).

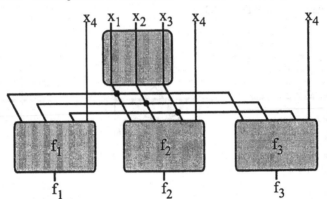

Example 4.1 (with respect to bound set $\{x_1, x_2, x_3\}$) looks like this:

$x_1 x_2 x_3$ \ x_4	0	1
0 0 0	(000)	(000)
0 0 1	(001)	(010)
0 1 0	(011)	(110)
0 1 1	(111)	(110)
1 0 0	(010)	(101)
1 0 1	(011)	(111)
1 1 0	(001)	(011)
1 1 1	(101)	(011)

As in the proof of Theorem 3.1 it is easy to see that the minimum number of decomposition functions of f in a decomposition of form $(\star)$ is $\lceil \log(nrp(\{x_1, \ldots, x_p\}, f)) \rceil$, where $nrp(\{x_1, \ldots, x_p\}, f)$ is the number of different row patterns in the decomposition matrix. However, the problem which occurs in decompositions of this kind lies in the fact that we are not so likely to find *nontrivial* decompositions (especially for large m). Likewise for the function in Example 4.1 there are 8 different row patterns and thus 3 decomposition functions. Each element of $\{0, 1\}^3$ has to be encoded by a unique code from $\{0, 1\}^3$. If the encoding is chosen as the identity function on $\{0, 1\}^3$, then the realization of Figure 4.5 results. α is the identity function, the composition functions g_i are equal to f_i ($i = 1, 2, 3$) and nothing is gained by this decomposition. (If another encoding were chosen, then the composition functions would possibly be even more complex than f_i.)

Because of the difficulty of finding nontrivial decompositions, decompositions of type $(\star)$ are advisable only for special cases (especially when m is small), and we will not use such decompositions for multi–output functions.

4.2.   Output partitioning and input partitioning

This section presents a method for partitioning the outputs of a multi–output function into groups and for choosing bound sets for decomposition of the functions of these groups. After applying this method a decomposition of the functions of these groups is performed computing common decomposition functions.

Here we assume that during the computation of an appropriate bound set the size of the bound set is not fixed (as in Section 3.7.2). A similar algorithm for fixed p (as in Section 3.7.1) can easily be derived from the algorithm in this section. The algorithms proceeds as follows:

In the first step the procedure from Section 3.7 for computing a good bound set is performed for each output function separately. The procedure for a fixed output function f_i computes for each variable order $<_{index}$ which is generated by the greedy approach and for each $2 \leq p \leq n - 1$ the estimated cost $est\_cost_p(<_{index})$.

Since the computation of common decomposition functions for several output functions requires that these output functions are decomposed with respect to the same bound set, the bound set which would result for a single–output function f_i from the greedy approach from Section 3.7.2 is *not* necessarily used for the decomposition of f_i. Instead, all variable orders (together with a $p \in \{2, \ldots, n - 1\}$) which occurred in the course of the procedure for some f_i are viewed as potential candidates for the decomposition of $f_1, \ldots, f_m$.

The second step tries to find bound sets which are appropriate *for all* functions $f_1, \ldots, f_m$. In this second step we use a greedy approach based on variable exchanges as in Section 3.7.2. The only difference lies in the fact that the estimated cost for a variable order $<_{index}$ and a bound set size $p \in \{2, \ldots, n-1\}$ is the *sum* of all estimated costs computed for the decomposition of the single–output functions. The variable orders generated in this second step are also added to the candidate set Φ.[3]

Now we choose — based on this set Φ of candidates — for each f_i a variable order $<_{index}$ from Φ and a $p \in \{2, \ldots, n-1\}$, such that f_i is decomposed with respect to $\{x_{index(1)}, \ldots, x_{index(p)}\}$. Here we try to decompose as many output functions as possible with respect to the same bound set — as long as this does

---

[3] If n is small enough, then Φ is chosen as the set of *all* possible variable orders, of course.

not lead to estimated costs which are too high. The partition of $f_1, \ldots, f_m$ into groups with a corresponding bound set is done by the following algorithm:

Input:

- A set $F = \{f_1, \ldots, f_m\}$ of functions of B_n ($n \geq 4$).

- A set Φ of variable orders. For each variable order $<_{index}$ of Φ, each $2 \leq p \leq n-1$ and each $1 \leq i \leq m$:
  Estimated cost $est\_cost_p^{(i)}(<_{index})$ for a decomposition of f_i with respect to $\{x_{index(1)}, \ldots, x_{index(p)}\}$.[4]

Output:

A partition of the functions of F into groups $G_1, \ldots, G_g$ with $\cup_{1 \leq i \leq g} G_i = F$ and $G_i \cap G_j = \emptyset$ for $i \neq j$.

For each group G_i a bound set $X_i^{(1)}$ for the decomposition of the functions in G_i.

Algorithm:

```
1     g := 0
2     ∀ fᵢ ∈ F : min_cost⁽ⁱ⁾ := min({est_costₚ⁽ⁱ⁾(<index) | <index ∈ Φ, 2 ≤ p ≤ n − 1})
3     while F ≠ ∅ do
4         g := g + 1
5         Choose fᵢ ∈ F with min_cost⁽ⁱ⁾ ≥ min_cost⁽ʲ⁾ ∀fⱼ ∈ F
6         Let <index_opt and p_opt with min_cost⁽ⁱ⁾ = est_cost_p_opt⁽ⁱ⁾(<index_opt)
7         Gg := {fᵢ}
8         F := F \ {fᵢ}
9         Xg⁽¹⁾ := {x_index_opt(1), ..., x_index_opt(p_opt)}
10        group_finished := false
11        while ((group_finished = false) and (F ≠ ∅)) do
12            /* Let Gg = {f_i₁, ..., f_iₖ} */
13            Determine f_iₖ₊₁ ∈ F, <index_new ∈ Φ and 2 ≤ p_new ≤ n − 1, such that
14                sharing(k + 1) (∑_{l=1}^{k+1} est_cost_p_new⁽ⁱˡ⁾(<index_new))
15                  − (sharing(k) (∑_{l=1}^{k} est_cost_p_opt⁽ⁱˡ⁾(<index_opt)) + min_cost⁽ⁱᵏ⁺¹⁾)
16            is minimal.
17            /* That is, maximize the estimated gain of including
18               a new function f_iₖ₊₁ in the set Gg. */
19            if (sharing(k + 1) (∑_{l=1}^{k+1} est_cost_p_new⁽ⁱˡ⁾(<index_new))
20                ≤ sharing(k) (∑_{l=1}^{k} est_cost_p_opt⁽ⁱˡ⁾(<index_opt)) + min_cost⁽ⁱᵏ⁺¹⁾)
21            /* If the estimated costs of a realization of the functions of Gg
22                together with f_iₖ₊₁ are not higher than the costs of a separate realization
23                of the functions of Gg (with the bound set optimized for Gg)
24                and f_iₖ₊₁ (with the bound set optimized for f_iₖ₊₁) */
25            then
26                <index_opt := <index_new
```

---

[4]Here we consider one-sided decompositions. A similar algorithm can be used for two-sided decompositions.

```
27                        p_opt := p_new
28                        G_g := G_g ∪ {f_{i_{k+1}}}
29                        F := F \ {f_{i_{k+1}}}
30                        X_g^(1) := {x_{index_opt(1)}, ..., x_{index_opt(p_opt)}}
31              else
32                        group_finished := true
33           fi
34      od
35  od
```

The algorithm looks at a current group G_g of functions which are supposed to be decomposed with respect to bound set $\{x_{index_{opt}(1)}, \ldots, x_{index_{opt}(p_{opt})}\}$, and it tries to add another function $f_{i_{k+1}}$ to G_g in the loop of lines 11–34. The function $f_{i_{k+1}}$, the new variable order $<_{index_{new}}$, and p_{new} are chosen in such a way that the estimated gain is maximized which is due to logic sharing in the decomposition (with respect to bound set $\{x_{index_{new}(1)}, \ldots, x_{index_{new}(p_{new})}\}$) of the functions of $G_g \cup \{f_{i_{k+1}}\}$ (compared to a separate decomposition of functions of G_g with respect to $\{x_{index_{opt}(1)}, \ldots, x_{index_{opt}(p_{opt})}\}$ and the function $f_{i_{k+1}}$ with respect to the best bound set for $f_{i_{k+1}}$) (lines 13–18 of the algorithm). (In this estimation *sharing* is — as in Section 3.7 — a monotonic decreasing function with $sharing(k) \leq 1$, e.g. $sharing(k) = (k)^c$ with $-1 < c \leq 0$.) If there is a function $f_{i_{k+1}}$, such that adding $f_{i_{k+1}}$ to G_g would result in a reduction of the estimated cost (lines 19–24), then $f_{i_{k+1}}$ is added to G_g and otherwise the determination of a new group is started.

When the algorithm terminates, all functions are assigned to a group G_i and all functions of a group G_i are decomposed with respect to the same bound set $X_i^{(1)}$.

It is important to note that the algorithm above chooses only *nontrivial* decompositions. The reason for this is that the cost estimation returns the value $+\infty$ for decompositions which are not nontrivial (see Section 3.7). Since there are nontrivial decompositions for each output function (at least for $p = n - 1$), only nontrivial decompositions are chosen.

The following two examples demonstrate how the heuristic method works:

EXAMPLE 4.2 Let $f = (f_0, f_1) \in B_{4,2}$ and $\forall (x_0, x_1, x_2, x_3) \in \{0,1\}^4$:

$$f_0(x_0, x_1, x_2, x_3) = x_0 \oplus x_1 \oplus x_2 \oplus x_3$$
$$f_1(x_0, x_1, x_2, x_3) = (x_0 \oplus x_2) \cdot (x_1 \oplus x_3)$$

Suppose we are looking for a two-sided decomposition of f. The partitioning of the output functions and the computation of variable partitions for the two-sided decomposition is done by an algorithm which is completely analogous to the heuristic algorithm above for one-sided decompositions.

We assume that the cost of a realization of a function with 2 inputs is estimated by $cost(2) = 1$ (one 2–input–gate); the cost of a realization of a function with 3 inputs is estimated by $cost(3) = 3$.

This results in the following numbers of decomposition functions and the following estimated cost:

- For f_0: For all variable partitions which are not balanced, we have 2 decomposition functions and the estimated costs are 4 (f_0 is totally symmetric). For all balanced decompositions we also have 2 decomposition functions and the estimated costs are 3.

- For f_1: For variable partition $\{\{x_0, x_2\}, \{x_1, x_3\}\}$ we have 2 decomposition functions and the estimated costs are 3. For all other variable partitions the estimated costs are higher.

The algorithm begins with the first group, which contains f_0 for instance. Also f_1 is included in this group, since for the variable partition $\{\{x_0, x_2\}, \{x_1, x_3\}\}$ the estimated costs of realizing f_0 and f_1 together (line 20 of the algorithm) are not higher than the estimated costs of a separate realization (even if we assume $sharing(n) = 1$). That means that the condition in line 20 is true anyway.

Both for f_0 and f_1, the only possible decomposition functions which depend on x_0 and x_2 are

$$\alpha_1(x_0, x_2) = x_0 \oplus x_2 \text{ and } \alpha_1'(x_0, x_2) = \overline{x_0 \oplus x_2}.$$

The only possible decomposition functions depending on x_1 and x_3 are

$$\beta_1(x_1, x_3) = x_1 \oplus x_3 \text{ and } \beta_1'(x_1, x_3) = \overline{x_1 \oplus x_3}.$$

If the decomposition functions are chosen in such a way that there are *common* decomposition functions for f_0 and f_1 (e.g. α_1 and β_1 both for f_0 and f_1), then the following realization results[5]:

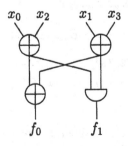

---

[5]More details on the choice of common decomposition functions for the general case can be found in the next section.

EXAMPLE 4.3 We consider a 2–bit–multiplier as another small example. The 2–bit–multiplier is a Boolean function $mult_2 \in B_{4,4}$ with

$$mult_2 : \{0,1\}^4 \to \{0,1\}^4, mult_2(a_1, a_0, b_1, b_0) = (r_3, r_2, r_1, r_0),$$

and

$$(2a_1 + a_0) \cdot (2b_1 + b_0) = \sum_{i=0}^{3} (r_i 2^i).$$

The single–output functions of $mult_2$ are denoted by $p_3, \ldots, p_0$, thus $mult_2 = (p_3, p_2, p_1, p_0)$.

To estimate the cost we assume again $cost(2) = 1$ and $cost(3) = 3$ as in the example above. The function $sharing$ to estimate the cost reduction by logic sharing is assumed as $sharing(n) = n^{-0.1}$.

We have the following numbers of decomposition functions and the following estimated cost:

- For p_3: For all non–balanced variable partitions, the number of decomposition functions is 2 (p_3 is totally symmetric) and the estimated costs are 4. For all balanced decompositions, the number of decomposition functions is also 2, the estimated costs are 3.

- For p_2: For the non–balanced variable partitions $\{\{a_1\}, \{a_0, b_0, b_1\}\}$ and $\{\{b_1\}, \{a_0, a_1, b_0\}\}$, we have exactly 1 decomposition function which depends on $\{a_0, b_0, b_1\}$ and $\{a_0, a_1, b_0\}$, respectively, such that the cost can be estimated by 4. For the two remaining non–balanced decompositions the number of decomposition functions which depend on the variable set with 3 variables is 2, which leads to estimated costs of $2 \cdot sharing(2) \cdot 3 + 3 \approx 8.598$. Both for variable partition $\{\{a_0, a_1\}, \{b_0, b_1\}\}$ and for $\{\{a_0, b_1\}, \{a_1, b_0\}\}$ the decomposition is not nontrivial, such that the costs are estimated by $+\infty$. For variable partition $\{\{a_0, b_0\}, \{a_1, b_1\}\}$ we have 2 decomposition functions and estimated costs of 3.

- For p_1: For all non–balanced variable partitions the estimated costs are ≈ 8.598 (2 decomposition functions, which depend on 3 variables). Both for $\{\{a_0, a_1\}, \{b_0, b_1\}\}$ and for $\{\{a_0, b_0\}, \{a_1, b_1\}\}$ the decomposition is not nontrivial, such that we have costs of $+\infty$. For variable partition $\{\{a_0, b_1\}, \{a_1, b_0\}\}$ there are 2 decomposition functions, such that the costs can be estimated by 3.

- For p_0: For all variable partitions the estimated costs are 1.

Suppose the heuristic method shown above begins the first group with p_3. After that p_2 is chosen with variable partition $\{\{a_0, b_0\}, \{a_1, b_1\}\}$ for p_3 and p_2. Then p_0 is added to this group; the variable partition remains $\{\{a_0, b_0\}, \{a_1, b_1\}\}$ for p_3, p_2 and p_0. At this point the variable partition chosen is still optimal for all functions in this group. In the next step, it is *not* possible to add p_1 to this group: Among all non–balanced variable partitions $\{\{a_1\}, \{a_0, b_0, b_1\}\}$ and $\{\{b_1\}, \{a_0, a_1, b_0\}\}$ are the candidates, where the sum of the estimated costs would be the smallest (cost 4 for p_3, cost 4 for p_2, cost ≈ 8.598 for p_1 and cost 1 for p_0). However, the cost of forcing all functions into the same group is too high in both cases, such that p_1 is not included in the group (line 18 of the algorithm): The estimated cost of realizing all output functions together is $> sharing(4) \cdot (4 + 4 + 8.598 + 1) > 15.32$, whereas the estimated cost of realizing p_3, p_2 and p_0 together is $sharing(3) \cdot (3 + 3 + 1) < 6.28$ and the estimated cost of realizing p_1 is 3 (assuming the optimal variable partition), such that the overall cost with a separate realization of p_1 can be estimated by 9.28. Thus the evaluation in line 18 of the algorithm gives a negative result and p_1 is put into an additional group with variable partition $\{\{a_0, b_1\}, \{a_1, b_0\}\}$.

Finally our implemented logic synthesis algorithm computes the following circuit:

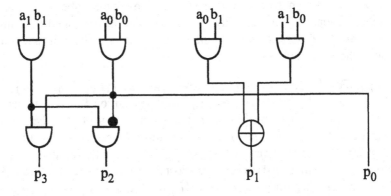

It is important to note that the decomposition function $a_1 \cdot b_1$ is used *both* in the decomposition of p_3 and p_2 and the decomposition function $a_0 \cdot b_0$ is even used in the decomposition of p_3, p_2, and p_0 as a common decomposition function.

4.3.   Computation of common decomposition functions

In this section we examine a solution for *computing* decomposition functions which can be used in the decomposition of several output functions. Here we assume that all output functions considered are decomposed with respect to the same variable set (bound set) or with respect to the same variable partition.

To work towards this goal we first prove a lemma, which gives us a criterion to decide, whether a certain set of functions $\alpha'_1, \ldots, \alpha'_h$ can be used as decomposition functions in a communication minimal decomposition of a function $f \in B_n$:

LEMMA 4.1 *Let $f \in B_n$ with input variables $x_1, \ldots, x_n$, let $X^{(1)} = \{x_1, \ldots, x_p\}$ and $r = \lceil \log(nrp(X^{(1)}, f)) \rceil$. Then there is a one-sided decomposition with respect to $X^{(1)}$ of form*[6]

$$f(\mathbf{x}^{(1)}, \mathbf{x}^{(2)}) = g(\alpha'_1(\mathbf{x}^{(1)}), \ldots, \alpha'_h(\mathbf{x}^{(1)}), \alpha_{h+1}(\mathbf{x}^{(1)}), \ldots, \alpha_r(\mathbf{x}^{(1)}), \mathbf{x}^{(2)}),$$

if and only if $nrp(X^{(1)}, f, \alpha') \leq 2^{r-h}$.

Here the notion $nrp(X^{(1)}, f, \alpha')$ is defined as follows:

DEFINITION 4.1 *Let $Z(X^{(1)}, f)$ be the decomposition matrix of f with respect to $X^{(1)}$ and let A be the image of $\alpha' = (\alpha'_1, \ldots, \alpha'_h)$. For an arbitrary $a \in A$*

$$nrp(X^{(1)}, f, a) := |\{f_{x_1^{\epsilon_1} \ldots x_p^{\epsilon_p}} \mid \alpha'(\epsilon_1, \ldots, \epsilon_p) = a\}|.$$

If all row indices of $Z(X^{(1)}, f)$ are considered which are assigned value a by α', then $nrp(X^{(1)}, f, a)$ is equal to the number of different row patterns occurring in these rows.
$nrp(X, f, \alpha')$ is defined as the maximum

$$nrp(X^{(1)}, f, \alpha') = \max_{a \in A}(nrp(X^{(1)}, f, a)).$$

Now we can prove Lemma 4.1:

PROOF: The proof is essentially analogous to the proof of Theorem 3.1. The basic idea is that $\alpha' = (\alpha'_1, \ldots, \alpha'_h)$ and $\alpha'' = (\alpha_{h+1}, \ldots, \alpha_r)$ have to 'distinguish between rows of $Z(X^{(1)}, f)$ with different row patterns'.

'$\Longrightarrow$': Consider a decomposition

$$f(\mathbf{x}^{(1)}, \mathbf{x}^{(2)}) =$$
$$g(\alpha'_1(\mathbf{x}^{(1)}), \ldots, \alpha'_h(\mathbf{x}^{(1)}), \alpha_{h+1}(\mathbf{x}^{(1)}), \ldots, \alpha_r(\mathbf{x}^{(1)}), \mathbf{x}^{(2)}).$$

Function $\alpha'' = (\alpha_{h+1}, \ldots, \alpha_r) \in B_{p,r-h}$ can have at most 2^{r-h} different function values.

---

[6]To simplify notations, we will abbreviate $x_1, \ldots, x_p$ by $\mathbf{x}^{(1)}$ and $x_{p+1}, \ldots, x_n$ by $\mathbf{x}^{(2)}$ in the following.

Suppose $nrp(X^{(1)}, f, \alpha') > 2^{r-h}$. Then α'' cannot produce enough function values to distinguish between different row patterns:

There is a $a \in \{0, 1\}^h$ in the image A of α', and there are $\epsilon^{(1)} = (\epsilon_1^{(1)}, \dots, \epsilon_p^{(1)})$ and $\epsilon^{(2)} = (\epsilon_1^{(2)}, \dots, \epsilon_p^{(2)}) \in \{0, 1\}^p$, such that

$$\alpha'(\epsilon^{(1)}) = \alpha'(\epsilon^{(2)}) = a,$$

$$\alpha''(\epsilon^{(1)}) = \alpha''(\epsilon^{(2)}),$$

but the row patterns for $\epsilon^{(1)}$, $\epsilon^{(2)}$ are different.
Thus there is no decomposition with decomposition functions

$$\alpha'_1, \dots, \alpha'_h, \alpha_{h+1}, \dots, \alpha_r.$$

'$\Longleftarrow$': Let

$$nrp(X^{(1)}, f, \alpha') \leq 2^{r-h}.$$

Consider an arbitrary value a in the image of α'.
Let $X_a^{(1)} \subseteq \{0, 1\}^p$ be the set of all $\epsilon \in \{0, 1\}^p$ with $\alpha'(\epsilon) = a$.
Then it is possible to define $\alpha''_a : X_a^{(1)} \to \{0, 1\}^{r-h}$, such that

$$\alpha''_a(\epsilon^{(1)}) \neq \alpha''_a(\epsilon^{(2)})$$

for all $\epsilon^{(1)}, \epsilon^{(2)} \in X_a^{(1)}$ with different row patterns in the decomposition matrix.
If for all a in the image of α' such a function α''_a is constructed, then these functions can be combined to a function α'' with domain $\{0, 1\}^p$ (since $X_a^{(1)} \cap X_{a'}^{(1)} = \emptyset$ for $a \neq a'$). Thus

$$(\alpha', \alpha'')(\epsilon^{(1)}) \neq (\alpha', \alpha'')(\epsilon^{(2)})$$

for all $\epsilon^{(1)}, \epsilon^{(2)} \in \{0, 1\}^p$ with different row patterns in the decomposition matrix.

$\square$

It is easy to generalize the criterion of Lemma 4.1 from single-output functions to multi-output functions:

LEMMA 4.2 *Let $f_1, \dots, f_m \in B_n$ with input variables $x_1, \dots, x_n$, $X^{(1)} = \{x_1, \dots, x_p\}$ and let $r_i = \lceil \log(nrp(X^{(1)}, f_i)) \rceil$ for all $1 \leq i \leq m$. Then $\alpha_1, \dots, \alpha_h$ can be used as decomposition functions in one-sided communication* *minimal decompositions of $f_1, \ldots, f_m$ with respect to $X^{(1)}$, i.e., there are one-sided decompositions of $f_1, \ldots, f_m$ with respect to $X^{(1)}$ of form*

$$f_1(\mathbf{x}^{(1)}, \mathbf{x}^{(2)}) =$$
$$g^{(1)}(\alpha_1(\mathbf{x}^{(1)}), \ldots, \alpha_h(\mathbf{x}^{(1)}), \alpha_{h+1}^{(1)}(\mathbf{x}^{(1)}), \ldots, \alpha_{r_1}^{(1)}(\mathbf{x}^{(1)}), \mathbf{x}^{(2)})$$

$$\vdots$$

$$f_m(\mathbf{x}^{(1)}, \mathbf{x}^{(2)}) =$$
$$g^{(m)}(\alpha_1(\mathbf{x}^{(1)}), \ldots, \alpha_h(\mathbf{x}^{(1)}), \alpha_{h+1}^{(m)}(\mathbf{x}^{(1)}), \ldots, \alpha_{r_m}^{(m)}(\mathbf{x}^{(1)}), \mathbf{x}^{(2)}),$$

if and only if for all $1 \le i \le m$

$$nrp(X^{(1)}, f_i, \alpha) \le 2^{r_i - h} \text{ with } \alpha = (\alpha_1, \ldots, \alpha_h).$$

When decomposition functions are computed step by step, we will often be faced with situations where a part of the decomposition functions for $f_1, \ldots, f_m$ has already been computed and the remaining decomposition functions should be chosen, such that as many functions as possible can be used as *common* decomposition functions of $f_1, \ldots, f_m$ (see Section 4.3.2). In this context we need Lemma 4.3, which can easily be concluded from Lemma 4.2. Before we formulate Lemma 4.3, we have to give the following notation:

NOTATION 4.1 *Let $f_i \in B_n$ be a function with input variables $x_1, \ldots, x_n$. Let $Z(X^{(1)}, f_i)$ be the decomposition matrix of f_i with respect to $X^{(1)} = \{x_1, \ldots, x_p\}$. Moreover let $\{0,1\}^p/_{\equiv_i} = \{K_1^{(i)}, \ldots, K_{nrp(X^{(1)}, f_i)}^{(i)}\}$ be the partition of $\{0,1\}^p$ into equivalence classes, which is induced by equality of row patterns in $Z(X^{(1)}, f_i)$. Let $\alpha_1^{(i)}, \ldots, \alpha_{k_i}^{(i)}, \alpha_1, \ldots, \alpha_h \in B_p$. In addition let $k_i + h \le r_i = \lceil \log(nrp(X^{(1)}, f_i)) \rceil$. For $a \in \{0,1\}^{k_i}$ from the image of $\alpha^{(i)}$ and $a' \in \{0,1\}^h$ from the image of α let $X_{aa'}^{(1)} \subseteq \{0,1\}^p$ be the set of all $\epsilon \in \{0,1\}^p$ with $(\alpha^{(i)}, \alpha)(\epsilon) = (a, a')$.*

Then $S_{aa'}^{(i)}$ is the following set:

$$S_{aa'}^{(i)} = \{1 \le j \le nrp(X^{(1)}, f_i) \mid K_j^{(i)} \cap X_{aa'}^{(1)} \ne \emptyset\}$$

For function f_i, $S_{aa'}^{(i)}$ contains all indices $j_1, \ldots, j_l$ of equivalence classes $K_{j_1}^{(i)}$, $\ldots, K_{j_l}^{(i)}$ with the property that at least one binary row index ϵ of $K_{j_k}^{(i)}$ is assigned (aa') by $(\alpha^{(i)}, \alpha)$ $(1 \le k \le l)$. (That is, $|S_{aa'}^{(i)}|$ is the number of *different* row patterns among the rows of $Z(X^{(1)}, f_i)$ with the property that $(\alpha^{(i)}, \alpha)$ assigns

(a, a') to the corresponding row index. Using the notions given above we have $|S_{aa'}^{(i)}| = nrp(X^{(1)}, f, (aa')).)$

Note that these sets $S_{aa'}^{(i)}$ are also used to derive an algorithm for computing common decomposition functions.

Using this notation we obtain the following Lemma 4.3:

LEMMA 4.3 *Let $f_1, \ldots, f_m$ be Boolean functions of B_n with input variables $x_1, \ldots, x_n$. For $1 \leq i \leq m$ let $Z(X^{(1)}, f_i)$ be the decomposition matrix of f_i with respect to $X^{(1)} = \{x_1, \ldots, x_p\}$ and $r_i = \lceil \log(nrp(X^{(1)}, f_i)) \rceil$. Suppose that for all $1 \leq i \leq m$ decomposition functions*

$$\alpha_1^{(i)}, \ldots, \alpha_{k_i}^{(i)} \in B_p$$

are preselected for the decomposition of f_i. Under these assumptions

$$\alpha_1, \ldots, \alpha_h \in B_p$$

can be used as common decomposition functions in a communication minimal decomposition of $f_1, \ldots, f_m$, i.e., there are one-sided decompositions of $f_1, \ldots, f_m$ with respect to $X^{(1)}$ of form

$$f_1(\mathbf{x}^{(1)}, \mathbf{x}^{(2)}) =$$
$$g^{(1)}(\alpha_1^{(1)}(\mathbf{x}^{(1)}), \ldots, \alpha_{k_1}^{(1)}(\mathbf{x}^{(1)}), \alpha_1(\mathbf{x}^{(1)}), \ldots, \alpha_h(\mathbf{x}^{(1)}),$$
$$\alpha_{k_1+h+1}^{(1)}(\mathbf{x}^{(1)}), \ldots, \alpha_{r_1}^{(1)}(\mathbf{x}^{(1)}), \mathbf{x}^{(2)})$$

$$\vdots$$

$$f_m(\mathbf{x}^{(1)}, \mathbf{x}^{(2)}) =$$
$$g^{(m)}(\alpha_1^{(m)}(\mathbf{x}^{(1)}), \ldots, \alpha_{k_m}^{(m)}(\mathbf{x}^{(1)}), \alpha_1(\mathbf{x}^{(1)}), \ldots, \alpha_h(\mathbf{x}^{(1)}),$$
$$\alpha_{k_m+h+1}^{(m)}(\mathbf{x}^{(1)}), \ldots, \alpha_{r_m}^{(m)}(\mathbf{x}^{(1)}), \mathbf{x}^{(2)}),$$

if and only if the following holds for all $1 \leq i \leq m$:
For all $a \in \{0, 1\}^{k_i}$ in the image of $\alpha^{(i)} = (\alpha_1^{(i)}, \ldots, \alpha_{k_i}^{(i)})$ and $a' \in \{0, 1\}^h$ in the image of $\alpha = (\alpha_1, \ldots, \alpha_h)$

$$|S_{aa'}^{(i)}| \leq 2^{r_i - k_i - h}.$$

Lemma 4.3 provides a necessary and sufficient condition for the existence of common decomposition functions in communication minimal decompositions of functions $f_1, \ldots, f_m$, where certain fixed decomposition functions may already be given.

Lemmas 4.1, 4.2 and 4.3 are fundamental for the computation of common decomposition functions. After looking into the complexity of this problem we will present an algorithm to solve the problem based on the lemmas proven above.

4.3.1   The problem CDF

The following problem has to be solved to compute common decomposition functions:

Problem CDF (Common Decomposition Functions)

Given: Functions $f_1, \ldots, f_m \in B_n$ with input variables $x_1, \ldots, x_n$ and a bound set $X^{(1)} = \{x_1, \ldots, x_p\}$. Let the number of decomposition functions which are needed in a communication minimal decomposition of f_i with respect to $X^{(1)}$, be $r_i = \lceil \log(nrp(X^{(1)}, f_i)) \rceil$ for $1 \leq i \leq m$. A natural number h with $h \leq r_i$ for all $1 \leq i \leq m$.

Find: Are there functions $\alpha_1, \ldots, \alpha_h \in B_p$, which can be used as common decomposition functions in one-sided decompositions of $f_1, \ldots, f_m$ with respect to $X^{(1)}$, i.e., are there one-sided decompositions of $f_1, \ldots, f_m$ with respect to $X^{(1)}$ of form

$$f_1(\mathbf{x}^{(1)}, \mathbf{x}^{(2)}) =$$
$$g^{(1)}(\alpha_1(\mathbf{x}^{(1)}), \ldots, \alpha_h(\mathbf{x}^{(1)}), \alpha_{h+1}^{(1)}(\mathbf{x}^{(1)}), \ldots, \alpha_{r_1}^{(1)}(\mathbf{x}^{(1)}), \mathbf{x}^{(2)})$$
$$\vdots$$
$$f_m(\mathbf{x}^{(1)}, \mathbf{x}^{(2)}) =$$
$$g^{(m)}(\alpha_1(\mathbf{x}^{(1)}), \ldots, \alpha_h(\mathbf{x}^{(1)}), \alpha_{h+1}^{(m)}(\mathbf{x}^{(1)}), \ldots, \alpha_{r_m}^{(m)}(\mathbf{x}^{(1)}), \mathbf{x}^{(2)}).$$

The following theorem gives a complexity result for **CDF**:

THEOREM 4.1 *The problem* **CDF** *is NP–hard.*

Theorem 4.1 is proved by a polynomial time transformation from the problem 3–PARTITION (see (Garey and Johnson, 1979)). The proof is given in Appendix H. Since the problem is NP–hard, even if $f_1, \ldots, f_m$ are given by function tables, it is also NP–hard, if $f_1, \ldots, f_m$ are given by m ROBDDs representing the functions.

Upon having a closer look at the proof in Appendix H, we can conclude the following two corollaries:

COROLLARY 4.1 *The problem* **CDF** *remains NP–hard, even if the number of Boolean functions m is equal to 2.*

COROLLARY 4.2 *The problem* **CDF** *remains NP–hard, even if the choice of decomposition functions is restricted to* strict *decomposition functions.*

Corollary 4.2 follows using Lemma 3.5 on page 95: For the functions f_1 and f_2 constructed in the proof of Theorem 4.1 (see Appendix H) $nrp(X^{(1)}, f_1)$ and $nrp(X^{(1)}, f_2)$ are powers of 2, such that the decomposition functions in a communication minimal decomposition are necessarily *strict*.

4.3.1.1   A preliminary solution to CDF

Based on the criterion of Lemma 4.3 we first give a branch–and–bound–algo-rithm, which represents a preliminary solution to the problem CDF. This algo-rithm is *not* used in the implementation of our logic synthesis procedure and is merely presented to lead to understanding of the modified branch–and–bound–algorithm of the next section, which is based on this preliminary algorithm. After presenting the preliminary algorithm we illustrate it by a detailed exam-ple (Example 4.4).

The algorithm constructs a function table of $\alpha = (\alpha_1, \ldots, \alpha_h)$ step by step and checks the initial part of the table so far constructed, whether the condition of Lemma 4.3 has (possibly) already been violated. If the condition of Lemma 4.3 has not been violated, then the function value for the next row in the function table is chosen; otherwise the value for the *current* row (and also for *previous* rows, if all possible function values have already been checked for the current row) is deleted. If an initial part of a function table for $\alpha = (\alpha_1, \ldots, \alpha_h)$ violates the condition of Lemma 4.3, then all function tables which continue this initial part will also violate the condition and need not to be considered in the following. A detailed description of the algorithm is as follows:

Input:

- Functions $f_1, \ldots, f_m$ of B_n with input variables $x_1, \ldots, x_n$.
- For $1 \leq i \leq m$: decomposition matrices $Z(X^{(1)}, f_i)$ of f_i with respect to $X^{(1)} = \{x_1, \ldots, x_p\}$ and $r_i = \lceil \log(nrp(X^{(1)}, f_i)) \rceil$.
- For $1 \leq i \leq m$: Let $\{K_1^{(i)}, \ldots, K_{nrp(X^{(1)}, f_i)}^{(i)}\}$ be the partition of $\{0, 1\}^p$ into equivalence classes, which is induced by equal row patterns in $Z(X^{(1)}, f_i)$. Let $clnr_i$ be a function, which assigns the index j to each v of an equivalence class $K_j^{(i)}$.

- For $1 \leq i \leq m$: 'Preselected' decomposition functions $\alpha_1^{(i)}, \ldots, \alpha_{k_i}^{(i)} \in B_p$ ($k_i < r_i$).

- A natural number h with $h \leq r_i - k_i$ for all $1 \leq i \leq m$. (The algorithm tries to find h common decomposition functions.)

Output:

- $\alpha_1, \ldots, \alpha_h \in B_p$, such that there are one-sided decompositions of f_1, $\ldots, f_m$ with respect to $X^{(1)}$ of form

$$f_1(x_1, \ldots, x_n) =$$
$$g^{(1)}(\alpha_1^{(1)}(\mathbf{x}^{(1)}), \ldots, \alpha_{k_1}^{(1)}(\mathbf{x}^{(1)}), \alpha_1(\mathbf{x}^{(1)}), \ldots, \alpha_h(\mathbf{x}^{(1)}),$$
$$\alpha_{k_1+h+1}^{(1)}(\mathbf{x}^{(1)}), \ldots, \alpha_{r_1}^{(1)}(\mathbf{x}^{(1)}), \mathbf{x}^{(2)})$$

$$\vdots$$

$$f_m(x_1, \ldots, x_n) =$$
$$g^{(m)}(\alpha_1^{(m)}(\mathbf{x}^{(1)}), \ldots, \alpha_{k_m}^{(m)}(\mathbf{x}^{(1)}), \alpha_1(\mathbf{x}^{(1)}), \ldots, \alpha_h(\mathbf{x}^{(1)}),$$
$$\alpha_{k_m+h+1}^{(m)}(\mathbf{x}^{(1)}), \ldots, \alpha_{r_m}^{(m)}(\mathbf{x}^{(1)}), \mathbf{x}^{(2)}),$$

if there are h functions in B_p with this property.

- Otherwise: The algorithm reports that it is not possible to find h functions of B_p with the property given above.

Algorithm: See Figure 4.6.

Remarks:
To simplify notations, the operations '$\cup$' and '$\backslash$' in lines 10, 12 and 19 are not given in absolute detail. If an entry for row v is removed from the function table of α, then for all $1 \leq i \leq m$ the sets $S_{\alpha^{(i)}(v)\alpha(v)}^{(i)}$ have to be corrected. However, the index $clnr_i(v)$ should only removed from $S_{\alpha^{(i)}(v)\alpha(v)}^{(i)}$, if there is no previous row v' in the function table with $\alpha^{(i)}(v') = \alpha^{(i)}(v)$, $\alpha(v') = \alpha(v)$ and $clnr_i(v) = clnr_i(v')$, such that $clnr_i(v)$ must remain in $S_{\alpha^{(i)}(v)\alpha(v)}^{(i)}$ because of this row. Index $clnr_i(v)$ is only allowed to be removed from set $S_{\alpha^{(i)}(v)\alpha(v)}^{(i)}$, if it has been 'removed' as often as it has been 'added' to the set. So operations '$\cup$' and '$\backslash$' of lines 10, 12 and 19 of the algorithms should be implemented as operations on multi–sets.

Some aspects of the algorithm are explained in more detail:

Figure 4.6. Preliminary algorithm to solve CDF.

1 For all $1 \leq i \leq m, a \in \{0,1\}^{k_i}, a' \in \{0,1\}^h : S_{aa'}^{(i)} = \emptyset$

2 $\forall v \in \{0,1\}^p : \alpha(v) := undef.$

3 $\alpha(0,\ldots,0) := (0,\ldots,0)$

4 $\forall 1 \leq i \leq m : S_{\alpha^{(i)}(0,\ldots,0)(0,\ldots,0)}^{(i)} = S_{\alpha^{(i)}(0,\ldots,0)(0,\ldots,0)}^{(i)} \cup \{clnr_i(0,\ldots,0)\}$

5 $v = (0,\ldots,0,1) \in \{0,1\}^p$

6 $a' = (0,\ldots,0) \in \{0,1\}^h$

7 **while** $(\exists v \in \{0,1\}^p$ with $\alpha(v) = undef.)$ **do**

8   /* Function table of α not computed for all $v \in \{0,1\}^p$ */

9 $\alpha(v) = a'$

10   **if** $(\forall 1 \leq i \leq m \,|'S_{\alpha^{(i)}(v)a'}^{(i)} \cup \{clnr_i(v)\}'| \leq 2^{r_i - k_i - h})$

11   **then**

12 $\forall 1 \leq i \leq m : 'S_{\alpha^{(i)}(v)a'}^{(i)} = S_{\alpha^{(i)}(v)a'}^{(i)} \cup \{clnr_i(v)\}'$

13   Increment v by 1.

14 $a' = (0,\ldots,0) \in \{0,1\}^h$

15   **else**

16   **while** $(\alpha(v) = (1,\ldots,1))$ **do**

17 $\alpha(v) = undef.$

18   Decrement v by 1.

19 $\forall 1 \leq i \leq m : 'S_{\alpha^{(i)}(v)\alpha(v)}^{(i)} = S_{\alpha^{(i)}(v)\alpha(v)}^{(i)} \setminus \{clnr_i(v)\}'$

20   **od**

21 $a' = \alpha(v)$

22   Increment a' by 1.

23 $\alpha(v) = undef.$

24   **fi**

25   **if** $(v = (0,\ldots,0))$

26   /* The procedure returns to the starting point without having found an

27   appropriate function α */

28   **then**

29   **return** 'No solution for α.'

30   **fi**

31 **od**

32 **return** α

Line 3: The function value for α in row $(0,\ldots,0)$ is chosen as $(0,\ldots,0)$ w.l.o.g.. (This can be done, since there is a decomposition with decomposition functions $\alpha'_1, \ldots, \overline{\alpha'_i}, \ldots, \alpha'_r$ whenever there is a decomposition with decomposition functions $\alpha'_1, \ldots, \alpha'_i, \ldots, \alpha'_r$.)

Lines 7–31: The function table of α is computed step by step.

If the initial part of the function table does *not* violate the condition of Lemma 4.3, then the algorithm tries to continue the initial part to a complete function table of α (lines 10–14).

If the condition of Lemma 4.3 is already violated by the current initial part of the function table, then all function tables which continue this initial part will also violate the condition. Thus we have to 'go to another branch in the branch–and–bound–algorithm' (lines 16–23).

At the beginning of the **while**–loop (lines 7–31) an initial part of the function table of α is already computed: The function values $\alpha(0,\ldots,0),\ldots,\alpha(v-1)$[7] are already determined and the current initial part of the function table does not violate the condition of Lemma 4.3 yet. If $a' > (0,\ldots,0)$, then in previous steps the attempts to choose function values for $\alpha(v)$ which are smaller than a' were not successful.

First we check whether the condition of Lemma 4.3 is already violated for decision $\alpha(v) = a'$. If this is not the case, then the assignment is accepted (for the time being), v is incremented by 1 for the next run of the **while**–loop and a' is reset to $(0,\ldots,0)$.

If the condition is violated, then a' $(= \alpha(v))$ is incremented by 1, the function value of $\alpha(v)$ is reset to 'undefined', and during the next run of the loop an assignment of $\alpha(v)$ by a new a' is tried. However, if a' $(= \alpha(v))$ is already equal to $(1,\ldots,1)$, then we have to backtrack and we are looking for the largest index v' in the previously defined function table of α, such that $\alpha(v')$ is not yet equal to $(1,\ldots,1)$. For v' and all larger indices v in the function table, the assignment of function values is discarded (including a correction of the sets $S^{(i)}_{\alpha^{(i)}(v)\alpha(v)}$) (see lines 16–20). After the **while**–loop has been finished, v has been set to this index v' for the next run. To check an assignment of $\alpha(v)$ by a value incremented by 1 in the next run of the outermost **while**–loop (lines 7–31), a' is set to the required value (lines 21–22).

For efficiency the branch–and–bound–algorithm computes only the first successful branch, i.e., at most *one* h–tuple of decomposition functions which solves the problem.

EXAMPLE 4.4 The preliminary algorithm is demonstrated with this example. Here we use the functions f_1 and f_2 from Example 4.1 (page 124) and we are looking for exactly one common decomposition function in the communication minimal decomposition of f_1 and f_2 with respect to $\{x_1, x_2, x_3\}$. We have $f_1(x_1,\ldots,x_4) = \overline{x_4}x_2x_3 + x_4(x_1 \oplus x_2)$ and $f_2(x_1,\ldots,x_4) = \overline{x_4}(x_1 \oplus x_2) + x_4(x_2 + x_3)$.

---

[7]Here for $v \in \{0,1\}^p$ $v-1$ means the element $v' \in \{0,1\}^p$ with $int(v') = int(v) - 1$. Also for v_1 and $v_2 \in \{0,1\}^p$ $v_1 < v_2$ if and only if $int(v_1) < int(v_2)$.

Based on equality of rows in the decomposition matrix $Z(\{x_1, x_2, x_3\}, f_1)$ of f_1, the following partition of $\{0, 1\}^3$ into equivalence classes results: $K_1^{(1)} = \{(000), (001), (110)\}$, $K_2^{(1)} = \{(010), (100), (101)\}$, $K_3^{(1)} = \{(111)\}$ and $K_4^{(1)} = \{(011)\}$. In the same way, we obtain the following partition into equivalence classes for f_2: $K_1^{(2)} = \{(000)\}$, $K_2^{(2)} = \{(001), (110), (111)\}$, $K_3^{(2)} = \{(100)\}$ and $K_4^{(2)} = \{(010), (011), (101)\}$.

In a communication minimal decomposition we need 2 decomposition functions for f_1 and f_2, respectively, i.e., $r_1 = r_2 = 2$.

We assume that no decomposition functions are preselected, i.e., $k_1 = k_2 = 0$.

Since we are looking for exactly one common decomposition function α_1 of f_1 and f_2 ($h = 1$), exactly one additional decomposition function remains both for f_1 and for f_2.

The sets which are considered within the condition of Lemma 4.3 are $S_0^{(1)}$ and $S_1^{(1)}$ for f_1 on the one hand and $S_0^{(2)}$ and $S_1^{(2)}$ for f_2 on the other hand. If we consider all row indices of the decomposition matrices $Z(\{x_1, x_2, x_3\}, f_i)$ ($i = 1, 2$) for which the function value of α_1 (or more precisely the function value of α_1 in the initial part of the function table constructed so far) is 0 (or 1), then the number of different row patterns in the corresponding rows must not be larger than 2 ($|S_0^{(i)}| \leq 2$ (and $|S_1^{(i)}| \leq 2$)). If this condition is fulfilled when the computation is finished, then we can make use of the remaining decomposition function $\alpha_2^{(i)}$ to ensure that $(\alpha_1, \alpha_2^{(i)})$ assigns *different* function values to the indices of rows with *different* row patterns.

Figure 4.7 illustrates how the branch–and–bound–algorithm to compute one common decomposition function of f_1 and f_2 works. The middle part of the illustration shows the part of the decision tree which is traversed by the algorithm. The assignments are numbered in the order they are performed by the algorithm. The path of arrows printed in bold face corresponds to the result computed for α. The remaining arrows correspond to decisions which were discarded because of a violation of the condition of Lemma 4.3. Finally we obtain a common decomposition function $\alpha_1 : \{0, 1\}^3 \to \{0, 1\}$ with $\alpha_1(v) = 1 \iff v \in \{(010), (011), (100), (101)\}$. The lower part of the figure shows the sets $S_0^{(1)}, S_1^{(1)}, S_0^{(2)}$ and $S_1^{(2)}$ for the different steps of the algorithm.

4.3.1.2   Solution of CDF for strict decomposition functions

The algorithm shown above has the disadvantage that it possibly leads to non–strict decomposition functions. In Section 3.8.1 we have seen that *strict* decomposition functions of a function f have the advantage that they preserve

Figure 4.7.   Example for branch–and–bound–algorithm.

f_1:

$x_1x_2x_3$ \ x_4	0 1	
000	0 0	row pattern 1
001	0 0	row pattern 1
010	0 1	row pattern 2
011	1 1	row pattern 4
100	0 1	row pattern 2
101	0 1	row pattern 2
110	0 0	row pattern 1
111	1 0	row pattern 3

$x_1x_2x_3$	α_1	
000	0	0 (1)
001	0	0 (2)
010	1	0 (3) 1 (4)
011	1	0 (5) 1 (6)
100	1	0 (7) 1 (8)
101	1	0 (9) 1 (10)
110	0	0 (11)
111	0	0 (12)

f_2:

$x_1x_2x_3$ \ x_4	0 1	
000	0 0	row pattern 1
001	0 1	row pattern 2
010	1 1	row pattern 4
011	1 1	row pattern 4
100	1 0	row pattern 3
101	1 1	row pattern 4
110	0 1	row pattern 2
111	0 1	row pattern 2

Sets $S_0^{(i)}$, $S_1^{(i)}$ during the run of the branch-and-bound-algorithm:

(1) $S_0^{(1)}=\{1\}$, $S_1^{(1)}=\{\}$ $S_0^{(2)}=\{1\}$, $S_1^{(2)}=\{\}$

(2) $S_0^{(1)}=\{1\}$, $S_1^{(1)}=\{\}$ $S_0^{(2)}=\{1,2\}$, $S_1^{(2)}=\{\}$

(3) $S_0^{(1)}=\{1,2\}$, $S_1^{(1)}=\{\}$ $S_0^{(2)}=\{1,2,4\}$, $S_1^{(2)}=\{\}$

(4) $S_0^{(1)}=\{1\}$, $S_1^{(1)}=\{2\}$ $S_0^{(2)}=\{1,2\}$, $S_1^{(2)}=\{4\}$

(5) $S_0^{(1)}=\{1,4\}$, $S_1^{(1)}=\{2\}$ $S_0^{(2)}=\{1,2,4\}$, $S_1^{(2)}=\{4\}$

(6) $S_0^{(1)}=\{1\}$, $S_1^{(1)}=\{2,4\}$ $S_0^{(2)}=\{1,2\}$, $S_1^{(2)}=\{4\}$

(7) $S_0^{(1)}=\{1,2\}$, $S_1^{(1)}=\{2,4\}$ $S_0^{(2)}=\{1,2,3\}$, $S_1^{(2)}=\{4\}$

(8) $S_0^{(1)}=\{1\}$, $S_1^{(1)}=\{2,4\}$ $S_0^{(2)}=\{1,2\}$, $S_1^{(2)}=\{3,4\}$

(9) $S_0^{(1)}=\{1,2\}$, $S_1^{(1)}=\{2,4\}$ $S_0^{(2)}=\{1,2,4\}$, $S_1^{(2)}=\{3,4\}$

(10) $S_0^{(1)}=\{1\}$, $S_1^{(1)}=\{2,4\}$ $S_0^{(2)}=\{1,2\}$, $S_1^{(2)}=\{3,4\}$

(11) $S_0^{(1)}=\{1\}$, $S_1^{(1)}=\{2,4\}$ $S_0^{(2)}=\{1,2\}$, $S_1^{(2)}=\{3,4\}$

(12) $S_0^{(1)}=\{1,3\}$, $S_1^{(1)}=\{2,4\}$ $S_0^{(2)}=\{1,2\}$, $S_1^{(2)}=\{3,4\}$

symmetry properties of f. In this section we modify the branch–and–bound–algorithm, such that it produces only strict decomposition functions.

The modified algorithm has the additional advantage that the restriction to common *strict* decomposition functions can speed up the computation greatly, since the efficiency of the algorithm is improved by preprocessing.

If we are looking for a common *strict* decomposition function α_k of two functions f_i and f_j (with equivalence classes $\{K_1^{(i)}, \ldots, K_{nrp(X^{(1)}, f_i)}^{(i)}\}$ for f_i and $\{K_1^{(j)}, \ldots, K_{nrp(X^{(1)}, f_j)}^{(j)}\}$ for f_j), then for ϵ and δ of $\{0,1\}^p$ $\alpha_k(\epsilon)$ and $\alpha_k(\delta)$ have to be equal, if $\epsilon, \delta \in K_l^{(i)}$ for an arbitrary $1 \leq l \leq nrp(X^{(1)}, f_i)$ or $\epsilon, \delta \in K_l^{(j)}$ for an arbitrary $1 \leq l \leq nrp(X^{(1)}, f_j)$. Thus, for the computation of common strict decomposition functions we will generally have larger subsets of $\{0,1\}^p$, which have the property that common strict decomposition functions have identical function values for *all* elements of these subsets. These subsets can be computed in a preprocessing step.

NOTATION 4.2 *Let $f \in B_{n,m}$ with input variables $x_1, \ldots, x_n$. f is decomposed with respect to $X^{(1)} = \{x_1, \ldots, x_p\}$.*

- *For all $1 \le i \le m$ let $\equiv_i$ be an equivalence relation on $\{0,1\}^p$ defined by*

$$\epsilon^{(1)} \equiv_i \epsilon^{(2)} \iff (f_i)_{x_1^{\epsilon_1^{(1)}} \ldots x_p^{\epsilon_p^{(1)}}} = (f_i)_{x_1^{\epsilon_1^{(2)}} \ldots x_p^{\epsilon_p^{(2)}}}$$

for all $\epsilon^{(1)}, \epsilon^{(2)} \in \{0,1\}^p$.
The corresponding partition into equivalence classes is denoted by

$$\{0,1\}^p /_{\equiv_i} = \{K_1^{(i)}, \ldots, K_{nrp(X^{(1)}, f_i)}^{(i)}\}.[8]$$

- *Let the relation $\sim'$ on $\{0,1\}^p$ be defined by*

$$\epsilon^{(1)} \sim' \epsilon^{(2)} \iff \exists 1 \le i \le m \text{ with } \epsilon^{(1)} \equiv_i \epsilon^{(2)}$$

for all $\epsilon^{(1)}, \epsilon^{(2)} \in \{0,1\}^p$.
Let $\sim$ be the transitive closure of $\sim'$.
The corresponding partition into equivalence classes is denoted by

$$\{0,1\}^p /_{\sim} = \{E_1, \ldots, E_l\}.$$

NOTATION 4.3 *A partition $P = \{P_1, \ldots, P_k\}$ is called a* coarsening *of a partition $Q = \{Q_1, \ldots, Q_l\}$ (or Q is called a* refinement *of P) if and only if $\cup_{i=1}^k P_i = \cup_{i=1}^l Q_i$ and for all Q_i $(1 \le i \le l)$ there is a $j \in \{1, \ldots, k\}$ with $Q_i \subseteq P_j$.*

REMARK 4.1 *The partition $\{0,1\}^p /_{\sim}$ is the 'finest' common coarsening of the partitions $\{0,1\}^p /_{\equiv_i}$ $(1 \le i \le m)$. If the partitions of $\{0,1\}^p$ are viewed as a lattice with $P \le Q$ iff P is a coarsening of Q, then $\{0,1\}^p /_{\sim}$ is simply the infimum of all $\{0,1\}^p /_{\equiv_i}$ $(1 \le i \le m)$.*

A strict decomposition function of f_i has to assign the same function value to all elements of a class $K_j^{(i)}$ from $\{0,1\}^p /_{\equiv_i}$. For this reason a *common strict* decomposition function of $f_1, \ldots, f_m$ has to assign the same function value to all elements of a class E_j from $\{0,1\}^p /_{\sim}$.

---

[8]See also Notation 2.2 on p. 26.

Figure 4.8. Computation of $\{0,1\}^p/_\sim$.

f_1: $x_1x_2x_3$	x_4: 0 1			f_2: $x_1x_2x_3$	x_4: 0 1	
0 0 0	0 0	row pattern 1	0 0 0	0 0 0	0 0	row pattern 1
0 0 1	0 0	row pattern 1	0 0 1	0 0 1	0 1	row pattern 2
0 1 0	0 1	row pattern 2	0 1 0	0 1 0	1 1	row pattern 4
0 1 1	1 1	row pattern 4	0 1 1	0 1 1	1 1	row pattern 4
1 0 0	0 1	row pattern 2	1 0 0	1 0 0	1 0	row pattern 3
1 0 1	0 1	row pattern 2	1 0 1	1 0 1	1 1	row pattern 4
1 1 0	0 0	row pattern 1	1 1 0	1 1 0	0 1	row pattern 2
1 1 1	1 0	row pattern 3	1 1 1	1 1 1	0 1	row pattern 2

The partition into equivalence classes $\{0,1\}^p/_\sim$ can easily be computed based on the decomposition matrices of $f_1, \ldots, f_m$. This is illustrated by the following example:

EXAMPLE 4.4 (CONTINUED):
Figure 4.8 demonstrates the computation of $\{0,1\}^3/_\sim$ for functions f_1 and f_2 from Example 4.4 (page 143), which are decomposed with respect to $\{x_1, x_2, x_3\}$. The elements of $\{0,1\}^3$ are represented by nodes of a graph and there is an edge between two nodes ϵ and δ iff $\epsilon \equiv_1 \delta$ (edges on the left hand side) or $\epsilon \equiv_2 \delta$ (edges on the right hand side). $\{0,1\}^3/_\sim$ results from the computation of the connected components of this graph. There are exactly two connected components (white and gray shaded nodes), such that

$$\{0,1\}^3/_\sim = \{\{(000),(001),(110),(111)\},\{(010),(011),(100),(101)\}\}.$$

When α_1 is a common strict decomposition function of f_1 and f_2, then the function value of α_1 for (000) directly implies the function value for (001), (110) and (111).

During the computation of common strict decomposition functions $\alpha_1, \ldots, \alpha_h$ of functions by a branch–and–bound–algorithm we can make use of the knowledge that $\alpha_1, \ldots, \alpha_h$ have to assign the same function value to *all* elements of a class E_j of $\{0,1\}^p/_\sim$. As shown in the example above, the partition of $\{0,1\}^p/_\sim$ into equivalence classes can easily be computed by the computation of the connected components of an (implicit) graph with node set $\{0,1\}^p$.

A modified branch–and–bound–algorithm does not construct the function table of $\alpha = (\alpha_1, \ldots, \alpha_h)$ row by row, but rather assigns function values to the *different equivalence classes* E_j *of* $\{0,1\}^p/\sim$. In this way the efficiency is improved to a great extent.

Before we give a detailed description of the modifications of the algorithm resulting from the use of *strict* decomposition functions, we have to define some notations. Finally the algorithm is illustrated by an example.

The preliminary branch–and–bound–algorithm computes common decomposition functions $\alpha_1, \ldots, \alpha_h$ for a set of functions f_i under the assumption of preselected decomposition functions $(\alpha_1^{(i)}, \ldots, \alpha_{k_i}^{(i)}) = \alpha^{(i)}$ and maintains sets $S_{aa'}^{(i)}$ to control the computation. If the function value of α for a new element $v \in \{0,1\}^p$ is set to a value a', e.g., then $S_{\alpha^{(i)}(v)a'}^{(i)}$ is updated by

$$`S_{\alpha^{(i)}(v)\alpha(v)}^{(i)} = S_{\alpha^{(i)}(v)\alpha(v)}^{(i)} \cup \{clnr_i(v)\}`$$

($clnr_i(v) = j$ for $v \in K_j^{(i)}$, cf. line 12 in Figure 4.6).

If we decide in the modified branch–and–bound–algorithm to choose $\alpha(E_j) = a'$ for a set E_j ($a' \in \{0,1\}^h$), then we have to perform operation

$$`S_{\alpha^{(i)}(v)a'}^{(i)} = S_{\alpha^{(i)}(v)a'}^{(i)} \cup \{clnr_i(v)\}` \qquad (\star)$$

for all $v \in E_j$ and all $1 \leq i \leq m$.

However, the following observations show that the maintenance of these sets to check the condition of Lemma 4.3 can be simplified for strict decomposition functions:

If we define for all a in the image $\alpha^{(i)}(E_j)$ of the preselected decomposition functions $(\alpha_1^{(i)}, \ldots, \alpha_{k_i}^{(i)})$ for f_i

$$CLNR_{aj}^{(i)} = \{k \mid K_k^{(i)} \subseteq E_j \text{ and } \alpha^{(i)}(K_k^{(i)}) = a\},$$

then, for the decision $\alpha(E_j) = a'$, $(\star)$ can be replaced by

$$`S_{aa'}^{(i)} = S_{aa'}^{(i)} \cup CLNR_{aj}^{(i)}`$$

for all $a \in \alpha^{(i)}(E_j)$. Thus $|CLNR_{aj}^{(i)}|$ tells us the number of different row patterns for all rows which have an index of E_j *and* are mapped to a by $\alpha^{(i)}$.

Finally we can observe for $1 \leq j_1, j_2 \leq l$ $j_1 \neq j_2$, $a_1 \in \alpha^{(i)}(E_{j_1})$, $a_2 \in \alpha^{(i)}(E_{j_2})$ that

$$CLNR_{a_1 j_1}^{(i)} \cap CLNR_{a_2 j_2}^{(i)} = \emptyset$$

holds. This can be directly concluded from the facts that

$$E_{j_1} = \bigcup_{K_j^{(i)} \subseteq E_{j_1}} K_j^{(i)},$$

$$E_{j_2} = \bigcup_{K_j^{(i)} \subseteq E_{j_2}} K_j^{(i)}$$

and that $\{0,1\}^p/_\sim$ is a partition. So the set unions of the algorithm

$$S_{aa'}^{(i)} = S_{aa'}^{(i)} \cup CLNR_{aj}^{(i)}$$

are always *disjoint* unions. Since we are interested only in the size of the sets $S_{aa'}^{(i)}$ when checking whether the condition of Lemma 4.3 is violated, it suffices to *add* the sizes of $CLNR_{aj}^{(i)}$ and $S_{aa'}^{(i)}$. The maintenance of *sets* $S_{aa'}^{(i)}$ can be replaced by the maintenance of *natural numbers*. Using the notations $SS_{aa'}^{(i)}$ for $|S_{aa'}^{(i)}|$ and $SCLNR_{aj}^{(i)}$ for $|CLNR_{aj}^{(i)}|$, we finally define the following modified branch–and–bound–algorithm:

Input:

- Functions $f_1, \ldots, f_m$ of B_n with input variables $x_1, \ldots, x_n$.
- For $1 \leq i \leq m$: Decomposition matrices $Z(X^{(1)}, f_i)$ of f_i with respect to $X^{(1)} = \{x_1, \ldots, x_p\}$ and $r_i = \lceil \log(nrp(X^{(1)}, f_i)) \rceil$.
- For $1 \leq i \leq m$: Decomposition functions $\alpha_1^{(i)}, \ldots, \alpha_{k_i}^{(i)} \in B_p$ ($k_i < r_i$).
- For $1 \leq i \leq m$: $\{0,1\}^p/_{\equiv_i} = \{K_1^{(i)}, \ldots, K_{nrp(X^{(1)}, f_i)}^{(i)}\}$ and $\{0,1\}^p/_\sim = \{E_1, \ldots, E_l\}$.
- For $1 \leq i \leq m, 1 \leq j \leq l, a \in \alpha^{(i)}(E_j)$:
  $CLNR_{aj}^{(i)} = \{k \mid K_k^{(i)} \subseteq E_j \text{ and } \alpha^{(i)}(K_k^{(i)}) = a\}$,
  $SCLNR_{aj}^{(i)} = |CLNR_{aj}^{(i)}|$.
- A natural number h with $h \leq r_i - k_i$ for all $1 \leq i \leq m$. (The algorithm tries to find h common *strict* decomposition functions.)

Output:

- $\alpha_1, \ldots, \alpha_h \in B_p$, such that there are one-sided decompositions of f_1, $\ldots, f_m$ with respect to $X^{(1)}$ of form

  $$f_1(x_1, \ldots, x_n) =$$

$$g^{(1)}(\alpha_1^{(1)}(\mathbf{x}^{(1)}), \ldots, \alpha_{k_1}^{(1)}(\mathbf{x}^{(1)}), \alpha_1(\mathbf{x}^{(1)}), \ldots, \alpha_h(\mathbf{x}^{(1)}),$$

$$\alpha_{k_1+h+1}^{(1)}(\mathbf{x}^{(1)}), \ldots, \alpha_{r_1}^{(1)}(\mathbf{x}^{(1)}), \mathbf{x}^{(2)})$$

$$\vdots$$

$$f_m(x_1, \ldots, x_n) =$$
$$g^{(m)}(\alpha_1^{(m)}(\mathbf{x}^{(1)}), \ldots, \alpha_{k_m}^{(m)}(\mathbf{x}^{(1)}), \alpha_1(\mathbf{x}^{(1)}), \ldots, \alpha_h(\mathbf{x}^{(1)}),$$

$$\alpha_{k_m+h+1}^{(m)}(\mathbf{x}^{(1)}), \ldots, \alpha_{r_m}^{(m)}(\mathbf{x}^{(1)}), \mathbf{x}^{(2)}),$$

where $\alpha_1 \ldots, \alpha_h$ are strict, if there are h functions in B_p with this property.

- Otherwise: The algorithm reports that it is not possible to find h functions of B_p with the property given above.

Algorithm: See Figure 4.9.

In many cases the preprocessing phase of computing $\{0, 1\}^p / \sim$ already results in large equivalence classes of elements of $\{0, 1\}^p$. Based on the knowledge that the function values of α have to be identical for all elements of these equivalence classes, the run time of the modified branch–and–bound–algorithms is reduced compared to the preliminary algorithm presented before.

If the modified algorithm is used for Example 4.4 (which was also used to illustrate the preliminary branch–and–bound–algorithm), then the problem is substantially simplified:

EXAMPLE 4.5
$\equiv_1$ induces the equivalence classes $K_1^{(1)} = \{(000), (001), (110)\}$, $K_2^{(1)} = \{(010), (100), (101)\}$, $K_3^{(1)} = \{(111)\}$ and $K_4^{(1)} = \{(011)\}$ and $\equiv_2$ induces the equivalence classes $K_1^{(2)} = \{(000)\}$, $K_2^{(2)} = \{(001), (110), (111)\}$, $K_3^{(2)} = \{(100)\}$ and $K_4^{(2)} = \{(010), (011), (101)\}$.

So we have $\{0, 1\}^3 / \sim = \{E_1, E_2\}$ with $E_1 = \{\{(000), (001), (110), (111)\}$ and $E_2 = \{(010), (011), (100), (101)\}\}$.

No decomposition functions are preselected and we are looking for one common *strict* decomposition function of f_1 and f_2.

These assumptions lead to $CLNR_1^{(1)} = \{1, 3\}$, $CLNR_2^{(1)} = \{2, 4\}$, $CLNR_1^{(2)} = \{1, 2\}$ and $CLNR_2^{(2)} = \{3, 4\}$ and thus $SCLNR_1^{(1)} = SCLNR_2^{(1)} = SCLNR_1^{(2)} = SCLNR_2^{(2)} = 2$.

Figure 4.9. Modified algorithm to solve CDF, when CDF is restricted to *strict* decomposition functions.

At first compute a 'compressed version' β of α. β is constructed for the equivalence classes of $\{0,1\}^p/_\sim$. α is then easily computed from β by $\alpha(E_j) = a' \iff \beta(j) = a'$ for all $1 \leq j \leq l$.

```
 1   if (l = 1)
 2     then
 3         return 'No solution for α.'
 4   fi
 5   For all 1 ≤ i ≤ m, a ∈ {0,1}^{k_i}, a' ∈ {0,1}^h : SS^{(i)}_{aa'} = 0
 6   ∀1 ≤ j ≤ l : β(j) := undef.
 7   β(1) := (0,...,0)
 8   ∀1 ≤ i ≤ m, ∀a ∈ α^{(i)}(E_1) :  SS^{(i)}_{a(0,...,0)} = SS^{(i)}_{a(0,...,0)} + SCLNR^{(i)}_{a1}
 9   if (∃1 ≤ i ≤ m, a ∈ α^{(i)}(E_1) with SS^{(i)}_{a(0,...,0)} > 2^{r_i - k_i - h})
10     then
11         return 'No solution for α.'
12   fi
13   j = 2
14   a' = (0,...,0) ∈ {0,1}^h
15   while ((j > 1) and (j ≤ l)) do
16       β(j) = a'
17       if (∀1 ≤ i ≤ m, ∀a ∈ α^{(i)}(E_j) : SS^{(i)}_{aa'} + SCLNR^{(i)}_{aj} ≤ 2^{r_i - k_i - h})
18         then
19             ∀1 ≤ i ≤ m, ∀a ∈ α^{(i)}(E_j) : SS^{(i)}_{aa'} = SS^{(i)}_{aa'} + SCLNR^{(i)}_{aj}
20             j = j + 1
21             a' = (0,...,0) ∈ {0,1}^h
22         else
23             while (β(j) = (1,...,1)) do
24                 β(j) = undef.
25                 j = j - 1
26                 a' = β(j)
27                 ∀1 ≤ i ≤ m, ∀a ∈ α^{(i)}(E_j) : SS^{(i)}_{aa'} = SS^{(i)}_{aa'} - SCLNR^{(i)}_{aj}
28             od
29             Increment a' by 1.
30             β(j) = undef.
31         fi
32   od
33   if (j = 1)
34       /* The procedure returns to the starting point without having found an
35       appropriate function β */
36     then
37         return 'No solution for α.'
38   fi
39   ∀1 ≤ j ≤ l, ∀v ∈ E_j : α(v) = β(j)
40   return α
```

After the assignment $\alpha(E_1) = 0$ we obtain $SS^{(1)}_0 = SS^{(2)}_0 = 2$ and $SS^{(1)}_1 = SS^{(2)}_1 = 0$.

The assignment $\alpha(E_2) = 0$ leads to a violation of the condition of Lemma 4.3 (in this case α would be a constant function) and it is necessary to set

$\alpha(E_2) = 1$. This implies $SS_0^{(1)} = SS_0^{(2)} = 2$ and $SS_1^{(1)} = SS_1^{(2)} = 2$ and a common strict decomposition of f_1 and f_2 is found.

If we are looking for a common strict decomposition function not for f_1 and f_2, but rather for f_1 and f_3 in Example 4.1 on page 124 ($f_1(x_1, \ldots, x_4) = \overline{x_4}x_2x_3 + x_4(x_1 \oplus x_2)$, $f_3(x_1, \ldots, x_4) = \overline{x_4}(x_2 + x_3) + x_1x_4$), then we notice already after the computation of $\{0, 1\}^3/_\sim$ that there is no common strict decomposition function in a communication minimal decomposition of f_1 and f_3: Because of $K_1^{(3)} = \{(000)\}$, $K_2^{(3)} = \{(001), (010), (011)\}$, $K_3^{(3)} = \{(100)\}$ and $K_4^{(3)} = \{(101), (110), (111)\}$ we obtain $\{0, 1\}^3/_\sim = \{\{0, 1\}^3\}$. That means that a common strict decomposition function of f_1 and f_3 would assign the same value to all elements of $\{0, 1\}^3$, and thus it would be a constant function. But constant functions can never occur as decomposition functions in communication minimal decompositions.

4.3.1.3   Solution of CDF based on ROBDD–representations

In the previous sections we considered a preliminary and an improved algorithm to compute common decomposition functions under the assumption that the Boolean functions are given as function tables or as decomposition matrices. However, for instances of realistic sizes we will not be able to represent the functions by function tables. Therefore we extend our algorithm to compute common strict decomposition functions also to the case of Boolean functions which are represented by ROBDDs.

This step is very important for the application of the logic synthesis algorithm to practical examples. When the functions to be decomposed are given by a compact ROBDD–representation, then the complete synthesis procedure can be performed based on compact ROBDD–representations.

The main task consists in generating the inputs of the modified branch–and–bound–algorithm in an efficient way: The modified branch–and–bound–algorithm computes the function values of common decomposition functions α_1, $\ldots, \alpha_h$ of functions $f_1, \ldots, f_m$ for elements $E_1, \ldots, E_l$ of $\{0, 1\}^p/_\sim$ and thus it computes also function values for classes $K_1^{(i)}, \ldots, K_{nrp(X^{(1)}, f_i)}^{(i)}$ ($1 \leq i \leq m$), since a class E_j ($1 \leq j \leq l$) is a union of classes of $\{0, 1\}^p/_{\equiv_i}$. In Section 3.6 a method was given, which computes ROBDDs for decomposition and composition functions of f_i based on an encoding of the equivalence classes $K_1^{(i)}, \ldots, K_{nrp(X^{(1)}, f_i)}^{(i)}$. So we just have to find an efficient method to generate the inputs of the modified branch–and–bound–algorithm based on ROBDDs for $f_1, \ldots, f_m$.

Basically we have the problem of obtaining representations of the classes $K_1^{(i)}, \ldots, K_{nrp(X^{(1)}, f_i)}^{(i)}$ $(1 \leq i \leq m)$ and the classes $E_1, \ldots, E_l$. These classes are represented by ROBDDs for their characteristic functions.

ROBDDs $bdd_j^{(i)}$ for the characteristic functions of classes $K_j^{(i)} \subseteq \{0, 1\}^p$ $(1 \leq j \leq nrp(X^{(1)}, f_i))$ can easily be computed from the ROBDD representing f_i. Here we can make use of the fact that each class $K_j^{(i)}$ has a one-to-one correspondence to a linking node n_j in the ROBDD of f_i (assuming a cut after the variables in $X^{(1)} = \{x_1, \ldots, x_p\}$). $K_j^{(i)}$ is just the set of all $(\epsilon_1, \ldots, \epsilon_p) \in \{0, 1\}^p$ with the property that the linking node n_j is reached by $(\epsilon_1, \ldots, \epsilon_p)$. So we obtain (as illustrated by Figure 4.10) an (possibly non–reduced) OBDD representing the characteristic function of $K_j^{(i)}$ by replacing the node n_j by the constant 1 and the other linking nodes by constant 0 (see also Section 3.6).

We still have to solve the problem of computing representations of the classes $E_1, \ldots, E_l$ of $\{0, 1\}^p/\sim$. In the last section these classes were computed as connected components in a graph whose nodes were elements of $\{0, 1\}^p$. Also when we start from ROBDDs, $E_1, \ldots, E_l$ are computed by a computation of connected components of an (implicitly defined) undirected graph $G_\sim = (V, E)$: The set V of nodes is given by the set of all ROBDDs $bdd_j^{(i)}$ $(1 \leq i \leq m, 1 \leq j \leq nrp(X^{(1)}, f_i))$ representing the equivalence classes $K_j^{(i)}$. There is an edge $\{bdd_{j_1}^{(i_1)}, bdd_{j_2}^{(i_2)}\}$ iff $bdd_{j_1}^{(i_1)} \wedge bdd_{j_2}^{(i_2)} \neq 0$, i.e., iff $K_{j_1}^{(i_1)} \cap K_{j_2}^{(i_2)} \neq \emptyset$ (see Figure 4.11).

Obviously for a class $K_j^{(i)}$ we obtain the equivalence class of $\{0, 1\}^p/\sim$ which contains $K_j^{(i)}$ by computing the connected component of $G_\sim$ which contains $bdd_j^{(i)}$. If for an arbitrary $k \in \{1, \ldots, m\}$ exactly the nodes $bdd_{j_1}^{(k)}, \ldots, bdd_{j_q}^{(k)}$ are contained in this connected component including $bdd_j^{(i)}$, then the corresponding equivalence class E_t in $\{0, 1\}^p/\sim$ is given by $\bigcup_{s=1}^q K_{j_s}^{(k)}$. The characteristic function for E_t is represented by the ROBDD for $\bigvee_{s=1}^q bdd_{j_s}^{(k)}$.

It was shown above that the computation of the equivalence classes of $\{0, 1\}^p/\sim$ can be reduced to the computation of connected components in a graph $G_\sim$. If we are prepared to construct the graph $G_\sim$ explicitly, then the problem is solved: For each pair $bdd_{j_1}^{(i_1)}, bdd_{j_2}^{(i_2)}$ of nodes in the node set of $G_\sim$ we merely have to check whether $bdd_{j_1}^{(i_1)} \wedge bdd_{j_2}^{(i_2)} \neq 0$. However, it turns out that it is not necessary to construct $G_\sim$ explicitly and it is not necessary to perform this *and*–operation for all pairs of nodes. Rather, the number of *apply* operations (*and*, *or*, ... of 2 ROBDDs (Bryant, 1986)) needed is linear in the number of *edges which really exist in $G_\sim$.*

Figure 4.10. Computation of OBDDs for equivalence classes $K_1^{(1)}$, $K_2^{(1)}$, $K_3^{(1)}$ and $K_4^{(1)}$ for function f_1 of Example 4.4 decomposed with respect to $\{x_1, x_2, x_3\}$. (OBDDs are not reduced.)

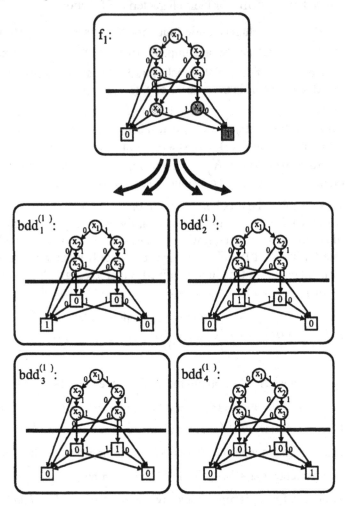

In Appendix I we present an algorithm which computes a representation of $\{0,1\}^p/_\sim$ and we prove that the total number of *apply*–operations needed by this algorithm is only linear in the number of edges in $G_\sim$.

Another important observation concerns the size of the ROBDDs computed during the algorithm of Appendix I: It is easy to see that the sizes of these ROBDDs cannot 'explode'. During the algorithm ROBDDs $cc^{(i)}$ are computed, which are disjunctions of ROBDDs $bdd_j^{(i)}$ ($1 \le j \le nrp(f_i, X^{(1)})$) and thus they could (in this special case) be computed by a replacement of linking nodes

Figure 4.11.   Computation of $\{0,1\}^p/_\sim$ for f_1 and f_2 of Example 4.4. The connected components of graph $G_\sim$ represent equivalence classes E_i.

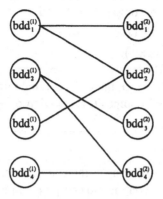

in the ROBDD of f_i by the constants 1 and 0, followed by a reduction of the resulting OBDD. For this reason they cannot be larger than the ROBDD for f_i.

The algorithm of Appendix I provides sets $CLNR_1^{(i)}, \ldots, CLNR_l^{(i)}$ (for all $1 \le i \le m$). The set $CLNR_j^{(i)}$ contains all indices s of classes $K_s^{(i)}$ which are subsets of class E_j, i.e., for all classes $E_1, \ldots, E_l$ of $\{0,1\}^p/_\sim$ the equation $E_j = \bigcup_{s \in CLNR_j^{(i)}} K_s^{(i)}$ holds. Now it is easy to generate the input for the modified branch–and–bound–algorithm of the previous section: Based on the sets $CLNR_1^{(i)}, \ldots, CLNR_l^{(i)}$ and on decomposition functions $\alpha_1^{(i)}, \ldots, \alpha_{k_i}^{(i)}$, which may be preselected for the decomposition of f_i, we can easily compute the sets $CLNR_{aj}^{(i)} = \{k \mid K_k^{(i)} \subseteq E_j$ and $\alpha^{(i)}(K_k^{(i)}) = a\}$ (for $1 \le i \le m$, $1 \le j \le l$, $a \in \alpha^{(i)}(E_j)$) and their sizes $SCLNR_{aj}^{(i)} = |CLNR_{aj}^{(i)}|$.

4.3.2   Applying a solution to CDF

This section shows how the solution to CDF is integrated into an overall method to compute common decomposition functions of as many output functions as possible.

The algorithm explained so far is able to compute h common strict decomposition functions for the decomposition of functions $f_1, \ldots, f_m$. Using a binary search with respect to h, the maximum number h_{max} of common strict decomposition functions of $f_1, \ldots, f_m$ can be computed ($1 \le h \le \min_{1 \le i \le m}(r_i - k_i)$, where r_i is the total number of decomposition functions in a communication minimal decomposition of f_i and k_i is the number of decomposition functions, which are already preselected for the decomposition of f_i).

After we have computed h_{max} common decomposition functions of $f_1, \ldots, f_m$, it is not possible to find additional common decomposition functions of $f_1, \ldots,$ f_m for communication minimal decompositions. However, it is still possible to find further common decomposition functions which can be used in the decomposition of *subsets* of $\{f_1, \ldots, f_m\}$.

EXAMPLE 4.1 (CONTINUED):
In Example 4.1 (page 124) there is no common decomposition function of f_1, f_2 and f_3 ($h_{max} = 0$). However, there is a common decomposition function of f_1 and f_2 and a common decomposition function of f_2 and f_3.

A naive approach would — as long as there are decomposition functions which remain to be computed — continue to compute common decomposition functions for subsets of $\{f_1, \ldots, f_m\}$ using the decomposition functions already computed as a preselection in the following computations. Of course it makes sense to begin with larger subsets of $\{f_1, \ldots, f_m\}$ and to finish with smaller subsets. However, such an approach would process an unnecessarily large number of subsets in general, including subsets containing functions f_{i_1} and f_{i_2} which cannot share any decomposition functions at all (at least not in communication minimal decompositions).

The remaining part of this section deals with a method of preventing unnecessary computations for subsets which cannot have common decomposition functions at all. This method has to process only a sub–problem of CDF which can be solved efficiently.

We have shown that the problem CDF remains NP–hard, even if the search for common decomposition functions is restricted to 2 functions and to strict decomposition functions (see Corollaries 4.1 and 4.2), but it can be solved efficiently, if

- it is restricted to 2 functions f_1 and f_2,

- it is restricted to strict decomposition functions,

- no decomposition functions $\alpha_1^{(1)}, \ldots, \alpha_{k_1}^{(1)}$ and $\alpha_1^{(2)}, \ldots, \alpha_{k_2}^{(2)}$ are preselected ($k_1 = k_2 = 0$) *and*

- we are looking for exactly one decomposition function ($h = 1$).

For this special class of instances of CDF it is possible to find an efficient solution using dynamic programming. How we apply solutions for such special cases of CDF to reduce the search space for the computation of common decomposition functions will be shown later. First we will present a dynamic program for solving this subclass of instances.

Input:

- Functions f_1 and f_2 of B_n with input variables $x_1, \ldots, x_n$.
- $r_1 = \lceil \log(nrp(X^{(1)}, f_1)) \rceil > 0$, $r_2 = \lceil \log(nrp(X^{(1)}, f_2)) \rceil > 0$ for a decomposition of f_1 and f_2 with respect to $X^{(1)} = \{x_1, \ldots, x_p\}$.
- For $i \in \{1, 2\}$: Partitions into equivalence classes $\{0, 1\}^p / \equiv_i = \{K_1^{(i)}, \ldots, K_{nrp(X^{(1)}, f_i)}^{(i)}\}$.
  Partition into equivalence classes $\{0, 1\}^p / \sim = \{E_1, \ldots, E_l\}$.
- For $i \in \{1, 2\}$, $1 \leq j \leq l$, $CLNR_j^{(i)} = \{k | K_k^{(i)} \subseteq E_j\}$, $SCLNR_j^{(i)} = |CLNR_j^{(i)}|$.

Output:

- $\alpha_1 \in B_p$, such that there is a one–sided decomposition of f_1 and f_2 with respect to $X^{(1)}$ of form

$$f_1(\mathbf{x}^{(1)}, \mathbf{x}^{(2)}) = g^{(1)}(\alpha_1(\mathbf{x}^{(1)}), \alpha_2^{(1)}(\mathbf{x}^{(1)}), \ldots, \alpha_{r_1}^{(1)}(\mathbf{x}^{(1)}), \mathbf{x}^{(2)}),$$
$$f_2(\mathbf{x}^{(1)}, \mathbf{x}^{(2)}) = g^{(2)}(\alpha_1(\mathbf{x}^{(1)}), \alpha_2^{(2)}(\mathbf{x}^{(1)}), \ldots, \alpha_{r_2}^{(2)}(\mathbf{x}^{(1)}), \mathbf{x}^{(2)}),$$

  where α_1 has to be strict, if there is a function of B_p with this property.
- Otherwise: The algorithm reports that it is not possible to find a function of B_p with the required property.

Algorithm:
  See Figure 4.12.

The algorithm is a 'two–dimensional generalization' of the well–known dynamic program for solving the Knapsack problem (see (Mehlhorn, 1984), p. 210, e.g.). It is based on the following considerations:

Using Lemma 4.2 it can be concluded that there is one common decomposition function α_1 with the properties given above if and only if there is a partition of $\{1, \ldots, l\}$ into 2 sets

$$D_0 = \{j \mid \alpha_1(x) = 0 \ \forall x \in E_j\} \text{ and } D_1 = \{j \mid \alpha_1(x) = 1 \ \forall x \in E_j\}$$

with

$$\sum_{k \in D_0} SCLNR_k^{(1)} \leq 2^{r_1 - 1} \text{ and } \sum_{k \in D_1} SCLNR_k^{(1)} \leq 2^{r_1 - 1} \text{ and}$$

Figure 4.12. Dynamic program to compute one common strict decomposition function of two functions f_1 and f_2.

```
 1  ∀0 ≤ l₁ ≤ 2^{r₁-1}, 0 ≤ l₂ ≤ 2^{r₂-1} :  B[l₁][l₂] = false, D[l₁][l₂] = ∅
 2  B[0][0] = true
 3  for j = 1 to l do
 4      for l₁ = 2^{r₁-1} downto SCLNR_j^{(1)} do
 5          for l₂ = 2^{r₂-1} downto SCLNR_j^{(2)} do
 6              if (B[l₁ − SCLNR_j^{(1)}][l₂ − SCLNR_j^{(2)}] = true)
 7                  then
 8                      B[l₁][l₂] = true
 9                      D[l₁][l₂] = D[l₁ − SCLNR_j^{(1)}][l₂ − SCLNR_j^{(2)}] ∪ {j}
10                      /* Invariant: There is exactly one subset D of {1,...,j} with
11                          ∑_{l∈D} SCLNR_l^{(1)} = l₁ and ∑_{l∈D} SCLNR_l^{(2)} = l₂, if B[l₁][l₂] = true.
12                          Then D = D[l₁][l₂] is a possible choice for D. */
13                  fi
14          od
15      od
16  od
17  if (∃l₁, l₂ with nrp(X, f₁) − 2^{r₁-1} ≤ l₁ ≤ 2^{r₁-1}
18      and nrp(X, f₂) − 2^{r₂-1} ≤ l₂ ≤ 2^{r₂-1} and B[l₁][l₂] = true)
19  then
20      for j = 1 to l do
21          if (j ∈ D[l₁][l₂])
22              then
23                  ∀x ∈ E_j : α₁(x) = 0
24              else
25                  ∀x ∈ E_j : α₁(x) = 1
26          fi
27      od
28  else
29      Report that there is no α₁ with required properties.
30  fi
```

$$\sum_{k \in D_0} SCLNR_k^{(2)} \le 2^{r_2-1} \quad \text{and} \quad \sum_{k \in D_1} SCLNR_k^{(2)} \le 2^{r_2-1}.$$

Since

$$\sum_{k=1}^{l} SCLNR_k^{(1)} = nrp(X^{(1)}, f_1),$$

we see that

$$\sum_{k \in D_1} SCLNR_k^{(1)} = \sum_{k=1}^{l} SCLNR_k^{(1)} - \sum_{k \in D_0} SCLNR_k^{(1)} \le 2^{r_1-1}.$$

is equivalent to

$$\sum_{k \in D_0} SCLNR_k^{(1)} \geq nrp(X^{(1)}, f_1) - 2^{r_1 - 1}.$$

Thus there is such a decomposition function α_1 if and only if there is a set

$$D_0 = \{ j \mid \alpha_1(x) = 0 \; \forall x \in E_j \}$$

with

$$nrp(X^{(1)}, f_1) - 2^{r_1 - 1} \leq \sum_{k \in D_0} SCLNR_k^{(1)} \leq 2^{r_1 - 1} \text{ and}$$

$$nrp(X^{(1)}, f_2) - 2^{r_2 - 1} \leq \sum_{k \in D_0} SCLNR_k^{(2)} \leq 2^{r_2 - 1}.$$

The dynamic program simply computes whether there is such a set D_0. To achieve this goal it uses two two–dimensional arrays B and D. The entries of B are initialized to **false**, the entries of D are initialized to $\emptyset$ (line 1). In a loop for $j = 1 \ldots, l$ (lines 3–16) B and D are changed. The invariant of lines 10–12 is essential for the correctness of the dynamic program: When the loop is executed for the jth time, then there is a subset D of $\{1, \ldots, j\}$ with $\sum_{l \in D} SCLNR_l^{(1)} = l_1$ and $\sum_{l \in D} SCLNR_l^{(2)} = l_2$, if and only if $B[l_1][l_2] =$ **true**. A possible choice for D is saved in $D[l_1][l_2]$. After l executions of the loop of lines 3–16 there is a set D_0 with

$$nrp(X^{(1)}, f_1) - 2^{r_1 - 1} \leq \sum_{k \in D_0} SCLNR_k^{(1)} \leq 2^{r_1 - 1} \text{ and}$$

$$nrp(X^{(1)}, f_2) - 2^{r_2 - 1} \leq \sum_{k \in D_0} SCLNR_k^{(2)} \leq 2^{r_2 - 1},$$

if and only if there are l_1 and l_2 with $B[l_1][l_2] =$ **true** and

$$nrp(X, f_1) - 2^{r_1 - 1} \leq l_1 \leq 2^{r_1 - 1} \text{ and}$$

$$nrp(X, f_2) - 2^{r_2 - 1} \leq l_2 \leq 2^{r_2 - 1}.$$

Then D_0 can be chosen as $D_0 = D[l_1][l_2]$.

The run time of the dynamic program is

$$O(2^{r_1 - 1} \cdot 2^{r_2 - 1} \cdot l) = O(nrp(X^{(1)}, f_1) \cdot nrp(X^{(1)}, f_2) \cdot l).$$

Using the dynamic program shown above, we can check, whether there are common strict decomposition functions of f_{i_1} and f_{i_2} for all pairs of functions f_{i_1} and f_{i_2}. Based on this information it is possible to define a graph $G_{CDF} =$

$(\{f_1, \ldots, f_m\}, E)$, where there is an edge between two nodes f_{i_1} and f_{i_2} if and only if there is a common strict decomposition function of f_{i_1} and f_{i_2}. If the branch–and–bound–algorithm for computing common strict decomposition functions is applied to some subset T of the functions $\{f_1, \ldots, f_m\}$, then this makes sense only if the functions of T form a clique in G_{CDF}.

EXAMPLE 4.6 Consider the binary addition of two 2^k–bit–numbers add_{2^k} :
$\{0,1\}^{2^{k+1}} \rightarrow \{0,1\}^{2^k}, add_{2^k}(a_{2^k-1}, \ldots, a_0, b_{2^k-1}, \ldots, b_0) = (r_{2^k-1}, \ldots, r_0)$, where

$$\sum_{i=0}^{2^k-1} r_i 2^i = (\sum_{i=0}^{2^k-1} a_i 2^i + \sum_{i=0}^{2^k-1} b_i 2^i) \bmod 2^{2^k}.$$

The single–output functions of add_{2^k} are denoted by $s_{2^k-1}, \ldots, s_0$, i.e., $add_{2^k} = (s_{2^k-1}, \ldots, s_0)$. If for add_{2^k} a two–sided and balanced decomposition is performed, then our implemented logic synthesis algorithm computes for all s_i ($0 \leq i \leq 2^k - 1$) the variable partition $A = \{\{a_{2^k-1}, \ldots, a_{2^{k-1}}, b_{2^k-1}, \ldots, b_{2^{k-1}}\} \{a_{2^{k-1}-1}, \ldots, a_0, b_{2^{k-1}-1}, \ldots, b_0\}\}$. It is easy to see that two–sided decompositions can be reduced to a series of two one–sided decompositions. In this special case we can reduce the two–sided decomposition of add_{2^k} with respect to A to two one–sided decompositions with respect to $A_1 = \{a_{2^k-1}, \ldots, a_{2^{k-1}}, b_{2^k-1}, \ldots, b_{2^{k-1}}\}$ and $A_0 = \{a_{2^{k-1}-1}, \ldots, a_0, b_{2^{k-1}-1}, \ldots, b_0\}$. A (communication minimal) decomposition with respect to A_0 has exactly one decomposition function for all s_i. The corresponding graph G_{CDF} has 2^{k-1} nodes $s_0, \ldots, s_{2^{k-1}-1}$ which are not the source of any edge, and 2^{k-1} additional nodes which form a clique in G_{CDF} (see left hand side of Figure 4.13 for add_{16}). In fact, there is one common decomposition function of $s_{2^k-1}, \ldots, s_{2^{k-1}}$ which is equal to the carry function for the addition of $(a_{2^{k-1}-1}, \ldots, a_0)$ and $(b_{2^{k-1}-1}, \ldots, b_0)$ (or equal to the negated carry function).

In a (communication minimal) decomposition with respect to A_1, there is no decomposition function for $s_0, \ldots, s_{2^{k-1}-1}$ (these functions are independent of $a_{2^k-1}, \ldots, a_{2^{k-1}}, b_{2^k-1}, \ldots, b_{2^{k-1}}$), 1 decomposition function for s_{2^k-1}, and there are 2 decomposition functions for $s_{2^{k-1}+1} \cdots, s_{2^k-1}$, respectively. In the graph G_{CDF} for this decomposition there is only one edge from s_{2^k-1} to $s_{2^{k-1}+1}$ (see right hand side of Figure 4.13 for add_{16}). It only makes sense to look for common decomposition functions of s_{2^k-1} and $s_{2^{k-1}+1}$. Because of G_{CDF}, other subsets of functions s_i need not be considered.

If the relations $\equiv_{i_1}$ and $\equiv_{i_2}$ for the decomposition of 2 functions with respect to a certain variable set are exactly identical, then we need not include both f_{i_1} and f_{i_2} in the graph G_{CDF}, since it is possible to choose the decomposition

Figure 4.13.   Graph G_{CDF} for function add_{16}.

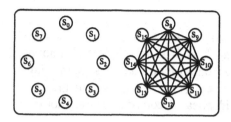
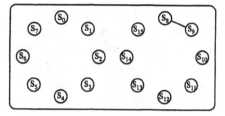

Figure 4.14.   Reduced graph G_{CDF} of add_{16}.

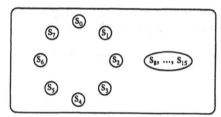
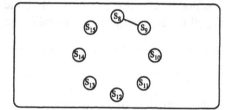

functions for f_{i_1} and f_{i_2} to be *all* identical. Of course we can also remove functions from G_{CDF}, which do not depend on the bound set and thus do not have any decomposition functions in a communication minimal decomposition.

EXAMPLE 4.6 (CONTINUED):

For function add_{2^k} using a decomposition with respect to A_0 the relations $\equiv_i$ of s_i are identical for all $i \in \{2^{k-1}, \ldots, 2^k - 1\}$. For this reason G_{CDF} can be reduced to the graph in Figure 4.14 (left hand side).

In graph G_{CDF} (right hand side of Figure 4.14) we can remove all nodes $s_0, \ldots, s_{2^{k-1}-1}$ for a decomposition with respect to A_1, since the corresponding functions do not have any decomposition functions in this decomposition.

For the computation of common decomposition functions it suffices to consider subsets of $\{f_1, \ldots, f_m\}$ which form a clique in the reduced graph G_{CDF}. The logic synthesis algorithm we implemented based on this begins with a maximal clique in G_{CDF} and computes a maximum number of common decomposition functions for the output functions of this clique. Common decomposition functions found so far are treated as preselected decomposition functions in subsequent applications of the branch–and–bound–algorithm. The method processes other cliques of G_{CDF} in order of decreasing size. Whenever all decomposition functions are computed for some output function, the corresponding node is removed from G_{CDF}.

EXAMPLE 4.6 (CONTINUED):
For function add_{2^k} and decomposition with respect to A_0, there are no edges in the *reduced* graph G_{CDF}. So the branch–and–bound–algorithm is not applied at all for the reduced graph. The decomposition functions for the single outputs are computed separately. Since $\equiv_{2^k-1}, \ldots, \equiv_{2^k-1}$ are identical, we know in advance that we can choose identical decomposition functions for $s_{2^k-1}, \ldots, s_{2^k-1}$.

For the decomposition with respect to A_1 there is exactly one clique which contains more than one function. A common decomposition function is computed for this clique, then s_{2^k-1} is removed from G_{CDF}, since the number of decomposition functions for s_{2^k-1} is only 1 and after that the remaining decomposition functions are computed for all other output functions separately.

4.4. Experiments

4.4.1 Example: A binary adder

We have evaluated the methods presented in this chapter using a series of examples, including adders of various bit widths. Here we use an 8–bit adder to illustrate the algorithm (and in particular the computation of common decomposition functions). In this example we use two–sided decompositions. Using the function of an 8–bit adder as an input, our algorithm produces the realization in Figure 4.15. In this figure, shaded boxes represent the respective results for different calls of the recursive decomposition procedure.

An analysis of the circuit structure reveals that this circuit has the same structure as a conditional sum adder (Slansky, 1960). (For comparison we show the 8–bit conditional sum adder in Figure 4.16.)

On the first level of the recursion, the variable partition into sets $\{x_0, y_0, x_1, y_1, x_2, y_2, x_3, y_3\}$ and $\{x_4, y_4, x_5, y_5, x_6, y_6, x_7, y_7\}$ was computed. In this decomposition the decomposition function depending on $\{x_0, y_0, x_1, y_1, x_2, y_2, x_3, y_3\}$ (i.e., the carry bit of the addition of $(x_3x_2x_1x_0)$ and $(y_3y_2y_1y_0)$) is *shared* between f_4, f_5, f_6 and f_7.

Unlike in the conditional sum scheme, our algorithm

1. *always* uses a minimum number of decomposition functions and

Figure 4.15.   Automatically generated circuit for an 8–bit adder.

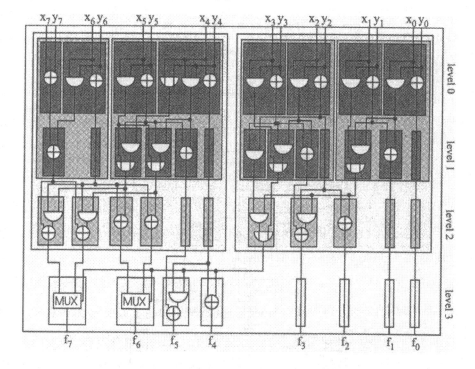

2. computes *common decomposition functions* whenever possible.

For f_4 this leads to only *one* decomposition function depending on the most significant bits $x_4, y_4, x_5, y_5, x_6, y_6, x_7, y_7$ on the first level of the recursion ('level 3' in Figure 4.15), whereas the conditional sum adder uses 2 functions here, which are the least significant bit of the sum of $(x_7 x_6 x_5 x_4)$ and $(y_7 y_6 y_5 y_4)$ and the least significant bit of the sum of $(x_7 x_6 x_5 x_4)$, $(y_7 y_6 y_5 y_4)$ and 1.

Depending on $\{x_4, y_4, x_5, y_5, x_6, y_6, x_7, y_7\}$, there are 2 decomposition functions for f_5, f_6 and f_7, both in the implementation in Figure 4.15 and the conditional sum adder, since 4 different alternatives for the information depending on the most significant bits are encoded by 2 bits. The conditional sum adder uses a fixed encoding for these 4 alternatives. The information is encoded by the corresponding bits of the sum and the sum + 1, respectively. However, for function f_5 the implementation in Figure 4.15 uses a different encoding for these 4 alternatives. The reason for this choice lies in the use of common decomposition functions: here the decomposition function computed for f_4 depending on variables $x_4, y_4, x_5, y_5, x_6, y_6, x_7, y_7$ can be *reused* for function f_5.

Figure 4.16.   Conditional sum adder with bit width 8.

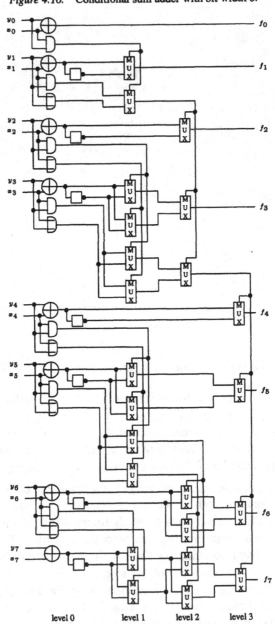

Table 4.1.   Comparison between our tool *mulopII* and *sis* with respect to layout size and signal delay.

circuit	layout size			signal delay		
	sis	*mulopII*	*ratio*	*sis*	*mulopII*	*ratio*
9symml	1194336	201400	5.93	27.6	13.6	2.03
C17	28800	25704	1.12	4.2	4.2	1.00
cm138a	103896	87480	1.19	5.8	6.8	0.85
cm151a	95312	140728	0.68	12.6	19.2	0.66
cm152a	85536	106704	0.80	10.0	13.2	0.76
cm162a	131976	178296	0.74	12.0	12.4	0.97
cm163a	144008	140728	1.02	13.0	9.4	1.38
cm82a	74784	61320	1.22	7.2	7.0	1.03
cm85a	165456	170288	0.97	10.2	13.0	0.78
cmb	204792	120624	1.70	9.4	6.6	1.42
decod	140448	119496	1.18	6.2	5.0	1.24
f51m	561184	251392	2.23	51.0	18.4	2.77
majority	42200	39168	1.08	7.8	6.6	1.18
parity	99408	96976	1.03	5.0	5.0	1.00
z4ml	156288	93800	1.67	16.2	9.8	1.65
$\sum$	3228K	1834K	1.76	198.2	150.2	1.32

These differences lead to an R_2–complexity of 89 for the automatically generated adder, whereas the conditional sum adder has a R_2–complexity of 106. The depth is the same in both cases (it is proportional to the logarithm of the number of inputs, when we consider adders with various bit widths).

For the synthesis of adders with higher bit widths (up to 128 inputs), we have already demonstrated the advantage of ROBDD based decompositions over function table based decompositions in Table 3.1 on page 91.

4.4.2   Benchmark results

Here the methods from Chapters 3 and 4 are applied to compute benchmark circuits using 2–input gates. Results for FPGA synthesis will be given later on in Chapter 5, after we have considered decompositions for incompletely specified functions.

For comparison we generated circuits with 2–input gates both using *sis* (Brayton et al., 1987; Sentovich et al., 1992) as well as our decomposition procedure. For the resulting circuits we then generated layouts using the tool *TimberWolf* (integrated into the system *octtools* from the University of California, Berkeley).

Table 4.1 shows results for layout sizes and signal delays both for our tool, which is called *mulopII* in the following, and *sis*. Although there are examples like *cm151a* and *cm162a*, which do not seem appropriate for disjoint decompositions, our tool *mulopII* outperforms *sis* in 11 out of 15 cases, when layout sizes are compared. On average the layout sizes of *sis* are about 76 % larger. Nevertheless, the signal delays of 10 out of 15 circuits generated by our tool *mulopII* are shorter than the signal delays for *sis* or equal to them.

One of the motivations for applying communication minimal decompositions is to minimize global wires with respect to the layout generation. We hope that the layout sizes can be further reduced using a layout tool which is able to make use of the hierarchical structures of the decomposed circuits. The minimization of global wires will become increasingly important as the transistor sizes in 'deep submicron' designs shrink, so that the wire delays will dominate the cell delays in future designs.

But even when using a layout tool not optimized for recursive decomposition, we can observe effects of the communication minimization: Figure 4.17 shows layouts of circuits generated for benchmark circuit *9symml*, where the largest improvement was observed for our tool. The circuit on the left hand side was generated by *sis* and the circuit on the right hand side by our tool. Not only is the number of cells much smaller in our design, but also the height of the routing channels.

Finally, we give results to further encourage the use of ROBDD based decompositions rather than function table based decompositions. In Table 4.2, the run times of our ROBDD based tool *mulopII* (Scholl and Molitor, 1994; Scholl and Molitor, 1995b; Scholl and Molitor, 1995a) are compared to the run times of a function table based version. (The CPU times were measured using a SPARC-station 10/30 with 64 MByte RAM.) As expected, Table 4.2 shows considerable improvement in run times for the ROBDD based tool.

The last column of Table 4.2 shows the percentage of run time which is used to solve the problem CDF of Section 4.3.1, i.e., which is used to compute common decomposition functions. The results show that the ROBDD based computation of common decomposition functions (see Section 4.3.1.3) is very efficient for the benchmark circuits.

4.5. Alternative methods

In this section we give an overview of other existing methods for the decomposition of multi–output functions.

Figure 4.17. Layouts for benchmark circuit *9symml*. The circuit on the left hand side was generated by *sis* and the circuit on the right hand side by our tool *mulopII*.

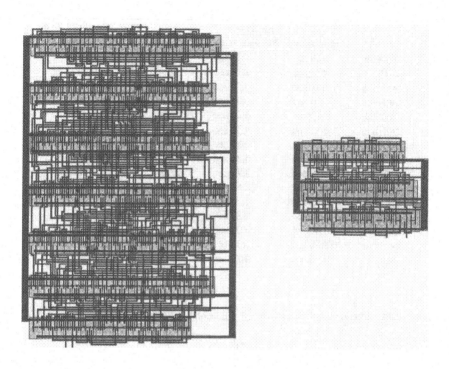

4.5.1 Interpreting multi–output functions as integer–valued functions

As already mentioned in Section 4.1, an alternative to a single–output decomposition with computation of common decomposition functions is the decomposition of a multi–output function *as a whole*, i.e., the decomposition of a multi–output function $f \in B_{n,m}$ into a decomposition function $\alpha \in B_{p,r}$ and a composition function $g \in B_{n-p+r,m}$ such that

$$f(x_1, \ldots, x_p, x_{p+1}, \ldots, x_n) =$$
$$g(\alpha_1(x_1, \ldots, x_p), \ldots, \alpha_r(x_1, \ldots, x_p), x_{p+1}, \ldots, x_n). \qquad (\star)$$

For these types of decompositions the minimum number of decomposition functions can also be determined using a decomposition matrix of f. Here the entries of the decomposition matrix are not elements of $\{0, 1\}$, but elements of $\{0, 1\}^m$. In Figure 4.18 you again see the decomposition matrix for the function from Example 4.1 on page 124 (with respect to bound set $\{x_1, x_2, x_3\}$).

Table 4.2. Comparison of run times for our ROBDD based tool *mulopII* and for a function table based version.

	run time			percentage of run time
circuit	table based	ROBDD based	ratio	for CDF (mulopII)
9symml	1.40 sec	1.23 sec	1.14	0.69%
C17	0.32 sec	0.15 sec	2.13	0.01%
cm138a	1.01 sec	0.18 sec	5.61	0.89%
cm151a	4.16 sec	1.09 sec	3.82	0.13%
cm152a	2.15 sec	0.50 sec	4.30	0.36%
cm162a	350.65 sec	3.32 sec	105.62	0.26%
cm163a	2923.31 sec	2.35 sec	1243.96	0.08%
cm82a	0.38 sec	0.21 sec	1.81	0.01%
cm85a	7.46 sec	3.73 sec	2.00	0.27%
cmb	1836.13 sec	2.52 sec	728.62	0.05%
decod	26.15 sec	2.56 sec	10.21	1.93%
f51m	3.14 sec	1.83 sec	1.71	0.28%
majority	0.44 sec	0.08 sec	5.50	0.01%
parity	111.06 sec	1.37 sec	81.07	0.00%
z4m1	0.66 sec	0.76 sec	0.87	0.16%

Figure 4.18. Decomposition matrix for f of Example 4.1 (all three outputs together).

x_4	0	1
$x_1 x_2 x_3$		
0 0 0	(000)	(000)
0 0 1	(001)	(010)
0 1 0	(011)	(110)
0 1 1	(111)	(110)
1 0 0	(010)	(101)
1 0 1	(011)	(111)
1 1 0	(001)	(011)
1 1 1	(101)	(011)

The minimum number of decomposition functions of f is equal to $\lceil \log(nrp(\{x_1, \ldots, x_p\}, f)) \rceil$, where $nrp(\{x_1, \ldots, x_p\}, f)$ is the number of different row patterns in the decomposition matrix.

Lai et al. (Lai et al., 1993b; Lai et al., 1994b) propose a very similar method for multi–output decomposition: They interpret a multi–output function function $f = (f_1, \ldots, f_m) : \{0, 1\}^n \rightarrow \{0, 1\}^m$ as an integer–valued function $\tilde{f} : \{0, 1\}^n \rightarrow \{0, \ldots, 2^m - 1\}$ with $\tilde{f}(x_1, \ldots, x_n) = \sum_{i=1}^{m} f_i(x_1, \ldots, x_n) \cdot 2^{i-1}$.

Figure 4.19.   Decomposition matrix for $\tilde{f}$ of Example 4.1 (integer–valued function).

$x_1x_2x_3$	x_4 0	1
0 0 0	0	0
0 0 1	4	2
0 1 0	6	3
0 1 1	7	3
1 0 0	2	5
1 0 1	6	7
1 1 0	4	6
1 1 1	5	6

Of course the number of row patterns in the decomposition matrix of $\tilde{f}$ is equal to the number of row patterns in the decomposition matrix of f (regarding the multi–output function as a whole). Figure 4.19 shows the decomposition matrix, which results from the decomposition matrix of Figure 4.18 by interpreting f as an integer–valued function $\tilde{f}$.

Lai et al. compute the number of row patterns in the decomposition matrix of $\tilde{f}$ by considering linking nodes not in ROBDDs, but in EVBDDs. EVBDDs are representations of integer–valued functions with a Boolean domain.

Unfortunately, we are not so likely to find nontrivial decompositions when multi–output functions are decomposed according to equation $(\star)$. Especially for multi–output functions with a large number m of outputs, the single–output functions are often nontrivially decomposable, whereas the multi–output function is not.[9] If an output partitioning into very small groups is used to avoid this problem, then the potential of logic sharing is not fully exploited, since many output functions which can potentially share logic are put into different groups. In our experience this kind of decomposition can only make sense when it is combined with other logic synthesis techniques. In that case other techniques like *sis* (Brayton et al., 1987; Sentovich et al., 1992) already compute a circuit with logic sharing and then multi–output decomposition is only performed for parts of this circuit, keeping the global structure fixed.

---

[9] As already mentioned in Section 4.1, for the function in Example 4.1 there is also no nontrivial multi–output decomposition with respect to $\{x_1, x_2, x_3\}$, although f_1, f_2 and f_3 are nontrivially decomposable. The decomposition matrices in Figures 4.18 and 4.19 have a maximum number of 8 row patterns.

4.5.2   Representing the set of all possible decomposition functions

The computation of common decomposition functions for several output functions, which was described in this chapter, is directly done by a branch–and–bound algorithm without the need to compute a set of candidate functions first. Other approaches, such as (Lai et al., 1994a) and (Wurth et al., 1995), compute a set of possible decomposition functions for each single–output function f_i and try to select common decomposition functions by identifying identical functions in these sets of candidates.

4.5.2.1   Explicit representation

Lai et al. (Lai et al., 1994a; Lai et al., 1996) proposed a method of computing common decomposition functions which is based on an explicit representation of sets of candidate functions for these decomposition functions.

First, they compute the set of all strict decomposition functions which can occur in the decomposition of f_i with respect to a bound set $X^{(1)} = \{x_1, \ldots, x_p\}$ for each single–output function f_i ($1 \leq i \leq m$). For this computation the partition $\{0,1\}^p/_{\equiv_i}$ of $\{0,1\}^p$ into equivalence classes resulting from identical row patterns in $Z(X^{(1)}, f_i)$ is determined based on the ROBDD for f_i. Suppose that the minimum number of decomposition functions is $r_i := \lceil \log(nrp(X^{(1)}, f_i)) \rceil$ and that $\{0,1\}^p/_{\equiv_i} = \{K_1^{(i)}, \ldots, K_{nrp(X^{(1)},f_i)}^{(i)}\}$. For the computation of candidates for strict decomposition functions, the following criterion can be derived from Lemma 4.1:

$\alpha \in B_p$ can be used as a strict decomposition function in the decomposition of f_i if and only if

$$|\{K_j^{(i)} \mid \alpha(K_j^{(i)}) = 0\}| \leq 2^{r_i-1} \text{ and } |\{K_j^{(i)} \mid \alpha(K_j^{(i)}) = 1\}| \leq 2^{r_i-1}.$$

Thus a strict decomposition function α of f_i corresponds to a partition of $\{K_1^{(i)}, \ldots, K_{nrp(X^{(1)},f_i)}^{(i)}\}$ into two sets

$$S_0 = \{K_j^{(i)} \mid \alpha(K_j^{(i)}) = 0\} \quad \text{and} \quad S_1 = \{K_j^{(i)} \mid \alpha(K_j^{(i)}) = 1\}$$
$$\text{with } |S_0| \leq 2^{r_i-1} \quad \text{and} \quad |S_1| \leq 2^{r_i-1}. \tag{4.4}$$

Thus the set of all partitions of $\{K_1^{(i)}, \ldots, K_{nrp(X^{(1)},f_i)}^{(i)}\}$ into two sets with condition (4.4) is a representation of all possible strict decomposition functions of f_i.[10]

---

[10]It it sufficient to consider *partitions* $\{S_0, S_1\}$ instead of ordered sets (S_0, S_1), since exchanging S_0 and S_1 means changing from α to $\overline{\alpha}$ and it will not be necessary to represent both α and $\overline{\alpha}$ in the candidate set.

In (Lai et al., 1994b) all possible partitions of $\{K_1^{(i)}, \ldots, K_{nrp(X^{(1)}, f_i)}^{(i)}\}$ into two sets fulfilling condition (4.4) are generated for each f_i. Then, for each partition, the corresponding decomposition function is computed. After that, for each candidate decomposition function α, the number of functions f_i is determined, such that α can be used in a communication minimal decomposition of f_i. Lai et al. select the candidate function which can be used for the largest number of output functions f_i.

For the selection of the second common decomposition function, we have to cope with preselected decomposition functions: Suppose in the first step we have selected the decomposition function α which corresponds to the partition $\{S_0, S_1\}$. If α has been selected for function f_i (among others), then we have to decide whether other candidate functions β of f_i can still be used in the decomposition of f_i *under the assumption that α is preselected*. Assuming a partition $\{T_0, T_1\}$ for β, this can be decided based on Lemma 4.1: Using α and β as decomposition functions of f_i, we assign $(0, 0)$ to the equivalence classes of $R_0 = S_0 \cap T_0$, $(0, 1)$ to the classes in $R_1 = S_0 \cap T_1$, $(1, 0)$ to the classes in $R_2 = S_1 \cap T_0$ and $(1, 1)$ to the classes in $R_3 = S_1 \cap T_1$. For the partition $\{R_0, \ldots, R_3\}$ representing decomposition functions α *and* β together we have (according to Lemma 4.1) the condition:

$$|R_j| \leq 2^{r_i - 2} \text{ for all } j \in \{0, \ldots, 3\}. \tag{4.5}$$

That means that we have to remove all candidate functions β which do not fulfill the condition of equation (4.5), i.e, which cannot used in a communication minimal decomposition with α as a preselected decomposition function, from the set of candidates for function f_i.

After removing decomposition functions from candidate sets, the second step selects the second candidate function. Again, the decomposition function with the most occurrences in candidate sets for functions f_i is selected. Then candidates are removed based on the current selection of a decomposition function, and so on. The removal of candidate functions is done by means of a trivial generalization of condition (4.5) based on Lemma 4.1. This process is repeated until r_i decomposition functions are selected for each output function f_i.

The problem with this approach lies in the potential size of the candidate sets: Suppose that $nrp(X^{(1)}, f_i) = 2^{r_i}$. Then there are $\frac{1}{2}\binom{2^{r_i}}{2^{r_i-1}} = \theta(2^{2^{r_i} - \frac{r_i}{2}})$ different partitions fulfilling the condition of equation (4.4). [11] This results in a huge number of candidate decomposition functions even for examples with a relatively small number of equivalence classes.

---

[11] We have to select exactly $2^{r_i - 1}$ out of 2^{r_i} equivalence classes to be included in S_0 and there is no difference between $\{S_0, S_1\}$ and $\{S_1, S_0\}$.

The problem of huge candidate sets is moderated in (Lai et al., 1994b) by a restriction to bound sets of size 5. For bound sets of size 5 there are at most $2^5 = 32$ equivalence classes and when there is a nontrivial decomposition, there are at most $2^4 = 16$ equivalence classes. (Nevertheless, the number of candidates for strict decomposition functions assuming 16 equivalence classes is equal to 6435.)

4.5.2.2   Implicit representation

Wurth et al. (Wurth et al., 1995) also decided to compute the sets of all candidate decomposition functions in order to select common decomposition functions. However, they do not represent the candidate functions explicitly, but rather *implicitly* using ROBDD representations. The common decomposition functions computed by Wurth et al. may also be non–strict.

- As in (Lai et al., 1994a) Wurth et al. compute decomposition functions step by step, always selecting single–output decomposition functions, which can be used for the decomposition of as many output functions as possible.

- In each step and for each output function f_i, they compute a set of functions which can be used as decomposition functions in a communication minimal decomposition assuming that the decomposition functions, which were already selected, remain fixed. These functions are called *assignable* functions for f_i. The set of these candidate functions is represented *implicitly* by a single ROBDD for each f_i (as described for (Legl et al., 1996b) in Section 3.8.4.2).

The algorithm is based on the computation of partitions $\{0,1\}^p/_{\equiv_i} = \{K_1^{(i)}, \ldots, K_{nrp(X^{(1)}, f_i)}^{(i)}\}$ for each function f_i. Again we define $r_i := \lceil \log(nrp(X^{(1)}, f_i)) \rceil$. Since Wurth et al. also use non–strict decomposition functions, the conclusion from Lemma 4.1 is not exactly the same as in Section 4.5.2.1. A function α can be used as a decomposition function in a communication minimal decomposition of f_i if and only if

$$|\{K_j^{(i)} \mid \exists \epsilon \in K_j^{(i)} \text{ with } \alpha(\epsilon) = 0\}| \leq 2^{r_i - 1}$$
$$\text{and}$$
$$|\{K_j^{(i)} \mid \exists \epsilon \in K_j^{(i)} \text{ with } \alpha(\epsilon) = 1\}| \leq 2^{r_i - 1}. \qquad (4.6)$$

It is easy to see that $|\{K_j^{(i)} \mid \exists \epsilon \in K_j^{(i)} \text{ with } \alpha(\epsilon) = 0\}| \leq 2^{r_i - 1}$ can be fulfilled if and only if at least $nrp(X^{(1)}, f_i) - 2^{r_i - 1}$ classes $K_j^{(i)}$ are *completely* mapped to 1 by α. This means that equation (4.6) is equivalent to

$$|\{K_j^{(i)} \mid \alpha(K_j^{(i)}) = \{1\}\}| \geq nrp(X^{(1)}, f_i) - 2^{r_i - 1} \qquad (4.7)$$

and

$$|\{K_j^{(i)} \mid \alpha(K_j^{(i)}) = \{0\}\}| \geq nrp(X^{(1)}, f_i) - 2^{r_i - 1} \qquad (4.8)$$

Equations (4.7) and (4.8) form the basis for the implicit computation of the set of all assignable functions for f_i.

To represent sets of functions a 'characteristic function' is used. For a set $S \subseteq B_p$ of functions, a characteristic function $\chi_S \in B_{2^p}$ could be used with the property that $\chi_S(\epsilon_0, \ldots, \epsilon_{2^p - 1}) = 1$ if and only if the function h_ϵ with function table $(\epsilon_0, \ldots, \epsilon_{2^p - 1})$ is in S. However, to represent functions $\chi_S \in B_{2^p}$ by ROBDDs, 2^p variables would be needed. For this reason Wurth et al. introduced the notion of 'global classes' to minimize the number of variables for characteristic functions:

NOTATION 4.4
Let $f = (f_1, \ldots, f_m) \in B_{n,m}$, $\{0,1\}^p/_{\equiv_i} = \{K_1^{(i)}, \ldots, K_{nrp(X^{(1)}, f_i)}^{(i)}\}$ for all $1 \leq i \leq m$. Then the smallest set $\mathcal{G} = \{G_1, \ldots, G_g\}$, such that $\mathcal{G}$ is a refinement[12] of the partitions $\{0,1\}^p/_{\equiv_1}, \ldots, \{0,1\}^p/_{\equiv_m}$ is called the set of global classes of f.

REMARK 4.2 *The size g of the set $\mathcal{G} = \{G_1, \ldots, G_g\}$ of global classes of f is equal to $g = nrp(X^{(1)}, f)$. For all $\epsilon \in G_j$ the corresponding row patterns of $Z(X^{(1)}, f_i)$ are equal for all $1 \leq i \leq m$ and thus the corresponding row patterns of $Z(X^{(1)}, f)$ are equal. A global class results from an intersection of classes $K_{j_i}^{(i)}$ for different $1 \leq i \leq m$.*

In (Wurth et al., 1995) a function $\alpha \in B_p$ is called *constructable with respect to* $\mathcal{G} = \{G_1, \ldots, G_g\}$ if and only if $|\alpha(G_j)| = 1$ for all $1 \leq j \leq g$, i.e., if and only if each global class G_j is either completely contained in the ON-set or in the OFF-set of α.

Wurth et al. observed that the sets of assignable functions can be restricted to *constructable* functions with respect to the set of global classes without reducing the chances of finding common decomposition functions. So for an assignable and constructable function it suffices to give the function values for the classes G_j, such that for a representation by characteristic functions we need only g variables and not 2^p variables (if $g < 2^p$).

---

[12]See Notation 4.3 on page 146.

Thus, sets of assignable (and constructable) functions $S \subseteq B_p$ can be represented by characteristic functions $\chi_S \in B_g$. $\chi_S(\epsilon_1, \ldots, \epsilon_g) = 1$ if and only if the function α with $\alpha(G_j) = \{\epsilon_j\}$ $\forall 1 \leq j \leq g$ is in S. χ_S is represented by an ROBDD with g variables $z_1, \ldots, z_g$.

Now we have to explain how the implicit representation of the set of assignable and constructable functions for a function f_i is computed. For ease of explanation we assume here that there is no preselected decomposition function from some previous step. We can use Equations (4.7) and (4.8) to compute the characteristic function.

For the time being, we introduce an auxiliary variable v_j for each equivalence class $K_j^{(i)}$ and we define a characteristic function based on these auxiliary variables $v_1, \ldots, v_{nrp(X^{(1)}, f_i)}$. The number of equivalence classes (and auxiliary variables) for f_i is abbreviated by $nrp_i = nrp(X^{(1)}, f_i)$ in the following. We consider the threshold function $s_{nrp_i - 2^{r_i - 1}}^{nrp_i}$ with $s_{nrp_i - 2^{r_i - 1}}^{nrp_i}(v_1, \ldots, v_{nrp_i}) = 1$ iff $\sum_{j=1}^{nrp_i} v_j \geq nrp_i - 2^{r_i - 1}$. An element $(v_1, \ldots, v_{nrp_i})$ of the ON–set of $s_{nrp_i - 2^{r_i - 1}}^{nrp_i}$ represents a function α with $\alpha(K_j^{(i)}) = \{v_j\}$. $(v_1, \ldots, v_{nrp_i})$ can only be a member of the ON–set of $s_{nrp_i - 2^{r_i - 1}}^{nrp_i}$, if the number of v_j which are equal to 1 is at least $nrp_i - 2^{r_i - 1}$. So $s_{nrp_i - 2^{r_i - 1}}^{nrp_i}$ represents the set of all functions α, such that α is 1 at least $nrp_i - 2^{r_i - 1}$ equivalence classes $K_j^{(i)}$ and therefore it represents the set of all functions α which fulfill Equation (4.7). To express $s_{nrp_i - 2^{r_i - 1}}^{nrp_i}$ not for equivalence classes of $\{0, 1\}^p / \equiv_i$, but rather for global classes, we merely need a simple manipulation: For each class $K_j^{(i)}$ there is a set $\{G_{j_1}, \ldots, G_{j_{l_j}}\}$ such that $K_j^{(i)} = \cup_{i=1}^{l_j} G_{j_i}$. We simply have to replace in $s_{nrp_i - 2^{r_i - 1}}^{nrp_i}$ all variables v_j by the corresponding conjunctions $z_{j_1} \cdot \ldots \cdot z_{j_{l_j}}$ and obtain a representation of the same set of functions, the only difference being that the characteristic function is now expressed based on the global classes. The result of the replacement of variables $v_1, \ldots, v_{nrp_i}$ by variables $z_1, \ldots, z_g$ is denoted by ψ_i^1. The characteristic function ψ_i^1 represents all constructable functions which fulfill Equation (4.7). In the same way we also construct a characteristic function ψ_i^0 which represents all constructable functions fulfilling equation (4.8). The set of all functions fulfilling *both* equations is represented by $\chi_i(z_1, \ldots, z_g) = \psi_i^0(z_1, \ldots, z_g) \cdot \psi_i^1(z_1, \ldots, z_g)$, such that χ_i represents all functions which are constructable and assignable for function f_i.

Still remaining is the task of selecting *common* decomposition functions from the representations $\chi_1, \ldots, \chi_m$. To find a decomposition function which is contained in as many candidate sets as possible, we have to look for a vector which is in the ON-set of as many functions χ_i as possible. This problem is solved implicitly using Kam's *Lmax* algorithm (Kam et al., 1994).

When a decomposition function is selected for a function f_i, then the characteristic function χ_i representing the set of assignable (and constructable) functions has to be recomputed. For the general case, Equations (4.7) and (4.8) have to be modified (using Lemma 4.1) to consider the preselected decomposition functions as well.

The algorithm stops when all decomposition functions are computed for all f_i.

4.5.3 Decomposition using support minimization

Sawada et al. (Sawada et al., 1995) used an approach for the computation of shared logic during decomposition, which completely differs from the computation of common decomposition functions. They just use decomposition methods for single–output functions, and logic sharing comes into play by Boolean resubstitution.

Boolean resubstitution (Brayton et al., 1987) is a technique for checking whether an existing function is useful for realizing other functions. In (Sawada et al., 1995) Boolean resubstitution is used to check whether decomposition functions of the decomposition of some function f_i are useful to simplify other functions f_j $(1 \leq j \leq m, i \neq j)$.

To check whether decomposition functions $\alpha_1^{(i)}, \ldots, \alpha_{r_i}^{(i)}$ of f_i are useful for simplifying another output function f_j, Sawada et al. use support minimization. For $\alpha_1^{(i)}, \ldots, \alpha_{r_i}^{(i)}$ they introduce r_i new variables $a_1, \ldots, a_{r_i}$. Suppose that $\{x_1, \ldots, x_n\}$ contains the support sets of $\alpha_1^{(i)}, \ldots, \alpha_{r_i}^{(i)}$ and f_j. Then the function f_j is viewed as a function of variables $x_1, \ldots, x_n$ and $a_1, \ldots, a_{r_i}$. Now

$$dc(x_1, \ldots, x_n, a_1, \ldots, a_{r_i}) = \bigvee_{k=1}^{r_i} a_k \oplus \alpha_k^{(i)}(x_1, \ldots, x_n)$$

is a characteristic function of the 'satisfiability don't care set' $ON(dc)$ for f_j. Since variables a_k correspond to functions $\alpha_k^{(i)}$, the vectors in $ON(dc)$ cannot applied to the inputs of f_j and thus they are don't cares for f_j. The don't care set $ON(dc)$ is used to form an incompletely specified function f_j' from f_j: f_j' has the don't care set $ON(dc)$ and the function values of f_j' are equal to f_j for all input vectors not in $ON(dc)$.

Now $f_j'(x_1, \ldots, x_n, a_1, \ldots, a_{r_i})$ is minimized with respect to some cost function, to check whether it makes sense to use $\alpha_1^{(i)}, \ldots, \alpha_{r_i}^{(i)}$ to realize f_j. If the completely specified result f_j'' of the minimization of f_j' depends essentially on some variable a_k, then it turns out that it makes sense to use the corresponding function $\alpha_k^{(i)}$ to realize f_j.

Since Sawada et al. synthesize look–up table FPGAs, they use the size of the support set of f_j'' as the cost function for the minimization, i.e., they consider a Boolean function to be simpler, if its support set is smaller. Thus they have to solve the problem of computing a completely specified extension f_j'' of f_j' which depends on a minimal number of variables.[13] In (Sawada et al., 1995) an exact solution to this problem is computed by a branch–and–bound algorithm.

Now that we have described the Boolean resubstitution method of (Sawada et al., 1995), we will see how this resubstitution is applied during decomposition. When a realization for functions $f_1, \ldots, f_m \in B_n$ has to be computed, all f_i are first decomposed separately as single–output functions. However, these decompositions are only intended to check which one of the m decompositions should really be applied. Only one function f_i is really decomposed and its decomposition functions are then resubstituted into $f_1, \ldots, f_{i-1}, f_{i+1}, \ldots, f_m$. Still remaining is the task of making a good choice for f_i.

Suppose that each f_i is decomposed with respect to a bound set $X_i^{(1)}$ and the decomposition functions for f_i are $\alpha_1^{(i)}, \ldots, \alpha_{r_i}^{(i)}$, the composition function for f_i is g_i. Now for each $1 \leq i \leq m$, $\alpha_1^{(i)}, \ldots, \alpha_{r_i}^{(i)}$ are resubstituted into $f_1, \ldots, f_m$. The results of the resubstitution of $\alpha_1^{(i)}, \ldots, \alpha_{r_i}^{(i)}$ into $f_1, \ldots f_m$ are functions $f_1^{(i)}, \ldots, f_m^{(i)}$.[14] After that, the cost of $f_1^{(i)}, \ldots, f_m^{(i)}$ is evaluated for each i, i.e., the sizes of the support sets are added by $\sum_{k=1}^{m} ess\_var(f_k^{(i)})$. The decomposition with the smallest sum $\sum_{k=1}^{m} ess\_var(f_k^{(b)})$ is identified as the best decomposition and only this decomposition is performed for function f_b together with the resubstitution of $\alpha_1^{(b)}, \ldots, \alpha_{r_b}^{(b)}$ into f_k ($1 \leq k \leq m, k \neq b$). Finally, the procedure is applied recursively to the composition function g_b of f_b and the functions $f_k^{(b)}$ ($1 \leq k \leq m, k \neq b$) after resubstitution.

---

[13] By definition f_j is one of the possible extensions of f_j', such that the result will never be worse than f_j.

[14] For index i, i.e., function f_i, the resubstitution procedure computes the composition function $f_i^{(i)} = g_i$.

Chapter 5

FUNCTIONAL DECOMPOSITION FOR INCOMPLETELY SPECIFIED FUNCTIONS

In many logic synthesis problems we have to deal with incompletely specified functions. Incompletely specified functions may represent subcircuits which are embedded into a larger system, such that we know that certain input vectors cannot be applied to the subcircuits. These input vectors are viewed as don't cares for the subcircuit, since the function represented by the overall circuit is independent of the function values assigned to these don't care inputs. Moreover, if changing the output of a subcircuit will not change the output of the overall system for certain input vectors, then these input vectors can be used as don't cares, too.

Incompletely specified functions also occur in logic synthesis by decomposition. Even if the original function is completely specified, it is possible that incompletely specified functions arise during the recursive decomposition: If for a one–sided decomposition of a function f with respect to some bound set $X^{(1)}$ the number of different row patterns $nrp(X^{(1)}, f)$ is not a power of 2, then it is not necessary that all possible combinations of values in $\{0, 1\}^{\lceil \log(nrp(X^{(1)},f)) \rceil}$ occur in the image of the decomposition functions (even if we use communication minimal decompositions). Then the composition function can be incompletely specified (and *really is* incompletely specified, if a strict decomposition is used). When the original function f is already incompletely specified, the decomposition functions can be incompletely specified even if the number of different classes of compatible rows $ncc(X^{(1)}, f)$ is a power of 2.

Using the results of Shannon (Shannon, 1949) (see Section 1.5) it is easy to see that for incompletely specified functions with large don't care sets, a random assignment of the don't cares will lead to functions which show a large complexity with high probability. Taking the considerations of Section 3.4 into account (especially Lemma 3.3) it is also easy to see that a random don't care assignment also will lead to functions which do not display good decomposability properties.

For these reasons it makes sense to look for good extensions of incompletely specified functions. Here we look for extensions which lead to good decompositions.

5.1. Decomposition and BDD minimization for incompletely specified functions

In Section 2.2 we presented methods which make use of don't cares to minimize ROBDDs for incompletely specified functions. Since we can expect that small ROBDDs will show good decomposition properties (cf. Section 3.6.3), it makes sense to first minimize the sizes of the ROBDDs representing incompletely specified functions before a decomposition is applied. As described in Section 2.2, don't cares can be used to find extensions with strong symmetries and afterwards symmetric sifting can be used to optimize the variable order. Since the communication minimization method in Section 2.2 is compatible with the method of generating strong symmetries, a (slightly modified) communication minimization method can be applied afterwards.

When symmetries have been found, then it is possible to modify the greedy approaches to find good bound sets (or variable partitions) for a decomposition (see Section 3.7) in such a way that not single variables are exchanged during the search, but rather blocks of symmetric variables (as in symmetric sifting). Following the computation of a good bound set, a decomposition is performed, similar to the case of completely specified functions. Here we make use of the communication minimization methods which we have already presented for ROBDD minimization.

5.2. One–sided decomposition of incompletely specified functions

We start with one–sided decompositions of incompletely specified functions. We can make use of results previously shown in Section 2.2.1 for the minimization of ROBDDs.

As in the case of completely specified functions, the decomposition of incompletely specified Boolean functions can be illustrated using decomposition matrices. In Definition 2.2 (page 25) we have already defined a decomposition matrix also for incompletely specified functions and, apart from 'equal row patterns' we have also defined 'compatible row patterns' for incompletely specified functions (Definition 2.11 on page 40). Two row patterns are compatible if and only if there is no column in the two rows where there is a 1 in the first row pattern and a 0 in the second row pattern or vice versa. Compatibility of rows in a decomposition matrix $Z(X^{(1)}, f)$ with $X^{(1)} = \{x_1, \ldots, x_p\}$ leads to a relation '$\sim$' on $\{0, 1\}^p$ which is *not* an equivalence relation.

Additionally we defined the notion $ncc(X^{(1)}, f)$ for the minimum number of sets in which $\{0, 1\}^p$ can be partitioned, such that every pair of elements of the same set is compatible with respect to $\sim$ (Notation 2.4 on page 41). The minimum number of decomposition functions of a function f which is decomposed with respect to $X^{(1)}$ can be computed for incompletely specified functions based on $ncc(X^{(1)}, f)$ (Karp, 1963):

THEOREM 5.1 *Let $f \in BP_n(D)$ be a function with input variables x_1, $\ldots, x_n$. Then there is a one–sided decomposition of f with respect to variable set $X^{(1)} = \{x_1, \ldots, x_p\}$, such that for all $(x_1, \ldots, x_n) \in D$*

$$f(x_1, \ldots, x_p, x_{p+1}, \ldots, x_n) =$$
$$g(\alpha_1(x_1, \ldots, x_p), \ldots, \alpha_r(x_1, \ldots, x_p), x_{p+1}, \ldots, x_n)$$

if and only if

$$r \geq \log(ncc(X^{(1)}, f)).$$

PROOF: The proof is exactly analogous to the proof of Theorem 3.1. We merely have to replace 'equality of row patterns' by 'compatibility of row patterns'. □

However, computing $ncc(X^{(1)}, f)$ based on $Z(X^{(1)}, f)$ means solving the problem **CM** of Section 2.2.1, which is NP–hard (Theorem 2.2 on page 42). This does not directly imply that computing $\lceil \log(ncc(X^{(1)}, f)) \rceil$ is likewise NP–hard, but having a closer look at the proof of Theorem 2.2 it is easy to see that the computation of $\lceil \log(ncc(X^{(1)}, f)) \rceil$, i.e., the computation of the minimum number of decomposition functions in a one–sided decomposition of f with respect to $X^{(1)}$ is NP–hard indeed (assuming that the input of the problem is $Z(X^{(1)}, f)$).

As already shown in Section 2.2.1.2, $ncc(X^{(1)}, f)$ can be approximated based on heuristics for the 'Partition into Cliques' problem. We have also shown that we do not need to compute decomposition matrices when the incompletely specified function f is represented by ROBDDs. The graph, for which the 'Partition into Cliques' problem must be solved, can be computed based on linking nodes of an ROBDD representing the function $ext(f)$ (see Section 2.2.1.3). This method reduces the search space for the original 'Partition into Cliques' problem.

After an approximate algorithm for Partition into Cliques has computed cc different classes $PK_1, \ldots, PK_{cc}$ of $\{0, 1\}^p$ with pairwise compatible elements, we can determine an extension f' of the original function f, such that

$\{PK_1, \ldots, PK_{cc}\}$ is the partition into equivalence classes which results from equal row patterns in $Z(X^{(1)}, f')$. f' is not necessarily completely specified and it can also be computed based on ROBDDs (again see Section 2.2.1.3). Now we can apply our methods for the computation of decomposition and composition functions which were presented for completely specified functions. Section 5.4 will show how we can compute decomposition and composition functions where the don't care sets are maximal.

5.3.   Two–sided decomposition of incompletely specified functions

Two-sided decompositions of a completely specified function f with respect to a variable partition $\{X^{(1)}, X^{(2)}\}$ can be performed as a series of two one–sided decompositions: We first decompose f with respect to the bound set $X^{(1)}$ and then we decompose the resulting composition function g with respect to bound set $X^{(2)}$. According to Theorem 3.2 (page 78) the total number of decomposition functions in the two–sided decomposition can be computed by $\lceil \log(nrp(X^{(1)}, f)) \rceil + \lceil \log(nrp(X^{(2)}, f)) \rceil$.

However, an analogous result is not true for incompletely specified functions: The minimum number of decomposition functions in a two–sided decomposition of a function f with respect to variable partition $\{X^{(1)}, X^{(2)}\}$ *may be larger than* $\lceil \log(ncc(X^{(1)}, f)) \rceil + \lceil \log(ncc(X^{(2)}, f)) \rceil$. The following small example shows the reason:

EXAMPLE 5.1

$$\begin{array}{|cc|} \hline * & 1 \\ 0 & * \\ \hline \end{array}$$

Both rows and columns of the matrix are compatible, such that $\lceil \log(ncc(X^{(1)}, f)) \rceil + \lceil \log(ncc(X^{(2)}, f)) \rceil$ is equal to 0. However, there is no replacement of the don't cares by elements of $\{0, 1\}$, such that *both* rows *and* columns are equal, and consequently, at least one decomposition function is needed in a two–sided decomposition of the corresponding function.

The example shows that the expression $\lceil \log(ncc(X^{(1)}, f)) \rceil + \lceil \log(ncc(X^{(2)}, f)) \rceil$ can underestimate the minimum number of decomposition functions in a communication minimal two–sided decomposition .

It is not very surprising that also the following problem for two–sided decompositions is NP–hard (Scholl and Molitor, 1993):

Problem 2S-CM (Two–sided Communication Minimization)

Given: Incompletely specified function $f \in BP_n(D)$ with input variables $x_1, \ldots, x_n$ and decomposition matrix $Z(X^{(1)}, f)$ with respect to variable partition $\{X^{(1)}, X^{(2)}\} = \{\{x_1, \ldots, x_p\}, \{x_{p+1}, \ldots, x_n\}\}$.

Find: Minimum number of decomposition functions in a two–sided decomposition of f with respect to $\{X^{(1)}, X^{(2)}\}$.

Moreover, an *exact* solution to problem **2S-CM** cannot simply be reduced to a series of two solutions to the problem **CM**.[1]

All the same we can reduce a two–sided decomposition of an incompletely specified function f with respect to a variable partition $\{X^{(1)}, X^{(2)}\}$ to a series of two one–sided decompositions by decomposing f with respect to the bound set $X^{(1)}$ and then decomposing the resulting composition function g with respect to bound set $X^{(2)}$. The only difference as compared to a completely specified function lies in the fact that the result (viewed as a two–sided decomposition) may contain a non–minimal number of decomposition functions, even if the number of decomposition functions in the two one–sided decompositions is minimal. This is due to flexibility concerning the choice of classes in the 'Partition into Cliques' solution: it is possible that in the decomposition of f with respect to $X^{(1)}$ there are different solutions with the same number of cliques, which lead to different numbers of decomposition functions in the subsequent decomposition of g with respect to $X^{(2)}$.[1]

5.4. Computation of *incompletely specified* decomposition and composition functions

After computing the minimum number of decomposition functions in a one–sided decomposition with respect to a bound set $\{x_1, \ldots, x_p\}$, the task remains to compute the decomposition and composition functions. Here we try to compute *incompletely specified* decomposition and composition functions, where the sizes of the don't care sets are maximized, since don't cares grant a degree of freedom which can be used to obtain good implementations.

The computation of decomposition and composition functions for the decomposition of an incompletely specified Boolean function $f \in BP_n(D_f)$ with respect to bound set $\{x_1, \ldots, x_p\}$ is based on a partition $PK = \{PK_1, \ldots, PK_v\}$ of $\{0, 1\}^p$. All sets PK_i $(1 \leq i \leq v)$ have the property

$$\epsilon \sim \delta \text{ for all } \epsilon, \delta \in PK_i, \qquad (\star)$$

---

[1]Examples and more details can be found in (Scholl and Molitor, 1993).

where

$$\epsilon \sim \delta \iff \nexists (y_1, \ldots, y_{n-p}) \in \{0,1\}^{n-p} \text{ with}$$
$$(\epsilon_1, \ldots, \epsilon_p, y_1, \ldots, y_{n-p}), (\delta_1, \ldots, \delta_p, y_1, \ldots, y_{n-p}) \in D_f,$$
$$f(\epsilon_1, \ldots, \epsilon_p, y_1, \ldots, y_{n-p}) \neq f(\delta_1, \ldots, \delta_p, y_1, \ldots, y_{n-p}).$$

The partition $PK = \{PK_1, \ldots, PK_v\}$ (resulting from a heuristic solution to 'Partition into Cliques') does not need to have minimum size $v = ncc(\{x_1, \ldots, x_p\}, f)$. However, in the following we assume that it shows at least the following 'local optimality' property: For all $i \neq j \in \{1, \ldots, v\}$: $\exists \epsilon \in PK_i, \delta \in PK_j$ with $\epsilon \not\sim \delta$, i.e., it is *not* possible to replace PK_i and PK_j by $PK_i \cup PK_j$ without violating property $(\star)$.

The ROBDDs bdd_i of functions $ext(com\_ext(\{f_{x_1^{\epsilon_1} \ldots x_p^{\epsilon_p}} | (\epsilon_1, \ldots, \epsilon_p) \in PK_i\}))$ $(1 \leq i \leq v)$ are computed based on the ROBDD for $ext(f)$ as described in Section 2.2.1.3.

As in the method for completely specified functions we now assign a unique code $(a_1^{(i)}, \ldots, a_r^{(i)}) \in \{0,1\}^r$ ($r = \lceil \log(v) \rceil$) to each set PK_i (i.e., to each bdd_i).

Basically, the composition function g is also obtained as in the case of completely specified functions (see Section 3.6.2). We start from ROBDDs $bdd_1, \ldots, bdd_v$ and construct a 'code tree', which forms the upper part of the resulting ROBDD. If $v < 2^r$, then not all codes from $\{0,1\}^r$ are used. If the code $(\epsilon_1, \ldots, \epsilon_r)$ is not used, then for all $(y_1, \ldots, y_{n-p})$ of $\{0,1\}^{n-p}$ $(\epsilon_1, \ldots, \epsilon_r, y_1, \ldots, y_{n-p})$ is in the don't care set of g. Thus in the ROBDD of $ext(g) = \overline{z} \cdot g_{off} + z \cdot g_{on}$ the node reached by $(\epsilon_1, \ldots, \epsilon_r)$ has to be the $\boxed{0}$-node (which represents a don't care in this representation).

The computation of ROBDDs for the decomposition functions $\alpha_1, \ldots, \alpha_r$ is also similar to the case of completely specified functions. The classes PK_i $(1 \leq i \leq v)$ are obtained as unions of classes $K_1, \ldots, K_{nrp(\{x_1, \ldots, x_p\}, f)}$, where classes K_j are the equivalence classes which result from equal row patterns in the decomposition matrix of $ext(f)$ with respect to $\{x_1, \ldots, x_p\}$. When codes are assigned to classes PK_i, then this also results in an assignment of codes to classes K_j. Note however that the code for different (compatible) classes K_j may be equal. There is a one–to–one correspondence between these classes K_j and 'linking nodes' below a cut after variables $x_1, \ldots, x_p$ in the ROBDD representing $ext(f)$. In the first step, we obtain the ROBDD of a decomposition function α_i' as in the case of completely specified functions: In the ROBDD for $ext(f)$ the linking nodes are replaced by the ith bits of the corresponding codes and then the ROBDD is reduced.

The only difference for incompletely specified functions lies in the fact that it may be possible for $(\alpha_1', \ldots, \alpha_r')$ to be replaced by an *incompletely specified* function $\alpha = (\alpha_1, \ldots, \alpha_r)$ with $(\alpha'_1, \ldots, \alpha'_r)$ as a completely specified extension. In this case some elements of $\{0,1\}^p$ can be included in the don't care set of α without violating the following decomposition condition for g and α:

$$f(\epsilon_1, \ldots, \epsilon_n) = g''(\alpha''(\epsilon_1, \ldots, \epsilon_p), \epsilon_{p+1}, \ldots, \epsilon_n) \qquad (\star)$$

for all $(\epsilon_1, \ldots, \epsilon_n) \in D_f$ and all completely specified extensions α'' and g'' of α and g.

In the following we analyze how the don't cares for α can be chosen. We have to differentiate between two cases:

Case 1: $v < 2^r$

Consider an $\epsilon \in \{0,1\}^p$ with $D(f_{x_1^{\epsilon_1} \ldots x_p^{\epsilon_p}}) = \emptyset$. (That is, the row in $Z(\{x_1, \ldots, x_p\}, f)$ with index ϵ only contains $\star$). We claim that ϵ with this property can be included in the don't care set of α. This is due to the reason that for all $(y_1, \ldots, y_{n-p}) \in \{0,1\}^{n-p}$, $(\epsilon_1, \ldots, \epsilon_p, y_1, \ldots, y_{n-p})$ is not in the domain D_f of f anyway and that condition $(\star)$ has to be fulfilled only for elements of D_f.

All $\epsilon \in \{0,1\}^p$ with $D(f_{x_1^{\epsilon_1} \ldots x_p^{\epsilon_p}}) = \emptyset$ (or with $ext(f_{x_1^{\epsilon_1} \ldots x_p^{\epsilon_p}}) = 0$) are contained in a single class K_i. If there are elements $\epsilon \in \{0,1\}^p$ with the property given above, then it is clear that the corresponding class K_i can be used as a don't care set of α.

Case 2: $v = 2^r$

In this case the choice of $\epsilon \in \{0,1\}^p$ which can be included in the don't care set of α is even less restricted:

We consider for $1 \le i \le v$ the functions

$$cof_{PK_i} := com\_ext(\{f_{x_1^{\delta_1} \ldots x_p^{\delta_p}} \mid (\delta_1, \ldots, \delta_p) \in PK_i\}).$$

We claim that an $\epsilon \in \{0,1\}^p$ can be included in the don't care set of α, if for all $1 \le i \le v$, cof_{PK_i} is an extension of $f_{x_1^{\epsilon_1} \ldots x_p^{\epsilon_p}}$. (Note that this is a weaker requirement for ϵ as in Case 1, since for an ϵ with $D(f_{x_1^{\epsilon_1} \ldots x_p^{\epsilon_p}}) = \emptyset$ it is clear in advance that all $cof_{PK_i} (1 \le i \le v)$ are extensions of $f_{x_1^{\epsilon_1} \ldots x_p^{\epsilon_p}}$.) The row in $Z(\{x_1, \ldots, x_p\}, f)$ with index ϵ has the following property: If, in this row, there is a column c containing a 1 (0), then the matrix which results from $Z(\{x_1, \ldots, x_p\}, f)$ by the computation of 'common extensions' (see Example 2.3 on page 47) has only 1 (0) in this column c.

To prove that it is possible to include an ϵ with the properties given above in the don't care set of α, it is important to note that for all $(y_1, \ldots, y_{n-p})$ with $(\epsilon_1, \ldots, \epsilon_p, y_1, \ldots, y_{n-p}) \in D_f$ and all $(a_1, \ldots, a_r) \in \{0,1\}^r$ the following holds:

1. $(a_1, \ldots, a_r, y_1, \ldots, y_{n-p}) \in D(g)$, since

   (a) $v = 2^r$ and thus for $(a_1, \ldots, a_r)$ there is a set PK_i with $\alpha'(PK_i) = (a_1, \ldots, a_r)$ and

   (b) for this set PK_i with $\alpha'(PK_i) = (a_1, \ldots, a_r)$, cof_{PK_i} is an extension of $f_{x_1^{\epsilon_1} \ldots x_p^{\epsilon_p}}$, such that we can conclude $(y_1, \ldots, y_{n-p}) \in D(cof_{PK_i})$ from $(\epsilon_1, \ldots, \epsilon_p, y_1, \ldots, y_{n-p}) \in D_f$ (i.e., $(y_1, \ldots, y_{n-p}) \in D(f_{x_1^{\epsilon_1} \ldots x_p^{\epsilon_p}})$) and thus we can conclude $(a_1, \ldots, a_r, y_1, \ldots, y_{n-p}) \in D(g)$.

2.

$$g(a_1, \ldots, a_r, y_1, \ldots, y_{n-p}) =$$
$$= cof_{PK_i}(y_1, \ldots, y_{n-p}) \quad (\alpha'(PK_i) = (a_1, \ldots, a_r))$$
$$= f_{x_1^{\epsilon_1} \ldots x_p^{\epsilon_p}}(y_1, \ldots, y_{n-p}), \quad \text{since } cof_{PK_i} \text{ is an extension}$$
$$\text{of } f_{x_1^{\epsilon_1} \ldots x_p^{\epsilon_p}}.$$
$$= f(\epsilon_1, \ldots, \epsilon_p, y_1, \ldots, y_{n-p}).$$

For all $(\epsilon_1, \ldots, \epsilon_p, y_1, \ldots, y_{n-p}) \in D_f$ we have $g(a_1, \ldots, a_r, y_1, \ldots, y_{n-p}) = f(\epsilon_1, \ldots, \epsilon_p, y_1, \ldots, y_{n-p})$ independently of $(a_1, \ldots, a_r)$, such that we are free to choose $\alpha(\epsilon_1, \ldots, \epsilon_p)$ and consequently we are allowed to include $(\epsilon_1, \ldots, \epsilon_p)$ in the don't care set.

The don't care set of α is equal to the set of all $(\epsilon_1, \ldots, \epsilon_p)$ with the property given above. We may compute this don't care set as follows: Let $\{K_1, \ldots, K_{nrp(\{x_1, \ldots, x_p\}, f)}\}$ be the partition of $\{0, 1\}^p$ into equivalence classes resulting from equal row patterns in $Z(\{x_1, \ldots, x_p\}, f)$. The don't care set dc_set of α is computed as the union of all classes K_i, such that *for all* $1 \leq j \leq v$, cof_{PK_j} is an extension of $f_{x_1^{\epsilon_1} \ldots x_p^{\epsilon_p}}$ for an arbitrary element $\epsilon \in K_i$ (and thus for all elements $\epsilon \in K_i$). We recognize these classes by checking all classes K_i $(1 \leq i \leq nrp(\{x_1, \ldots, x_p\}, f))$, whether for all $1 \leq j \leq v$

$$(f_{x_1^{\epsilon_1} \ldots x_p^{\epsilon_p}})_{on} \cdot \overline{(cof_{PK_j})_{on}} = 0$$

and

$$(f_{x_1^{\epsilon_1} \ldots x_p^{\epsilon_p}})_{off} \cdot \overline{(cof_{PK_j})_{off}} = 0$$

for an arbitrary element $\epsilon \in K_i$.

In logic synthesis it is desirable to have functions with large don't care sets. Now the question arises whether the definition of decomposition and composition functions given above leads to a maximum number of don't cares. We can prove that the don't care sets given above are maximal in the following sense: If the don't care set of the decomposition function α or of the composition function g is extended by an arbitrary element, then there are extensions α'' and g'' of α and g, such that

$$g''(\alpha''(\epsilon_1, \ldots, \epsilon_p), \epsilon_{p+1}, \ldots, \epsilon_n) \neq f(\epsilon_1, \ldots, \epsilon_n)$$

for some $(\epsilon_1, \ldots, \epsilon_n) \in D_f$.

This means that we are *not* allowed to add further elements to the don't care sets. The corresponding Theorem J.3 (together with its proof) is given in Appendix J.

REMARK 5.1 *Theorem J.3 claims that the don't care sets of decomposition and composition functions as defined above cannot be increased. However, in Theorem J.3 α is defined as a multi–output function, such that for all α_i ($1 \le i \le r$) we use the same domain $D(\alpha_i) = D(\alpha)$. If we regard α_i as single–output functions with separate don't care sets, then it is possible that the don't care sets can be further increased.*

EXAMPLE 5.2 Figure 5.1 shows a decomposition matrix with respect to $\{x_1, x_2, x_3\}$. A possible solution to the 'Partition into Cliques' problem is

$$\underbrace{\{\{(000), (001)\}}_{PK_1}, \underbrace{\{(010), (011)\}}_{PK_2}, \underbrace{\{(100), (101)\}}_{PK_3}, \underbrace{\{(110), (111)\}\}}_{PK_4}.$$

We encode the classes PK_i ($1 \le i \le 4$) by $\alpha(PK_1) = (00)$, $\alpha(PK_2) = (01)$, $\alpha(PK_3) = (10)$, and $\alpha(PK_4) = (11)$. According to Theorem J.3 the don't care set of $\alpha = (\alpha_1, \alpha_2)$ has to be empty. However, it is easy to see that it is possible to define either for α_1 or for α_2 $D(\alpha_i) = \{0, 1\}^3 \setminus \{(000)\}$ ($i = 1, 2$). Including (000) in the don't care set of α_1, e.g., means that the code for (000) is possibly changed from (00) to (10). (10) has been originally assigned to class P_3. The function table of cof_{PK_3} is $\boxed{10 \star 0}$, which is an extension of $\boxed{\star \star \star 0}$.

Yet it is *not* possible to use (000) as a don't care both for α_1 *and* α_2 *at the same time*. Including (000) both in the don't care set of α_1 *and* α_2 means that the code for (000) can be changed to (11), for instance. (11) has been assigned to class PK_4. However, the function table $\boxed{11 \star 1}$ of cof_{PK_4} is *not* an extension of $\boxed{\star \star \star 0}$ and $g(1111) = 1 \neq f(00011)$.

5.5. Multi–output functions

Finally the question arises, how methods for decomposing multi–output functions and methods for decomposing incompletely specified functions can be combined.

Figure 5.1. Decomposition matrix for Example 5.2.

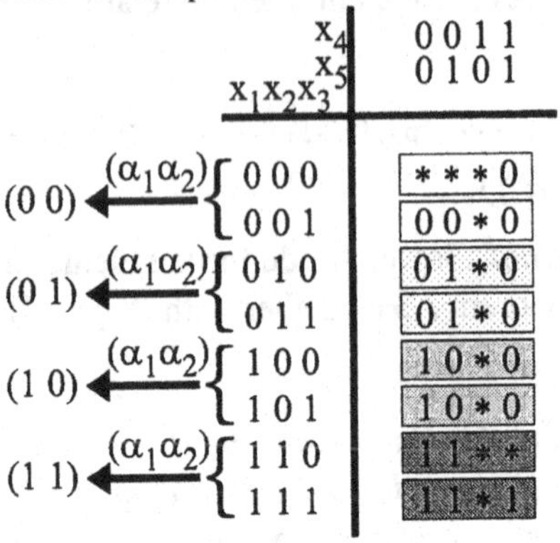

One method for the decomposition of incompletely specified functions $f_1, \ldots, f_m$ with respect to a bound set $X^{(1)} = \{x_1, \ldots, x_p\}$ would consist of a don't care assignment for all functions f_i to minimize the number of decomposition functions in the decomposition of f_i, followed by the method of computing common decomposition functions during the decomposition of the functions f_i.

On the other hand it would be desirable to assign don't cares already *with respect to a multi-output decomposition* of $f_1, \ldots, f_m$. This don't care assignment should take into account that $f_1, \ldots, f_m$ are decomposed by computing common decomposition functions for several output functions.

To achieve this goal we propose minimizing a lower bound on the *total number* of decomposition functions for $f_1, \ldots, f_m$ (Scholl, 1996; Scholl, 1997; Scholl, 1998). The don't care assignment for multi-output functions proceeds in three steps:

1. First we assign don't cares to obtain as many symmetries as possible (cf. Section 5.1),

2. then we perform a don't care assignment to minimize the lower bound on the total number of decomposition functions to prepare the computation of common decomposition functions (this method will be described in the following),

3. and finally we assign don't cares in the single–output functions to minimize the number of decomposition functions for the single–output functions f_i.

5.5.1 Don't care assignment with respect to logic sharing

Motivation. The don't care assignment with respect to logic sharing mentioned above works as follows: To minimize the total number of decomposition functions for $f_1, \ldots, f_m$ in order to obtain as much logic sharing as possible, we minimize a lower bound on this total number. The lower bound is obtained by regarding the decomposition of $f = (f_1, \ldots, f_m)$ *as a whole*. First let us consider a completely specified function $f \in B_{n,m}$. If $f \in B_{n,m}$ is decomposed as a whole, i.e., in the form of

$$f(\mathbf{x}^{(1)}, \mathbf{x}^{(2)}) = g(\alpha_1(\mathbf{x}^{(1)}), \ldots, \alpha_r(\mathbf{x}^{(1)}), \mathbf{x}^{(2)}) \qquad (\star)$$

with functions $\alpha \in B_{p,r}$ and $g \in B_{n-p+r,m}$, then the minimum number r of decomposition functions can be determined as in the case of single–output functions using the decomposition matrix $Z(X^{(1)}, f)$ — with the only difference being that the entries of the decomposition matrix are now elements of $\{0,1\}^m$, not only 0 or 1. As in the proof of Theorem 3.1 (page 75) it can be shown that the minimum number r of decomposition functions in a decomposition of type $(\star)$ is equal to $\lceil \log(nrp(X^{(1)}, f)) \rceil$, where $nrp(X^{(1)}, f)$ is the number of different row patterns in $Z(X^{(1)}, f)$.

Since $\lceil \log(nrp(X^{(1)}, f)) \rceil$ is a lower bound on the number of decomposition functions in decompositions of type $(\star)$, also for decompositions of type

$$f_1(\mathbf{x}^{(1)}, \mathbf{x}^{(2)}) = g^{(1)}(\alpha_1^{(1)}(\mathbf{x}^{(1)}), \ldots, \alpha_{r_1}^{(1)}(\mathbf{x}^{(1)}), \mathbf{x}^{(2)})$$

$$\vdots$$

$$f_m(\mathbf{x}^{(1)}, \mathbf{x}^{(2)}) = g^{(m)}(\alpha_1^{(m)}(\mathbf{x}^{(1)}), \ldots, \alpha_{r_m}^{(m)}(\mathbf{x}^{(1)}), \mathbf{x}^{(2)}),$$

with $r_i = \lceil \log(nrp(X^{(1)}, f_i)) \rceil$ we have:

$$\left| \bigcup_{i=1}^{m} \{\alpha_1^{(i)}, \ldots, \alpha_{r_i}^{(i)}\} \right| \geq \lceil \log(nrp(X^{(1)}, f)) \rceil.$$

This means that $\lceil \log(nrp(X^{(1)}, f)) \rceil$ is not only a lower bound on $\sum_{i=1}^{m} \lceil \log(nrp(X^{(1)}, f_i)) \rceil$, but it also provides an estimation of the extent to which we can expect to find common decomposition functions in the decomposition of $f = (f_1, \ldots, f_m)$. If $\lceil \log(nrp(X^{(1)}, f)) \rceil$ is small and $\sum_{i=1}^{m} r_i$ is large, then we can hope that there is a large potential for sharing decomposition functions in the decomposition of the single–output functions f_i.[2]

---

[2]Of course the value $nrp(X^{(1)}, f)$ will only be significant for us if it is significantly smaller than 2^p ($p = |X^{(1)}|$).

Minimizing the lower bound with respect to logic sharing. If we have incompletely specified functions $f_1, \ldots, f_m$, then we can assign don't cares to minimize this lower bound: We are looking for extensions $f'_1, \ldots, f'_m$ of $f_1, \ldots, f_m$, such that $nrp(X^{(1)}, f')$ is minimized. Similarly to single–output functions we also partition row indices of the decomposition matrix $Z(X^{(1)}, f)$ into classes of 'compatible' rows.

To this end we extend the definition of cofactors to vectors of several incompletely specified functions (which can have different don't care sets):

DEFINITION 5.1 *Let* $f_i \in BP_n(D_i)$ *for* $1 \leq i \leq m$ $(m > 1)$. *The* cofactor *of* $(f_1, \ldots, f_m)$ *with respect to* $x_1^{\epsilon_1} \ldots x_p^{\epsilon_p}$ *is a function* $f_{x_1^{\epsilon_1} \ldots x_p^{\epsilon_p}} : \{0,1\}^{n-p}$ $\rightarrow \{0, 1, \star\}^m$, *where for all* $(y_1, \ldots, y_{n-p}) \in \{0,1\}^{n-p}$

$$f_{x_1^{\epsilon_1} \ldots x_p^{\epsilon_p}}(y_1, \ldots, y_{n-p}) = (\delta_1, \ldots, \delta_m)$$

with

$$\delta_i = \begin{cases} \star, & \text{if } (\epsilon_1, \ldots, \epsilon_p, y_1, \ldots, y_{n-p}) \notin D_i \\ f_i(\epsilon_1, \ldots, \epsilon_p, y_1, \ldots, y_{n-p}), & \text{if } (\epsilon_1, \ldots, \epsilon_p, y_1, \ldots, y_{n-p}) \in D_i \end{cases}$$

REMARK 5.2 *If we define for cofactor* $f_{x_1^{\epsilon_1} \ldots x_p^{\epsilon_p}}$ *of* $(f_1, \ldots, f_m)$

$$D'_i = \{(y_1, \ldots, y_{n-p}) \mid f_{x_1^{\epsilon_1} \ldots x_p^{\epsilon_p}}(y_1, \ldots, y_{n-p}) = (\delta_1, \ldots, \delta_m) \text{ with } \delta_i \neq \star\}$$

and $h : D'_i \rightarrow \{0, 1\}$, *such that for* $(y_1, \ldots, y_{n-p}) \in D'_i$

$$h(y_1, \ldots, y_{n-p}) = \delta_i \text{ with } f_{x_1^{\epsilon_1} \ldots x_p^{\epsilon_p}}(y_1, \ldots, y_{n-p}) = (\delta_1, \ldots, \delta_m),$$

then h *is the cofactor of* f_i *with respect to* $x_1^{\epsilon_1} \ldots x_p^{\epsilon_p}$ *as defined in Definition 1.5.*

Two elements $\epsilon^{(1)}$ and $\epsilon^{(2)}$ of $\{0,1\}^p$ are called *compatible* if and only if there is no $(y_1, \ldots, y_{n-p}) \in \{0,1\}^{n-p}$ with $f_{x_1^{\epsilon_1^{(1)}} \ldots x_p^{\epsilon_p^{(1)}}}(y_1, \ldots, y_{n-p}) = (\delta_1^{(1)}, \ldots, \delta_m^{(1)})$, $f_{x_1^{\epsilon_1^{(2)}} \ldots x_p^{\epsilon_p^{(2)}}}(y_1, \ldots, y_{n-p}) = (\delta_1^{(2)}, \ldots, \delta_m^{(2)})$ and $\delta_i^{(1)} = 0, \delta_i^{(2)} = 1$ (or $\delta_i^{(1)} = 1, \delta_i^{(2)} = 0$) for some $1 \leq i \leq m$.

REMARK 5.3 *This notion of compatibility is a straightforward generalization of compatibility defined for incompletely specified single–output functions in Definition 2.11 on page 40.* $\epsilon^{(1)}$ *is compatible to* $\epsilon^{(2)}$ *with respect to compatibility defined for* $(f_1, \ldots, f_m)$ *if and only if for all* $1 \leq i \leq m$, $\epsilon^{(1)}$ *is compatible to* $\epsilon^{(2)}$ *with respect to compatibility defined for* f_i.

EXAMPLE 5.3 Consider the following decomposition matrix of (f_1, f_2, f_3, f_4):

$x_1 \ldots x_p$	$\begin{matrix} x_{p+1} \\ x_{p+2} \end{matrix}$ $\begin{matrix} 0 \\ 0 \end{matrix}$	$\begin{matrix} 0 \\ 1 \end{matrix}$	$\begin{matrix} 1 \\ 0 \end{matrix}$	$\begin{matrix} 1 \\ 1 \end{matrix}$
$\vdots$	$\vdots$	$\vdots$	$\vdots$	$\vdots$
$\epsilon_1^{(1)} \ldots \epsilon_p^{(1)}$	(1100)	$(\star 011)$	$(0 \star 10)$	$(\star\star\star\star)$
$\vdots$	$\vdots$	$\vdots$	$\vdots$	$\vdots$
$\epsilon_1^{(2)} \ldots \epsilon_p^{(2)}$	$(\star 100)$	(0011)	(0010)	$(\star\star\star\star)$
$\vdots$	$\vdots$	$\vdots$	$\vdots$	$\vdots$
$\epsilon_1^{(3)} \ldots \epsilon_p^{(3)}$	$(0 \star 00)$	$(0 \star 11)$	$(\star 0 \star 0)$	$(\star\star\star\star)$
$\vdots$	$\vdots$	$\vdots$	$\vdots$	$\vdots$

A row with index ϵ represents the function table of cofactor $f_{x_1^{\epsilon_1} \ldots x_p^{\epsilon_p}}$. The pairs of row indices $(\epsilon^{(1)}, \epsilon^{(2)})$ and $(\epsilon^{(2)}, \epsilon^{(3)})$ are compatible; the pair $(\epsilon^{(1)}, \epsilon^{(3)})$ is not compatible.

If two row indices are compatible, then it is possible to assign values to the don't cares, such that the corresponding row patterns will be equal. For example, the compatible rows

$$\boxed{(1100) \quad (\star 011) \quad (0 \star 10) \quad (\star\star\star\star)}$$

and

$$\boxed{(\star 100) \quad (0011) \quad (0010) \quad (\star\star\star\star)}$$

change into

$$\boxed{(1100) \quad (0011) \quad (0010) \quad (\star\star\star\star)},$$

when a minimum number of don't care assignments is used.

As for single–output functions here, too, we need a common extension of compatible cofactors:

NOTATION 5.1 *Let for* $1 \le i \le m$ $(m > 1)$ $f_i \in BP_n(D_i)$. *Let* $cof_1 = f_{x_1^{\epsilon_1^{(1)}} \ldots x_p^{\epsilon_p^{(1)}}}, \ldots, cof_l = f_{x_1^{\epsilon_1^{(l)}} \ldots x_p^{\epsilon_p^{(l)}}}$ *be cofactors of* $f = (f_1, \ldots, f_m)$ *and let* $\epsilon^{(1)}, \ldots, \epsilon^{(l)}$ *be pairwise compatible. Then the function*

$$com\_ext(\{cof_1, \ldots, cof_l\}) : \{0, 1\}^{n-p} \to \{0, 1, \star\}$$

with

$$com\_ext(\{cof_1, \ldots, cof_l\})(y_1, \ldots, y_{n-p}) = (\delta_1, \ldots, \delta_m),$$

and

$$\delta_i = \begin{cases} 0 & \text{if there is } j \in \{1, \ldots, l\} \text{ with} \\ & cof_j(y_1, \ldots, y_{n-p}) = (\epsilon_1, \ldots, \epsilon_m) \text{ and } \epsilon_i = 0, \\ 1 & \text{if there is } j \in \{1, \ldots, l\} \text{ with} \\ & cof_j(y_1, \ldots, y_{n-p}) = (\epsilon_1, \ldots, \epsilon_m) \text{ and } \epsilon_i = 1, \\ \star & \text{otherwise} \end{cases}$$

is called common extension *of* $cof_1, \ldots, cof_l$.

To minimize the number of different row patterns in the decomposition matrix $Z(X^{(1)}, f)$, $\{0, 1\}^p$ is partitioned into classes $PK_1, \ldots PK_l$, such that the elements of such a class PK_i ($1 \le i \le l$) are pairwise compatible. To obtain a minimum number l of classes we solve an instance of the problem 'Partition into Cliques' as for incompletely specified single–output functions (the only difference lies in a different definition of compatibility for multi–output functions). If $\{0, 1\}^p/_{\equiv} = \{K_1, \ldots, K_{nrp(\{x_1, \ldots, x_p\}, f)}\}$ is the partition of $\{0, 1\}^p$ into equivalence classes which results from equal row patterns in $Z(\{x_1, \ldots, x_p\}, f)$, then for multi–output functions, too, we can choose $\{PK_1, \ldots, PK_l\}$, such that each set PK_i is the union of certain sets K_j, i.e., such that all indices of rows in $Z(\{x_1, \ldots, x_p\}, f)$ with the same row pattern are in the same compatibility class. For all $i \in \{1, \ldots, l\}$ the cofactors with respect to elements in PK_i are replaced by their common extension. This leads to functions $f'_1, \ldots, f'_m$, such that the decomposition matrix $Z(X^{(1)}, f')$ of f' has exactly l different row patterns.

Solution for ROBDD representations. Finally we give a brief description of the method applied to incompletely specified functions which are given by ROBDDs for $ext(f_1), \ldots, ext(f_m)$. We assume that the variables of $X^{(1)} = \{x_1, \ldots, x_p\}$ are located before the variables of $\{x_{p+1}, \ldots, x_n\}$ in the variable order and the additional z–variable of the $ext(.)$–representation is also after the variables of $X^{(1)}$ in the order. Then the partitions into equivalence classes with respect to $\equiv_i$ for functions f_i can easily be computed using a cut of the ROBDD of $ext(f_i)$ after the variables of $X^{(1)}$ (see Section 2.2.1.3). $\equiv_i$ is the equivalence relation which results from equal row patterns in the decomposition matrix $Z(X^{(1)}, f_i)$ of f_i. For each f_i we obtain the partition $\{0, 1\}^p/_{\equiv_i} = \{K_1^{(i)}, \ldots, K_{nrp(\{x_1, \ldots, x_p\}, f_i)}^{(i)}\}$. However, for the method to minimize the lower bound on the number of decomposition functions for $f_1, \ldots, f_m$, we need the partition of $\{0, 1\}^p$ into equivalence classes which results from equal row patterns in the decomposition matrix of $f = (f_1, \ldots, f_m)$. Since two row patterns in the decomposition matrix of f are equal if and only if the corresponding row patterns of *all* decomposition matrices for f_i are equal ($1 \le i \le m$), $\{0,1\}^p/_\equiv$ can simply be computed as follows:

$$\{0,1\}^p/_\equiv = \{\bigcap_{i=1}^m K_{j_i}^{(i)} \mid 1 \le j_i \le nrp(\{x_1,\ldots,x_p\},f_i) \text{ and } \bigcap_{i=1}^m K_{j_i}^{(i)} \ne \emptyset\}.$$

Fortunately it is not necessary to compute all intersections $\bigcap_{i=1}^m K_{j_i}^{(i)}$ to obtain $\{0,1\}^p/_\equiv$. It is possible to compute $\{0,1\}^p/_\equiv = \{K_1,\ldots,K_{nrp(\{x_1,\ldots,x_p\},f)}\}$ directly using ROBDD techniques:[3]

The ROBDDs for $ext(f_i)$ are cut after the variables of $X^{(1)}$. For $ext(f_i)$ we obtain $nrp(\{x_1,\ldots,x_p\},f_i)$ different linking nodes $n_1^{(i)},\ldots,n_{nrp(\{x_1,\ldots,x_p\},f_i)}^{(i)}$. The node $n_j^{(i)}$ is reached exactly by the elements of $K_j^{(i)}$. Now we replace the linking nodes $n_j^{(i)}$ by ROBDDs for new, unique variables $v_j^{(i)}$. In this way we obtain, from the ROBDD of $ext(f_i)$, an ROBDD of

$$kl^{(i)} = \bigvee_{j=1}^{nrp(X^{(1)},f_i)} v_j^{(i)} \cdot \chi_{K_j^{(i)}},$$

where $\chi_{K_j^{(i)}}$ is the characteristic function of $K_j^{(i)}$. The classes K_i result from the computation of an ROBDD for $kl = \bigwedge_{i=1}^m kl^{(i)}$:

$$
\begin{aligned}
kl &= \bigwedge_{i=1}^m kl^{(i)} \\
&= \bigwedge_{i=1}^m \left(\bigvee_{j=1}^{nrp(X^{(1)},f_i)} v_j^{(i)} \cdot \chi_{K_j^{(i)}} \right) \\
&= \bigvee_{\substack{1 \le j_1 \le nrp(X^{(1)},f_1) \\ \vdots \\ 1 \le j_m \le nrp(X^{(1)},f_m)}} (v_{j_1}^{(1)} \cdot \ldots \cdot v_{j_m}^{(m)}) \cdot (\chi_{K_{j_1}^{(1)}} \cdot \ldots \cdot \chi_{K_{j_m}^{(m)}}) \\
&= \bigvee_{\substack{1 \le j_1 \le nrp(X^{(1)},f_1) \\ \vdots \\ 1 \le j_m \le nrp(X^{(1)},f_m)}} (v_{j_1}^{(1)} \cdot \ldots \cdot v_{j_m}^{(m)}) \cdot (\chi_{K_{j_1}^{(1)} \cap \ldots \cap K_{j_m}^{(m)}})
\end{aligned}
$$

---

[3] Here we use the same technique as Wurth et al. in (Wurth et al., 1995) for the computation of their 'global classes'.

$$= \bigvee_{\substack{i=1 \\ K_{i_1}^{(1)} \cap \ldots \cap K_{i_m}^{(m)} = K_i}}^{nrp(X^{(1)},f)} (v_{i_1}^{(1)} \cdot \ldots \cdot v_{i_m}^{(m)}) \cdot \chi_{K_i}$$

When the ROBDD of kl is cut after the variables of $X^{(1)}$, then we get $nrp(X^{(1)}, f)$ linking nodes $n_1, \ldots, n_{nrp(X^{(1)},f)}$ immediately below the cut line. The set of all elements of $\{0,1\}^p$ by which node n_i is reached is equal to K_i. (The ROBDD with root n_i defines a function $v_{i_1}^{(1)} \cdot \ldots \cdot v_{i_m}^{(m)}$ with $K_i = K_{i_1}^{(1)} \cap \ldots \cap K_{i_m}^{(m)}$.)

Two classes $K_j = K_{j_1}^{(1)} \cap \ldots \cap K_{j_m}^{(m)}$ and $K_k = K_{k_1}^{(1)} \cap \ldots \cap K_{k_m}^{(m)}$ are compatible if and only if for all $1 \leq i \leq m$ the classes $K_{j_i}^{(i)}$ and $K_{k_i}^{(i)}$ are compatible (cf. Remark 5.3). Compatibility of classes $K_{j_i}^{(i)}$ and $K_{k_i}^{(i)}$ is checked as given in Section 2.2.1.3.

After solving an instance of 'Partition into Cliques' we obtain a partition $\{PK_1, \ldots, PK_l\}$ of $\{0,1\}^p$. The classes PK_j are unions of equivalence classes K_{j_k}. Suppose $PK_j = \bigcup_{k=1}^{s_j} K_{j_k}$. Then the ROBDD of functions $ext(f_i')$ is computed in a completely analogous manner to the method in Section 2.2.1.3:

The ROBDD for $ext(f_i')$ is computed replacing sub–ROBDDs in the ROBDD representing kl. If class $PK_j = \bigcup_{k=1}^{s_j} K_{j_k}$, then the linking nodes n_{j_k} in the ROBDD for kl (which are reached by vectors of K_{j_k}) are replaced by a common extension of certain cofactors of $ext(f_i)$. The don't care assignment has to replace the cofactors for all elements of PK_j with a common extension.

The common extension needed for $PK_j = \bigcup_{k=1}^{s_j} K_{j_k}$ is computed as follows: For each k we obtain a cofactor of $ext(f_i)$ with respect to some element $\epsilon^{(j_k)}$ of K_{j_k} (and thus with respect to *all* elements of K_{j_k}, since K_{j_k} is a subset of some element of $\{0,1\}^p/_{\equiv_i}$) simply by determining the node reached by $\epsilon^{(j_k)}$ in the ROBDD for $ext(f_i)$. Suppose the node reached by $\epsilon^{(j_k)}$ is $n_{j_k}^{(i)}$ and the ROBDD rooted by $n_{j_k}^{(i)}$ is $cof_{j_k}^{(i)}$.

Now the common extension needed for PK_j is exactly the common extension of cofactors $cof_{j_k}^{(i)}$ $(1 \leq k \leq s_j)$. The computation of common extensions $com\_ext(\{cof_{j_1}^{(i)}, \ldots, cof_{j_{s_j}}^{(i)}\})$ is done as described in Section 2.2.1.3.

In the ROBDD for kl, the linking nodes n_{j_k} (reached by vectors of K_{j_k}) have to be replaced by the common extension of $cof_{j_k}^{(i)}$ $(1 \leq k \leq s_j)$. When this replacement has been done for all $PK_1, \ldots, PK_l$ and the resulting OBDD has been reduced, an ROBDD for $ext(f_i')$ has been constructed.

In terms of decomposition matrices, the construction ensures that for all indices in sets PK_j and all $1 \leq i \leq m$ the row patterns of $Z(X^{(1)}, ext(f'_i))$ are identical, and thus the row patterns of $Z(X^{(1)}, f')$ are identical for all indices in sets PK_j.

5.5.2 Compatibility with communication minimization for single–output functions

The method described computes extensions $f'_1, \ldots, f'_m$ of $f_1, \ldots, f_m$, such that the number of different row patterns of the decomposition matrix $Z(X^{(1)}, f')$ of f' is minimized. $f'_1, \ldots, f'_m$ are not necessarily *completely specified* functions. Usually there is some degree of freedom in assigning the remaining don't cares.

The remaining don't cares are used to minimize the number of decomposition functions for single–output functions f'_i. However, we have to be careful not to increase $nrp(X^{(1)}, f')$ again. Fortunately, we can easily prove that the number of row patterns of $Z(X^{(1)}, f')$ cannot increase when we assign don't cares to minimize the number of decomposition functions for the single–output functions f'_i using the 'communication minimization approach' of Sections 2.2.1.3 and 5.2.

The reason for this is simple: When rows in $Z(X^{(1)}, f')$ have the same row pattern, then the corresponding rows in $Z(X^{(1)}, f'_i)$ have the same row pattern, too. Our claim follows from the fact that the 'communication minimization approach' for f'_i guarantees that equal row patterns remain equal.

Consequently, we have proven the following lemma:

LEMMA 5.1 *Let $f = (f_1, \ldots f_m)$ be a vector of incompletely specified functions with $f_i \in BP_n(D_i)$ $(1 \leq i \leq m)$ and let $X^{(1)} \subseteq \{0, 1\}^n$. $\tilde{f}_i$ results from f_i $(1 \leq i \leq m)$ using the communication minimization method from Sections 2.2.1.3 and 5.2 with respect to $X^{(1)}$. Then we have*

$$nrp(X^{(1)}, f) \geq nrp(X^{(1)}, \tilde{f}).$$

This means, we have two methods which exploit don't cares: The method just described, which minimizes the lower bound $\lceil \log(nrp(X^{(1)}, f)) \rceil$ on the total number of decomposition functions in the decomposition of $f_1, \ldots, f_m$ and the method from Sections 2.2.1.3 and 5.2, which minimizes the number of decomposition functions for single–output functions f_i separately. Both methods are compatible in the sense that it is possible to use the method to minimize the lower bound (with respect to a decomposition of $f'_1, \ldots, f'_m$ with computation of common decomposition functions) and afterwards to use the remaining don't cares in the method from Sections 2.2.1.3 and 5.2 without increasing the lower bound previously computed.

Table 5.1. Comparison of CLB counts for the XC3000 device with and without don't care exploitation.

			Number of CLBs	
Circuit	in	out	mulopII	mulop–dc
5xp1	7	10	9	9
9sym	9	1	7	7
alu2	10	6	51	33
apex7	49	37	45	41
b9	41	21	30	28
C499	41	32	65	50
C880	60	26	87	71
clip	9	5	14	13
count	35	16	26	26
duke2	22	29	114	108
e64	65	65	55	55
f51m	8	8	8	8
misex1	8	7	9	8
misex2	25	18	24	24
rd73	7	3	5	5
rd84	8	4	8	8
rot	135	107	146	135
sao2	10	4	20	18
vg2	25	8	18	18
z4ml	7	4	4	4
$\sum$			745	669

5.6. Experimental results

5.6.1 FPGA synthesis

We applied the decomposition procedure to several MCNC and ISCAS benchmarks to compute FPGA realizations for the Xilinx XC3000 device (see Section 1.4). This FPGA device is based on lookup tables with 5 inputs, such that the recursive decomposition is performed up to functions depending on at most 5 inputs. The lookup tables are combined into Configurable Logic Blocks (CLBs) which can realize either one 5-input function or two 4-input functions which depend on at most 5 different variables. To evaluate the effect of don't care assignments, we compare the numbers of CLBs in Table 5.1 for two versions of the algorithm: In the version *mulopII* (Scholl and Molitor, 1995a) we didn't use any don't care assignment procedures (all don't cares were assigned to 0) and in *mulop–dc* (Scholl, 1998) we use the don't care assignment as described in this section.

Table 5.2. Comparison of CLB counts for the XC3000 device between *mulop–dcII*, *FGMap*, *mis–pga (new)*, *IMODEC* and *FGSyn*

Circuit	in	out	mulop–dcII	FGMap	mis–pga(new)	IMODEC	FGSyn
					Number of CLBs		
5xp1	7	10	9	15	13	9	9
9sym	9	1	7	7	7	7	7
alu2	10	6	33	53	96	46	52
apex7	49	37	39	47	43	41	45
b9	41	21	28	27	32	-	29
C499	41	32	50	49	66	50	54
C880	60	26	71	74	72	81	88
clip	9	5	13	20	23	12	18
count	35	16	24	24	30	26	24
duke2	22	29	101	178	94	122	94
e64	65	65	50	55	56	55	56
f51m	8	8	8	11	15	8	9
misex1	8	7	8	8	9	9	9
misex2	25	18	23	21	25	21	22
rd73	7	3	5	7	5	5	5
rd84	8	4	8	12	9	8	8
rot	135	107	123	194	143	127	144
sao2	10	4	18	27	28	17	20
vg2	25	8	18	23	18	19	20
z4ml	7	4	4	5	4	4	4
$\sum$ (subtot.)			612	830	756	667	688
$\sum$ (total)			640	857	788	-	717

The results of Table 5.1 show a considerable reduction in CLB counts with our new algorithm. There are reductions in CLB counts of up to 35% for *alu2* and the overall reduction is more than 10%. It is important to note that the benchmark functions are all completely specified functions, such that don't cares can occur only at higher levels of the recursion. For this reason it is clear that improvements can be obtained only for larger benchmarks.

A second table (Table 5.2) shows a comparison between our tool *mulop–dcII* (Scholl, 1998), *FGMap* (Lai et al., 1993a), *mis–pga (new)* (Murgai et al., 1991; Sangiovanni-Vincentelli et al., 1993), *IMODEC* (Wurth et al., 1995) and *FGSyn* (Lai et al., 1994a) demonstrating the advantages of our procedure.[4]

---

[4] *mulop–dcII* is similar to *mulop–dc*; the only difference is a modified procedure for merging lookup tables into CLBs (the merging problem is formulated as a maximum cardinality matching problem, as proposed in (Murgai et al., 1990)).

5.6.2   Example: A partial multiplier

In this section we consider the synthesis of a partial multiplier. With this example we particularly demonstrate the role of symmetries and the role of don't cares during decomposition. The example also shows that the automatic decomposition tool is able to produce competitive designs for arithmetic functions which have already been intensively studied using human intelligence.

A partial multiplier of bit width n is a Boolean function $pm_n : \{0,1\}^{n^2} \to \{0,1\}^{2n}$. The inputs are given by the bits of the n partial products of a n–bit multiplier; the outputs are the product bits of the n–bit multiplier. If $(a_{n-1}, \ldots, a_0)$ and $(b_{n-1}, \ldots, b_0)$ are the operands which have to be multiplied, and the result of the multiplication is a binary number $(r_{2n-1}, \ldots, r_0)$, then $(r_{2n-1}, \ldots, r_0)$ is the sum of n partial products $2^j \cdot (a_{n-1}b_j, \ldots, a_0b_j)$ $(0 \leq j \leq n-1)$. The inputs of the *partial* multiplier are the bits $p_{i,j} = a_i b_j$ $(0 \leq i,j \leq n-1)$. pm_n is defined as $pm_n : \{0,1\}^{n^2} \to \{0,1\}^{2n}$, $pm_n(p_{n-1,n-1}, \ldots, p_{0,n-1}, \ldots, p_{n-1,0}, \ldots, p_{0,0}) = (r_{2n-1}, \ldots, r_0)$, with

$$\sum_{i=0}^{2n-1} r_i 2^i = \sum_{j=0}^{n-1} \left[\left(\sum_{i=0}^{n-1} p_{i,j} \cdot 2^i \right) \cdot 2^j \right].$$

During the decomposition of the partial multiplier we use the strategy of decomposing with respect to symmetry sets of the functions, as long as there are symmetry sets of size 3 or larger. If there are l symmetry sets ($P_{\sim esym} = \{\mu_1, \ldots, \mu_l\}$) then we use l–sided decompositions. l–sided decompositions can be 'simulated' by l one–sided decompositions. When there are only symmetry sets of size two or one, then decompositions with respect to a symmetry set are not guaranteed to be nontrivial. In this case we prefer two–sided decompositions (which in turn can be 'simulated' by two one–sided decompositions).

Figure 5.2 shows the circuit for a partial 4–bit multiplier.

The circuit can be interpreted in the following way:

The inputs of the circuit are bits $p_{3,3}, p_{3,2}, \ldots, p_{0,0}$ of the partial products. In the first level of the recursion, pm_4 is decomposed with respect to the symmetry sets $\{p_{3,3}\}$, $\{p_{3,2}, p_{2,3}\}$, $\{p_{3,1}, p_{2,2}, p_{1,3}\}$, $\{p_{3,0}, p_{2,1}, p_{1,2}, p_{0,3}\}$, $\{p_{2,0}, p_{1,1}, p_{0,2}\}$, $\{p_{1,0}, p_{0,1}\}$ and $\{p_{0,0}\}$.

The decomposition functions which are computed for the symmetry sets (see Figure 5.2, 'Decomposition functions, 1st level of recursion') can be interpreted as the respective sums of the bits in the symmetry sets. (For example, $p_{3,1}, p_{2,2}$ and $p_{1,3}$ are added by a full adder, resulting in a 2–bit number $(y_1^{(4)}, y_0^{(4)})$ etc., see Figure 5.2.) Consequently, 7 different numbers are computed, such that the sum of all these numbers is equal to the final result $(r_7, \ldots, r_0)$ of the

Figure 5.2. Automatically generated circuit for a partial 4-bit multiplier.

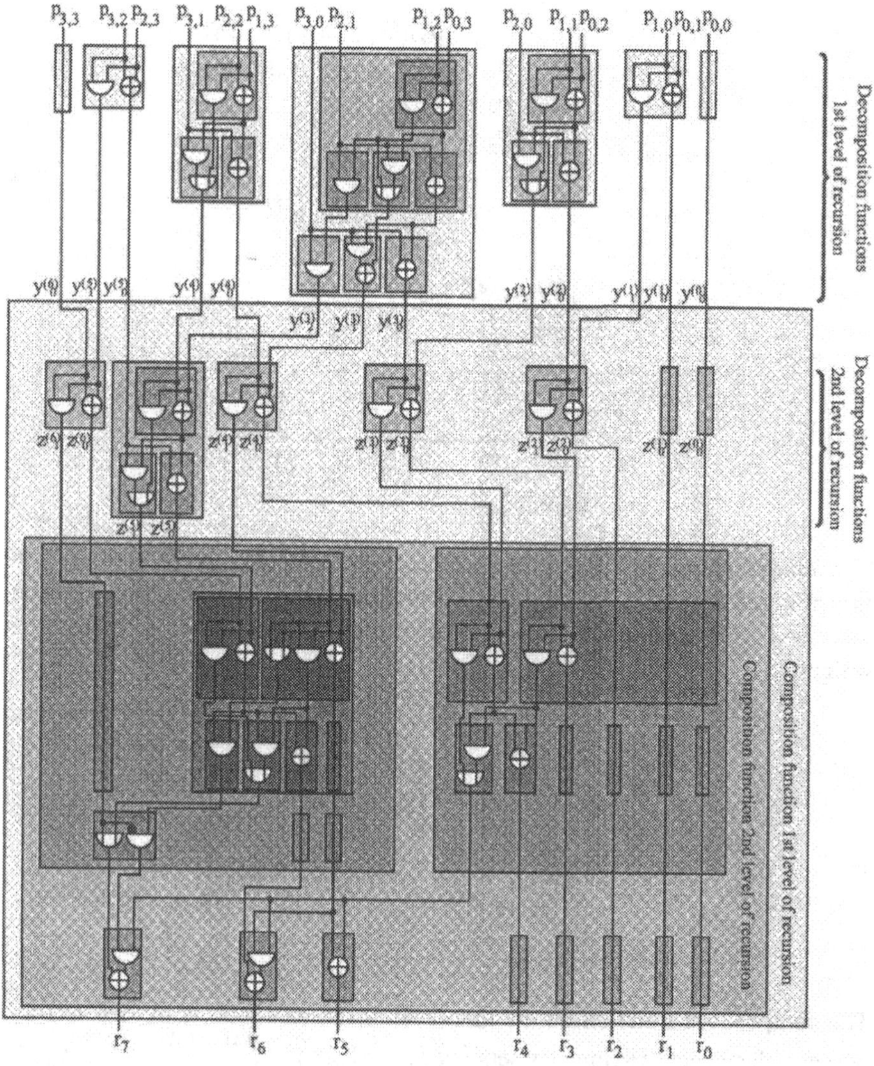

multiplication. The following addition is performed on the 1st level of the recursion (bits of symmetry sets μ_i shown in column i are added resulting in numbers $(y_l^{(i)}, \ldots, y_0^{(i)})$ shown in row i of the result):

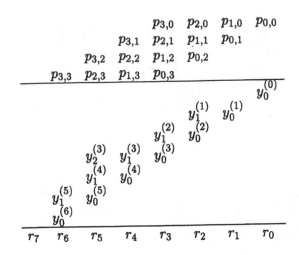

The remaining task of adding these numbers (implemented by the composition function, see Figure 5.2, 'Composition function, 1st level of recursion') can be described as the addition of three numbers (after 'shifting' of bits with the same weight):

$$
\begin{array}{ccccccc}
y_1^{(5)} & y_2^{(3)} & y_1^{(3)} & y_1^{(2)} & y_1^{(1)} & y_0^{(1)} & y_0^{(0)} \\
y_0^{(6)} & y_1^{(4)} & y_0^{(4)} & y_0^{(3)} & y_0^{(2)} & & \\
 & y_0^{(5)} & & & & & \\
\hline
r_7 & r_6 & r_5 & r_4 & r_3 & r_2 & r_1 \quad r_0
\end{array}
$$

The composition function on the first level of the recursion is further decomposed with respect to symmetry sets $\{y_1^{(5)}, y_0^{(6)}\}$, $\{y_2^{(3)}, y_1^{(4)}, y_0^{(5)}\}$, $\{y_1^{(3)}, y_0^{(4)}\}$, $\{y_1^{(2)}, y_0^{(3)}\}$, $\{y_1^{(1)}, y_0^{(2)}\}$, $\{y_1^{(1)}\}$ and $\{y_0^{(0)}\}$. Here the decomposition functions which are computed for the symmetry sets (see Figure 5.2, 'Decomposition functions, 2nd level of recursion'), can also be interpreted as sums of the bits in the symmetry sets. (For example, $y_2^{(3)}$, $y_1^{(4)}$ and $y_0^{(5)}$ are added by a full adder, resulting in a 2-bit number $(z_1^{(5)}, z_0^{(5)})$.) Again, 7 different numbers are computed, such that the sum of all these numbers is equal to the final result $(r_7, \ldots, r_0)$ of the multiplication. The following addition is performed on the second level of the recursion:

$$
\begin{array}{ccccccc}
y_1^{(5)} & y_2^{(3)} & y_1^{(3)} & y_1^{(2)} & y_1^{(1)} & y_0^{(1)} & y_0^{(0)} \\
y_0^{(6)} & y_1^{(4)} & y_0^{(4)} & y_0^{(3)} & y_0^{(2)} & & \\
& y_0^{(5)} & & & & &
\end{array}
$$

$$
\begin{array}{cccccccc}
 & & & & & & & z_0^{(0)} \\
 & & & & & & z_0^{(1)} & \\
 & & & & z_1^{(2)} & z_0^{(2)} & & \\
 & & & z_1^{(3)} & z_0^{(3)} & & & \\
 & & z_1^{(4)} & z_0^{(4)} & & & & \\
 & z_1^{(5)} & z_0^{(5)} & & & & & \\
z_1^{(6)} & z_0^{(6)} & & & & & & \\
\hline
r_7 & r_6 & r_5 & r_4 & r_3 & r_2 & r_1 & r_0
\end{array}
$$

Again, after 'shifting' bits with the same weight, the remaining task can be described as the addition of two numbers:

$$
\begin{array}{cccccccc}
z_1^{(6)} & z_1^{(5)} & z_1^{(4)} & z_1^{(3)} & z_1^{(2)} & z_0^{(2)} & z_0^{(1)} & z_0^{(0)} \\
 & z_0^{(6)} & z_0^{(5)} & z_0^{(4)} & z_0^{(3)} & & & \\
\hline
r_7 & r_6 & r_5 & r_4 & r_3 & r_2 & r_1 & r_0
\end{array}
$$

The composition function of the second level of the recursion (Figure 5.2) implements the addition of these two numbers. Since here there are no symmetry sets of size larger than 2, we perform a two–sided decomposition. The composition function of the second level of the recursion is decomposed with respect to $\{\{z_1^{(6)}, z_0^{(6)}, z_1^{(5)}, z_0^{(5)}, z_1^{(4)}\}, \{z_0^{(4)}, z_1^{(3)}, z_0^{(3)}, z_1^{(2)}, z_0^{(2)}, z_0^{(1)}, z_0^{(0)}\}\}$. The realization of the composition function of the second level of the recursion *basically* corresponds to an adder following the conditional–sum scheme.

An important issue concerning the decomposition of the partial multiplier has not been mentioned yet:

The composition functions used are *incompletely specified* functions. For example, for the addition of $p_{3,0}, p_{2,1}, p_{1,2}$ and $p_{0,3}$ on the first level, there are only 5 possible values for the sum. However, we need 3 decomposition functions, such that 3 out of 8 values cannot occur as outputs of the addition leading to don't cares in the composition function.

The partial multiplier is a good example demonstrating the importance of the don't care assignment methods presented. Our don't care assignment concept is essential for the result: Number and size of the symmetry sets essentially depend on the don't care assignment. Also, the use of don't care assignment methods of this chapter for the composition function of the second level is responsible for the quality of the result. Whereas the implementation found here has a R_2–complexity of 110, we have an increase in cost by 75% to an R_2–complexity of 194, if we do not use any exploitation of don't cares and assign 0 for all don't cares, for instance.

The interpretation of the composition function on the second level as the addition of a 7–bit number and a 4–bit number is only *one* possibility with a very special assignment to the don't cares occurring in the composition function. When we have a closer look at the realization of Figure 5.2, it becomes clear that the extension chosen by our procedure is not exactly equal to an adder function. Function r_7, which computes the most significant bit, differs from the corresponding output function of an adder. This difference even leads to a cost reduction as compared to the adder: The implementation for the composition function computed here has an R_2–complexity of 41. When the don't cares are assigned such that the composition function on the second level of the recursion is equal to an adder, then the decomposition method computes a realization following the conditional sum scheme (cf. Section 4.4.1) with an R_2–complexity of 45.

Based on an analysis of the circuit computed automatically for a partial multiplier with a fixed bit width (as shown above), we can even find a general construction scheme for multipliers with variable bit width: A generalization of the principle to variable bit widths n (i.e., to the design of partial multipliers pm_n) leads to the following result (Scholl, 1996): The B_2–complexity for the partial multiplier (without the adder of the last level of the recursion) can be estimated by

$$8\frac{1}{3} \cdot n^2 + O(n \log^2 n).$$

The B_2–depth of this circuit is bounded by

$$5.13 \cdot \log n + O(\log^* n \log \log n)\P$$

By comparison, the B_2–complexity of a partial multiplier with bit width n ($n = 2^k$) according to the Wallace tree principle (Wallace, 1964) amounts to $10n^2 - 20n$ and the corresponding depth to $5 \log n - 5$ (also without the final adder for the last two binary numbers).

$\P\log^* n := \min\{m \mid \log^{(m)}(n) \le 1\}$ with $\log^{(0)}(n) := n$ and $\log^{(i)}(n) := \log(\log^{(i-1)}(n))$ for $i \in \mathbb{N}$.

Chapter 6

NON–DISJOINT DECOMPOSITIONS

Non–disjoint decomposition were already defined by Roth and Karp in (Roth and Karp, 1962). However, to the best of our knowledge they did not do a systematic study of non–disjoint decompositions. Fortunately, non–disjoint decompositions can easily be integrated into our decomposition procedure of Chapters 3, 4 and 5. Another proposal for the use of non–disjoint decompositions was given in (Lai et al., 1993b).

Non–disjoint decompositions are defined as follows:

DEFINITION 6.1 (ONE–SIDED NON–DISJOINT DECOMPOSITION)
A **one–sided non–disjoint decomposition** *of a Boolean function* $f \in B_n$ *with input variables* $x_1, \ldots, x_n$ *with respect to the pair* $(X^{(1)}, X^{(2)})$ *of variable sets with* $X^{(1)} = \{x_1, \ldots, x_p\}$, $X^{(2)} = \{x_h, \ldots, x_n\}$, $1 < p < n$ *and* $1 < h < p + 1$ *is a representation of f of form*

$$f(x_1, \ldots, x_n) = g(\alpha_1(x_1, \ldots, x_p), \ldots, \alpha_r(x_1, \ldots, x_p), x_h, \ldots, x_n).$$

DEFINITION 6.2 (TWO–SIDED NON–DISJOINT DECOMPOSITION)
A **two–sided non–disjoint decomposition** *of a Boolean function* $f \in B_n$ *with input variables* $x_1, \ldots, x_n$ *with respect to the pair* $(X^{(1)}, X^{(2)})$ *of variables sets with* $X^{(1)} = \{x_1, \ldots, x_p\}$, $X^{(2)} = \{x_h, \ldots, x_n\}$, $1 < p < n$ *and* $1 < h < p + 1$ *is a representation of f of form*

$$f(x_1, \ldots, x_n) =$$
$$g(\alpha_1(x_1, \ldots, x_p), \ldots, \alpha_r(x_1, \ldots, x_p),$$
$$\beta_1(x_h, \ldots, x_n), \ldots, \beta_s(x_h, \ldots, x_n)).$$

6.1.   Motivation

Example 6.1 demonstrates how the number of decomposition functions can be reduced using non–disjoint decompositions:

EXAMPLE 6.1   Consider the Boolean function $sel\_add_{2^k} : \{0, 1\}^{2^{k+2}+1} \rightarrow \{0, 1\}^{2^k}$, $sel\_add_{2^k}(os, a_{2^k-1}, \ldots, a_0, b_{2^k-1}, \ldots, b_0, c_{2^k-1}, \ldots, c_0, d_{2^k-1}, \ldots, d_0) = (r_{2^k-1}, \ldots, r_0)$, where

$$\sum_{i=0}^{2^k-1} r_i 2^i = \begin{cases} (\sum_{i=0}^{2^k-1} a_i 2^i + \sum_{i=0}^{2^k-1} b_i 2^i) \bmod 2^{2^k}, & \text{if } os = 0 \\ (\sum_{i=0}^{2^k-1} c_i 2^i + \sum_{i=0}^{2^k-1} d_i 2^i) \bmod 2^{2^k}, & \text{if } os = 1 \end{cases}$$

The single–output functions of $sel\_add_{2^k}$ are denoted $s_{2^k-1}, \ldots, s_0$, thus $sel\_add_{2^k} = (s_{2^k-1}, \ldots, s_0)$. $sel\_add_{2^k}$ computes the binary addition of two 2^k–bit numbers. Input os selects the numbers which have to be added: If $os = 0$, then the *first* two operands $(a_{2^k-1}, \ldots, a_0)$ and $(b_{2^k-1}, \ldots, b_0)$ are added, if $os = 1$, then the *last* two operands $(c_{2^k-1}, \ldots, c_0)$ and $(d_{2^k-1}, \ldots, d_0)$ are added.

Figure 6.1 shows one possible two–sided balanced and disjoint decomposition of all s_i $(0 \le i \le 2^k - 1)$ with respect to the partition $A = \{X^{(1)}, X^{(2)}\}$ with $X^{(1)} = \{os, a_{2^k-1}, \ldots, a_{2^{k-1}}, b_{2^k-1}, \ldots, b_{2^{k-1}}, c_{2^k-1}, \ldots, c_{2^{k-1}}, d_{2^k-1}, \ldots, d_{2^{k-1}}\}$ and $X^{(2)} = \{a_{2^{k-1}-1}, \ldots, a_0, b_{2^{k-1}-1}, \ldots, b_0, c_{2^{k-1}-1}, \ldots, c_0, d_{2^{k-1}-1}, \ldots, d_0\}$.*

For each of the functions $s_0, \ldots, s_{2^{k-1}-1}$ we need two decomposition functions which depend on the variables in $X^{(2)}$ in this *disjoint* decomposition, since the value of os is not known for the computation of these decomposition functions ($os \notin X^{(2)}$). Both the corresponding sum bits of the addition of $(a_{2^{k-1}-1}, \ldots, a_0)$ and $(b_{2^{k-1}-1}, \ldots, b_0)$ and the corresponding sum bits of the addition of $(c_{2^{k-1}-1}, \ldots, c_0)$ and $(d_{2^{k-1}-1}, \ldots, d_0)$ are computed ($slab_i$ and $slcd_i$). In an analogous manner, for the functions $s_{2^{k-1}}, \ldots, s_{2^k-1}$, there are two decomposition functions depending on $X^{(2)}$: Both the carry bit of the addition of $(a_{2^{k-1}-1}, \ldots, a_0)$ and $(b_{2^{k-1}-1}, \ldots, b_0)$ and the carry bit of the addition of $(c_{2^{k-1}-1}, \ldots, c_0)$ and $(d_{2^{k-1}-1}, \ldots, d_0)$ ($carryab$ and $carrycd$). For the decomposition of $s_0, \ldots, s_{2^{k-1}-1}$ we need exactly one decomposition function which depends on $X^{(1)}$, namely the identity function os. For the decomposition of $s_{2^{k-1}+2}, \ldots, s_{2^k-1}$, however, we need 3 decomposition functions depending on the variables of $X^{(1)}$: The first decomposition function is the corresponding bit sh_i of the sum of $(a_{2^k-1}, \ldots, a_{2^{k-1}})$ and $(b_{2^k-1}, \ldots, b_{2^{k-1}})$ or the sum of

---

*An implementation of function add_{2^k} (addition of two operands) is presented in Section 4.4. This implementation is very similar to the implementation of Figure 6.1 for $sel\_add_{2^k}$.

$(c_{2^k-1}, \ldots, c_{2^k-1})$ and $(d_{2^k-1}, \ldots, d_{2^k-1})$ (depending on the value of os), the second decomposition function is the bit $shp1_i$ of the sum plus 1 (which is used for the case that the addition of the least significant bits of the operands generates a carry), and the third decomposition function is os, since (depending on os) we have to select between $carryab$ and $carrycd$ generated by the addition of the least significant bits[2].

The cost of the implementation can be reduced, when os, which is obviously important both for the computation of the most significant bits as well as the least significant bits, is included in *both* variable sets of the decomposition, resulting in a non–disjoint decomposition. When we perform a non–disjoint decomposition with respect to $X^{(1)}$ and $X^{(2)}$ with $X^{(1)} = \{os, a_{2^k-1}, \ldots, a_{2^k-1}, b_{2^k-1}, \ldots, b_{2^k-1}, c_{2^k-1}, \ldots, c_{2^k-1}, d_{2^k-1}, \ldots, d_{2^k-1}\}$ and $X^{(2)} = \{os, a_{2^{k-1}-1}, \ldots, a_0, b_{2^{k-1}-1}, \ldots, b_0, c_{2^{k-1}-1}, \ldots, c_0, d_{2^{k-1}-1}, \ldots, d_0\}$, then we need fewer decomposition functions (see Figure 6.2).

Since os is known in the decomposition of $s_0, \ldots, s_{2^{k-1}-1}$ with respect to $X^{(2)}$, we need only one decomposition function which is equal to the resulting function s_i. For functions $s_{2^k-1}, \ldots, s_{2^k-1}$ we need to compute only one carry bit based on $X^{(2)}$, since os is already contained in $X^{(2)}$ and thus we know during the computation based on variables of $X^{(2)}$, whether the operands $(a_{2^{k-1}-1}, \ldots, a_0)$ and $(b_{2^{k-1}-1}, \ldots, b_0)$ or $(c_{2^{k-1}-1}, \ldots, c_0)$ and $(d_{2^{k-1}-1}, \ldots, d_0)$ have to be added. For this reason we do not need os as a decomposition function in decompositions with respect to $X^{(1)}$, either.

The example shows that it may sometimes be more natural to include some variables in both variable sets of a two–sided decomposition.

REMARK 6.1 *We can even construct more extreme cases than Example 6.1: Let $X^{(1)}$ and $X^{(2)}$ be two disjoint variable sets with $|X^{(2)}| = |X^{(1)}|$, c an additional variable not in $X^{(1)} \cup X^{(2)}$. Then we can easily construct functions $f \in B_n$ ($n = |X^{(1)}| + |X^{(2)}| + 1$) with variable sets $X^{(1)} \cup X^{(2)} \cup \{c\}$ such that*

- *In a two–sided disjoint decomposition of f with respect to $\{X^{(1)}, X^{(2)} \cup \{c\}\}$ the minimum number of decomposition functions depending on $X^{(1)}$ is $|X^{(1)}|$ and the minimum number of decomposition functions depending on $X^{(2)} \cup \{c\}$ is $\frac{1}{2}|X^{(2)}| + 1$.*

---

[2]The decomposition of s_{2^k-1} and $s_{2^{k-1}+1}$ is a special case, since apart from os we need only one decomposition function depending on $X^{(1)}$ for s_{2^k-1} and this decomposition function can also be used in the decomposition of $s_{2^{k-1}+1}$ (see also add_{2^k} in Section 4.4).

Figure 6.1. Disjoint decomposition of $sel\_add_{2^k}$.

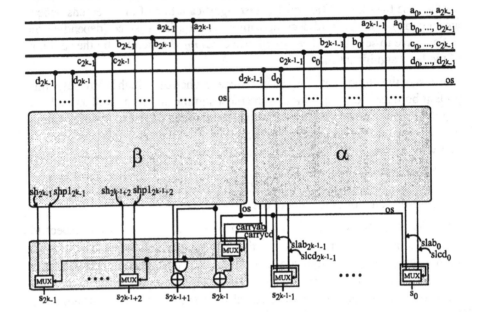

- In a two-sided disjoint decomposition of f with respect to $\{X^{(1)} \cup \{c\}, X^{(2)}\}$ the minimum number of decomposition functions depending on $X^{(2)}$ is $|X^{(2)}|$ and the minimum number of decomposition functions depending on $X^{(1)} \cup \{c\}$ is $\frac{1}{2}|X^{(1)}| + 1$.

- In a two-sided non-disjoint decomposition of f with respect to $\{X^{(1)} \cup \{c\}, X^{(2)} \cup \{c\}\}$ we need only $\frac{1}{2}|X^{(1)}| + \frac{1}{2}|X^{(2)}|$ decomposition functions.

An intuitive explanation for this situation is that c is 'essential' both for the computation of immediate results based on $X^{(1)}$ as well as the computation of immediate results based on $X^{(2)}$: If c is removed from $X^{(1)} \cup \{c\}$ (from $X^{(2)} \cup \{c\}$) to obtain a non-disjoint decomposition, then the number of decomposition functions depending on $X^{(1)}$ ($X^{(2)}$) is doubled. Both for the case $c = 0$ as well as the case $c = 1$, different immediate results depending on $X^{(1)}$ ($X^{(2)}$) have to be computed.

It is clear from Example 6.1 and from Remark 6.1 that there are cases where it indeed makes sense to use non-disjoint decompositions.

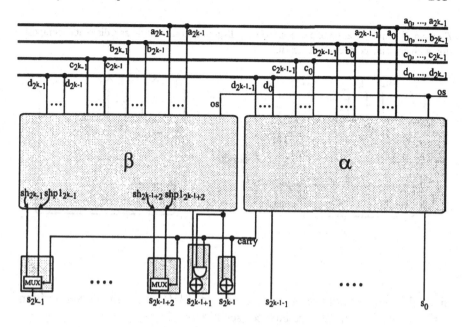

Figure 6.2. Non–disjoint decomposition of $sel\_add_{2^k}$.

6.2. Integration of non–disjoint into disjoint decomposition

Non–disjoint decompositions can easily be integrated into our decomposition procedure of Chapters 3, 4 and 5. Here we propose two different methods of integrating non–disjoint decompositions: The first method is based on the property that our decomposition procedure is able to compute decomposition functions one after another, and the second method is based on incompletely specified functions.

6.2.1 Preselected decomposition functions

During the computation of common decomposition functions we did not compute all decomposition functions for some output function at once, but rather the decomposition functions were computed step by step, assuming decomposition functions computed in earlier steps to be 'preselected' decomposition functions in later steps. Using preselected decomposition functions we can easily integrate non–disjoint decompositions: A one–sided non–disjoint decomposition with respect to the pair $(X^{(1)}, X^{(2)})$ of variable sets with $X^{(1)} = \{x_1, \ldots, x_p\}$, $X^{(2)} = \{x_h, \ldots, x_n\}$, $0 < p < n$ and $1 < h < p + 1$ can be viewed as a one–

Figure 6.3.   Interpretation of a one–sided non–disjoint decomposition as a disjoint decomposition with preselected decomposition functions.

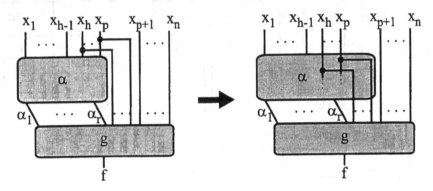

sided disjoint decomposition with respect to $X^{(1)}$, where the decomposition functions $x_h, \ldots, x_p$ are preselected (see Figure 6.3).

In Chapters 3 and 4 the computation of decomposition functions for $f \in B_n$ with respect to a bound set $X^{(1)}$ is based on a partition of $\{0, 1\}^p$ into equivalence classes $K_1, \ldots, K_{nrp(X^{(1)}, f)}$. In a strict encoding these equivalence classes are encoded with codes of minimum lengths $\lceil \log(nrp(X^{(1)}, f)) \rceil$. Chapter 4 handles preselected decomposition functions $\alpha'_1, \ldots, \alpha'_h$ in a natural way (see Lemmas 4.1, 4.2 and 4.3): $r - h$ additional decomposition functions are sufficient to decompose f if and only if for all a the number of equivalence classes K_j with $(\alpha'_1, \ldots, \alpha'_h)(K_j) = a$ is at most 2^{r-h} $(nrp(X^{(1)}, f, \alpha') \leq 2^{r-h}$ in Lemma 4.1).

The only difference for non–disjoint decompositions compared to Chapter 4 lies in the fact that preselected 'variable functions' (see Figure 6.3) may be non–strict decomposition functions. If we have one preselected decomposition function $\alpha'_1 = x_p$, e.g., there may be some classes K_j with $\alpha'_1(K_j) = \{0, 1\}$, such that $\alpha'_1 = x_p$ is not a strict decomposition function of f. But this requires only a minor modification: For the computation of the decomposition functions we need, for each code produced by the decomposition functions, the class of elements of $\{0, 1\}^p$ which are assigned this code. For this reason we might possibly have to divide some classes into two: If $\alpha'_1(K_j) = \{0, 1\}$ $(\alpha'_1 = x_p)$, then we have to compute the subclass $K_j^{(1)}$ of K_j, which is assigned 1 by α'_1, and the subclass $K_j^{(0)}$, which is assigned 0. However, using the characteristic function χ_{K_j} of K_j we simply have to compute the cofactors

$\chi_{K_j^{(1)}} = \chi_{K_j}|_{x_p^1}$ and $\chi_{K_j^{(0)}} = \chi_{K_j}|_{x_p^0}$ with respect to the variable x_p.[3] These operations can easily be performed based on the ROBDD representations of the characteristic function χ_{K_j} for K_j. Apart from a possible class splitting due to non–strict decomposition functions, we need no other changes to the ROBDD based algorithm.

In (Lai et al., 1993b) Lai et al. presented a method for one–sided non–disjoint decompositions of a function f with respect to $\{X^{(1)}, X^{(2)}\} := \{\{x_1, \dots, x_p\}, \{x_h, \dots, x_n\}\}$, which is based on a ROBDD representation of f with a particular variable order: It is assumed that the variables of set $\{x_1, \dots, x_{h-1}\}$ are in the variable order before the variables in $\{x_h, \dots, x_p\}$, and the variables in $\{x_{p+1}, \dots, x_n\}$ are the last variables in the order. Lai et al. compute not only linking nodes with respect to $\{x_1, \dots, x_p\}$ by following paths in the ROBDD, but they also compute, for each of the 2^{p-h+1} fixed assignments to the variables in $\{x_h, \dots, x_p\}$, the set of linking nodes (with respect to $\{x_1, \dots, x_p\}$) which can be reached from the root by this fixed assignment to the variables in $\{x_h, \dots, x_p\}$.

If for each such fixed assignment to the variables in $\{x_h, \dots, x_p\}$ the number of different linking nodes is not larger than 2^l, then a decomposition with l decomposition functions can be used. (Our notion is that they compute $nrp(X^{(1)}, f, \alpha')$ with $\alpha' = (x_h, \dots, x_p)$ in this way and $nrp(X^{(1)}, f, \alpha')$ must not be larger than 2^l, see also Lemma 4.1.)

6.2.2 Variable duplication with don't cares

A second possibility for integrating non–disjoint decompositions is to use decompositions of incompletely specified functions. Here in the decomposition of a function $f \in B_n$ with respect to the pair $(X^{(1)}, X^{(2)})$ of variable sets $(X^{(1)} = \{x_1, \dots, x_p\}, X^{(2)} = \{x_h, \dots, x_n\}, 1 < p < n$ and $1 < h < p+1)$, the input variables which occur both in $X^{(1)}$ and in $X^{(2)}$ are duplicated: In addition to the variables $x_h, \dots, x_p$, new variables $x'_h, \dots, x'_p$ are introduced and function f is replaced by a new function f' with

$$f'(\epsilon_1, \dots, \epsilon_{h-1}, \epsilon_h, \epsilon'_h, \dots, \epsilon_p, \epsilon'_p, \epsilon_{p+1}, \dots, \epsilon_n) =$$
$$f(\epsilon_1, \dots, \epsilon_{h-1}, \epsilon_h, \dots, \epsilon_p, \epsilon_{p+1}, \dots, \epsilon_n)$$

for all $(\epsilon_1, \dots, \epsilon_{h-1}, \epsilon_h, \epsilon'_h, \dots, \epsilon_p, \epsilon'_p, \epsilon_{p+1}, \dots, \epsilon_n) \in \{0, 1\}^{n+p-h+1}$ with $\epsilon_h = \epsilon'_h, \dots, \epsilon_p = \epsilon'_p$. f' is an incompletely specified function. The char-

---

[3] Of course the linking node now belonging to subclasses $K_j^{(1)}$ and $K_j^{(0)}$ is the same and it is equal to the linking node n_j originally belonging to K_j.

Figure 6.4. Non–disjoint decomposition by variable duplication and disjoint decomposition with don't care exploitation.

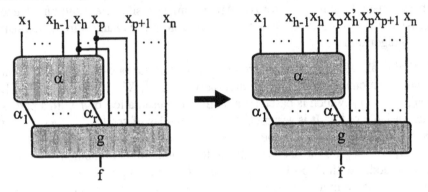

acteristic function of the don't care set is given by

$$dc(x_h, x_h', \ldots, x_p, x_p') = \bigvee_{j=h}^{p} x_j \oplus x_j'.$$

It is clear that a communication minimal two–sided disjoint decomposition of this incompletely specified function f' with respect to $\{Y^{(1)}, Y^{(2)}\}$ with $Y^{(1)} = \{x_1, \ldots, x_p\}$ and $Y^{(2)} = \{x_h', \ldots, x_p', x_{p+1}, \ldots, x_n\}$ corresponds to a communication minimal two–sided non–disjoint decomposition of f with respect to $(X^{(1)}, X^{(2)})$ and that a communication minimal one–sided disjoint decomposition of f' with respect to $X^{(1)}$ corresponds to a communication minimal one–sided non–disjoint decomposition of f with respect to $(X^{(1)}, X^{(2)})$ (see Figure 6.4).

6.3.   Computing non–disjoint variable sets for decomposition

Now the question remains as to how to choose non–disjoint variable sets for decomposition. To decide whether it makes sense to use non–disjoint decompositions it is necessary to have a method for computing or estimating the number of decomposition functions in non–disjoint decompositions.

One–sided decompositions.   We start with one–sided decompositions.

Let $\{X^{(1)}, X^{(2)} \cup \{c\}\}$ be a partition of the input variables. As mentioned in Section 6.2.1 we can see, using preselected decomposition functions, that the minimum number of decomposition functions in a one–sided non–disjoint decomposition with respect to $(X^{(1)} \cup \{c\}, X^{(2)} \cup \{c\})$ is equal to $\lceil \log(nrp(X^{(1)} \cup \{c\}, f, c)) \rceil$.[4]

$nrp(X^{(1)} \cup \{c\}, f, c)$ can easily be computed using cofactors f_c and $f_{\bar{c}}$:[5]

LEMMA 6.1 *Let $f \in B_n$, $\{X^{(1)}, X^{(2)} \cup \{c\}\}$ be a partition of the input variables of f. Then*

$$nrp(X^{(1)} \cup \{c\}, f, c) = \max(nrp(X^{(1)}, f_{\bar{c}}), nrp(X^{(1)}, f_c))$$

PROOF: Let $X^{(1)} = \{x_1, \ldots, x_p\}$ and $Z(X^{(1)} \cup \{c\}, f)$ be the decomposition matrix of f with respect to $X^{(1)} \cup \{c\}$. Then according to Definition 4.1, $nrp(X^{(1)} \cup \{c\}, f, 0)$ is defined as follows: Consider all rows of $Z(X^{(1)} \cup \{c\}, f)$ with the property that 'function c assigns value 0 to the row index', i.e., all rows having index $(x_1, \ldots, x_p, c)$ with $c = 0$. $nrp(X^{(1)} \cup \{c\}, f, 0)$ is equal to the number of different row patterns in this set of rows. It is clear that the rows considered are equal to the rows of decomposition matrix $Z(X^{(1)}, f_{\bar{c}})$ of the cofactor $f_{\bar{c}}$ and thus $nrp(X^{(1)} \cup \{c\}, f, 0) = nrp(X^{(1)}, f_{\bar{c}})$ Using an analogous argument we can conclude $nrp(X^{(1)} \cup \{c\}, f, 1) = nrp(X^{(1)}, f_c)$. Now $nrp(X^{(1)} \cup \{c\}, f, c) = \max(nrp(X^{(1)} \cup \{c\}, f, 0), nrp(X^{(1)} \cup \{c\}, f, 1))$ proves the claim. $\square$

The gain in non–disjoint decompositions (in the current decomposition step) can be much larger for two–sided decompositions than for one–sided decompositions: For two–sided disjoint decomposition there are examples where the number of decomposition functions depending on $X^{(2)}$ is higher by a factor of two in a disjoint decomposition with respect to $(X^{(1)} \cup \{c\}, X^{(2)})$ compared to a non–disjoint decomposition with respect to $(X^{(1)} \cup \{c\}, X^{(2)} \cup \{c\})$ (cf. Remark 6.1). However, the number of decomposition functions of a one–sided disjoint decomposition with respect to $X^{(1)} \cup \{c\}$ will not be much higher than the number of decomposition functions in a one–sided non–disjoint decomposition with respect to $(X^{(1)} \cup \{c\}, X^{(2)} \cup \{c\})$: Using the interpretation of one–sided non–disjoint decompositions as disjoint decompositions with preselected decomposition functions, we can see that there are only two cases:

1. $\lceil \log(nrp(X^{(1)} \cup \{c\}, f, c)) \rceil = \lceil \log(nrp(X^{(1)} \cup \{c\}, f)) \rceil - 1$, if c can be used as a decomposition function in a communication minimal disjoint decomposition of f with respect to $X^{(1)} \cup \{c\}$.

---

[4] $nrp(X^{(1)} \cup \{c\}, f, c) = \max(nrp(X^{(1)} \cup \{c\}, f, 0), nrp(X^{(1)} \cup \{c\}, f, 1))$ as defined in Definition 4.1 (page 135).

[5] For simplicity we confine ourselves to a single common variable c of $X^{(1)}$ and $X^{(2)}$. The generalization to more than one common variable is straightforward.

2. $\lceil \log(nrp(X^{(1)} \cup \{c\}, f, c)) \rceil = \lceil \log(nrp(X^{(1)} \cup \{c\}, f)) \rceil$ otherwise (when we cannot profit from the preselected decomposition function c).

Although the potential gain in non–disjoint decompositions (immediately observed in the current decomposition step) cannot be so high for one–sided decompositions, it makes sense to check, during an algorithm computing possible bound sets $Y^{(1)}$ for disjoint decompositions, whether there is a $c \in Y^{(1)}$, such that a non–disjoint decomposition with c both in bound and free set makes sense, i.e., to check whether there is a $c \in Y^{(1)}$, such that function c can be used as a decomposition function in a communication minimal decomposition with respect to $Y^{(1)}$ (case 1 in the enumeration above). This has the obvious advantage that we do not need any cells to realize function c, so that we have selected one decomposition function which is free of cost.

On the other hand we can say that it makes sense to include a variable c both in the bound set and the free set, if c is 'essential for the computation of intermediate results on both sets'. So another possibility is to also consider two–sided decompositions during the selection of one–sided decompositions[6]: Thus, as an additional criterion to move from a disjoint decomposition with respect to $X^{(1)} \cup \{c\}$ to a one–sided non–disjoint decomposition with respect to $(X^{(1)} \cup \{c\}, X^{(2)} \cup \{c\})$, we can use the number of decomposition functions in a two–sided decomposition with respect to $(X^{(1)} \cup \{c\}, X^{(2)} \cup \{c\})$ compared to the number of decomposition functions in a two–sided decomposition with respect to $\{X^{(1)} \cup \{c\}, X^{(2)}\}$.

Two–sided decompositions. To compute the number of decomposition functions for a non–disjoint two–sided decomposition with respect to $(X^{(1)}, X^{(2)})$, we could use an exact solution to problem **2S-CM** on p. 181 for the incompletely specified function defined by Figure 6.4. However, problem **2S-CM** is NP–hard. Fortunately, the incompletely specified functions resulting from the construction in Figure 6.4 have special properties which allow an efficient approximation without having to solve any 'partition into cliques' problems. More details can be found in (Scholl, 1996).

The search for appropriate non–disjoint two–sided decompositions can be integrated into the search for good disjoint decompositions (see Sections 3.7 and 4.2): For variable partitions $A = \{X^{(1)}, X^{(2)}\}$ found during the partitioning algorithm and for each input variable c, approximations for the minimum numbers of decomposition functions in a non–disjoint two–sided decomposition with respect to $(X^{(1)} \cup \{c\}, X^{(2)})$ (if $c \in X^{(2)}$) or with respect to $(X^{(1)}, X^{(2)} \cup \{c\})$ (if $c \in X^{(1)}$) are computed. Based on the minimum numbers of decomposition

---

[6]Having subsequent recursive decompositions in mind.

functions, estimated costs can also be computed for non–disjoint decompositions. If the estimated costs are minimal for a non–disjoint decomposition, then we check whether further variables should be included both into bound and free sets.

Chapter 7

LARGE CIRCUITS

Of course decomposition based methods cannot compute implementations for designs of arbitrary complexity. The size of the examples to be processed by decomposition is limited by the need to represent the corresponding Boolean functions in a form which is appropriate for the computations of a decomposition tool. Though the use of ROBDDs instead of function tables and decomposition matrices has led to a breakthrough concerning the sizes of functions which can be processed by decomposition, it is clear that there are limits for the use of ROBDDs. At the least, when it is not possible to represent a Boolean function by an ROBDD using the memory resources available, the current decomposition methods cannot be applied.

Fortunately, logic synthesis problems are scalable in the sense that the circuits to be optimized can be partitioned into (larger or smaller) clusters, which are then handled by the logic synthesis procedure. When only small clusters are processed by the logic synthesis procedure, then the potential for optimization will be small, but typically the amounts of resources needed for logic synthesis will also be small in this case. So there is a trade-off between the quality of the synthesis result and the resources needed to obtain it. Partitions of the specifying circuit into clusters are often given in the specification or they may be computed by the tool.

Partitioning a specifying circuit. Here we assume that the specification of the logic synthesis problem can be interpreted as a circuit, and that this specification does not contain any useful hierarchy or clustering information. There are several possibilities to partition the specifying circuit into clusters appropriate for a decomposition procedure:

- When we have a specifying circuit which is too large to apply decomposition methods directly, then we can apply *mincut* methods (e.g. (Schweikert and Kernighan, 1972; Breuer, 1979)) to find clusters of appropriate size. A *mincut* algorithm partitions a circuit into subcircuits (or clusters) with the goal of minimizing the wires which are cut by the clustering algorithm. By minimizing the numbers of wires to be cut, the numbers of inputs and outputs of the clusters are (heuristically) minimized. There is the hope that clusters with a small number of inputs represent simple logic synthesis problems and that the corresponding Boolean functions can be represented by small ROBDDs. Moreover, following the decomposition of the clusters, the layout generation for the overall circuit is simplified, since the number of global wires *between* clusters is minimized.

- If the specifying circuit consists of cells which are not too small, then it can also make sense to combine cells 'with similar sets of inputs', i.e., similar supports, into one cluster. This is done until the sizes of the clusters (measured by numbers of inputs and outputs) exceed predefined thresholds. Functions for cells with similar supports are grouped together, since they are more likely to share decomposition functions in a subsequent decomposition of the clusters.

  In *sis* (Brayton et al., 1987; Sentovich et al., 1992), for example, the internal data structure consists of single output nodes (or cells), whose functions are described by sums-of-products (cf. Definition 1.14). If the node functions are too small, then the circuit can be partially collapsed before combining nodes with similar support. When the circuit is partially collapsed, nodes are combined with their predecessors into new nodes and the Boolean function for a new node is computed by function composition. After collapsing, multi–output functions result from a combination of nodes with similar support.

- The third possibility for partitioning circuits is to use the *sizes of* ROBDD *representations* for clusters as a direct criterion in introducing cuts. Buch et al. (Buch et al., 1997) used such a method for the synthesis of Pass Transistor circuits. Cuts are determined during the ROBDD construction for a circuit: A conventional procedure to compute ROBDDs for the outputs of a circuit computes the ROBDD for gates of the circuit recursively based on the ROBDDs representing the functions of their predecessor gates. Buch et al. however use an ROBDD limit, which introduces auxiliary variables whenever the size of an ROBDD would grow too large during this process. An auxiliary variable represents an internal signal of the circuit, where a cut is introduced. The result of Buch's method is a clustered circuit, where each cluster is associated with an ROBDD depending on primary inputs and auxiliary variables for the internal signals, where cuts were introduced.

Figure 7.1. 16–bit divider designed using CADIC.

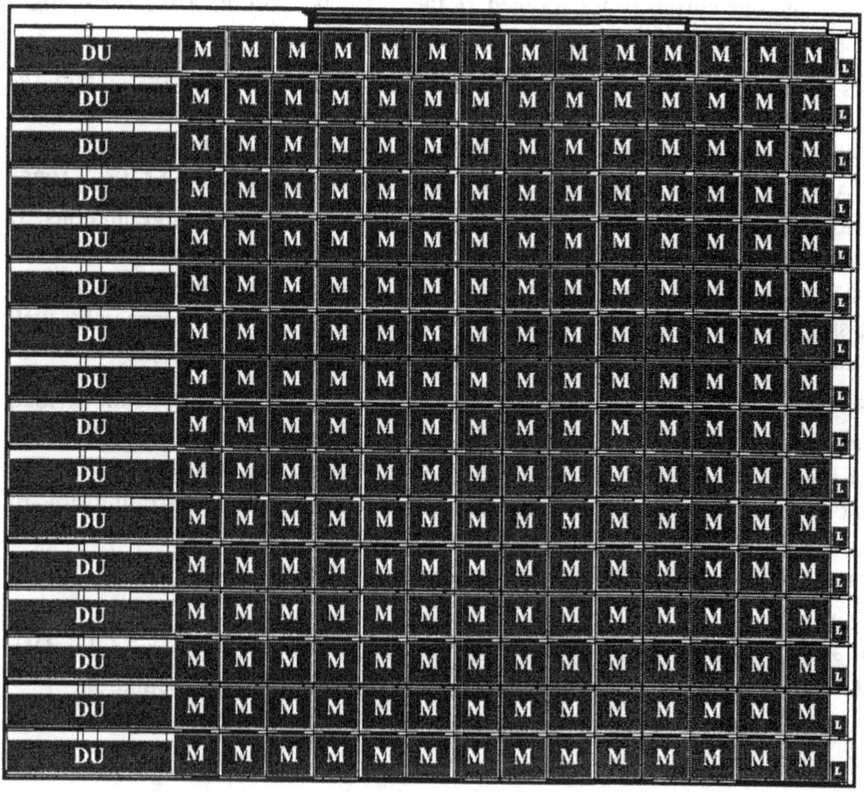

This method has the advantage that the size of the ROBDDs for the clusters can be strictly controlled by an ROBDD limit. The partitioning is guided by ROBDD sizes. Both quality and efficiency of ROBDD based methods for functional decomposition are based on the sizes of ROBDDs representing the functions to be decomposed.

Using hierarchy information for partitioning. In many cases there will already be a natural partition of the overall circuit given by a hierarchical specification. When the parts of the overall circuit given by this hierarchical specification are of appropriate size for the logic synthesis algorithm, then it will be applied directly to these parts, otherwise these parts can be clustered again using the approaches mentioned above.

Note that in the specification there may be 'hard macros', where an optimization across hierarchy boundaries is not allowed, i.e., where a combination of several parts into one cluster is not allowed, or there may be 'soft macros', where logic optimization is allowed to cross hierarchy boundaries.

As an example, we show the design of a fault–tolerant divider (Takagi and Yajima, 1988) using the design system CADIC (Becker et al., 1987a; Becker et al., 1987b; Becker et al., 1990) and our decomposition tool, which is integrated into CADIC. CADIC is based on a calculus developed in (Hotz, 1965; Molitor, 1988) and allows a specification of circuits by recursive, parameterized equations. If the parameters of the equations are fixed to constants, then the equations can be 'expanded' to a circuit of fixed size. A system of equations defines not just a single circuit, but rather a family of circuits. So CADIC is especially well suited to give a compact hierarchical description of regular circuits.

Figure 7.1 shows the expansion of a recursively defined fault–tolerant divider, here expanded to form a 16–bit divider. The expansion is done up to the level of basic modules DU, M and L defined in (Takagi and Yajima, 1988). The modules were specified by tables that were not conducive to a straightforward synthesis by hand. In CADIC the modules were defined as 'hard macros', such that an automatic logic synthesis by decomposition was done for DU, M and L separately.

Figure 7.2 shows the divider (also for bit width 16), where the modules DU, M and L are replaced by their automatically generated implementation.

Figure 7.2. 16–bit divider (modules *DU*, *M* and *L* are replaced by their automatically generated implementation).

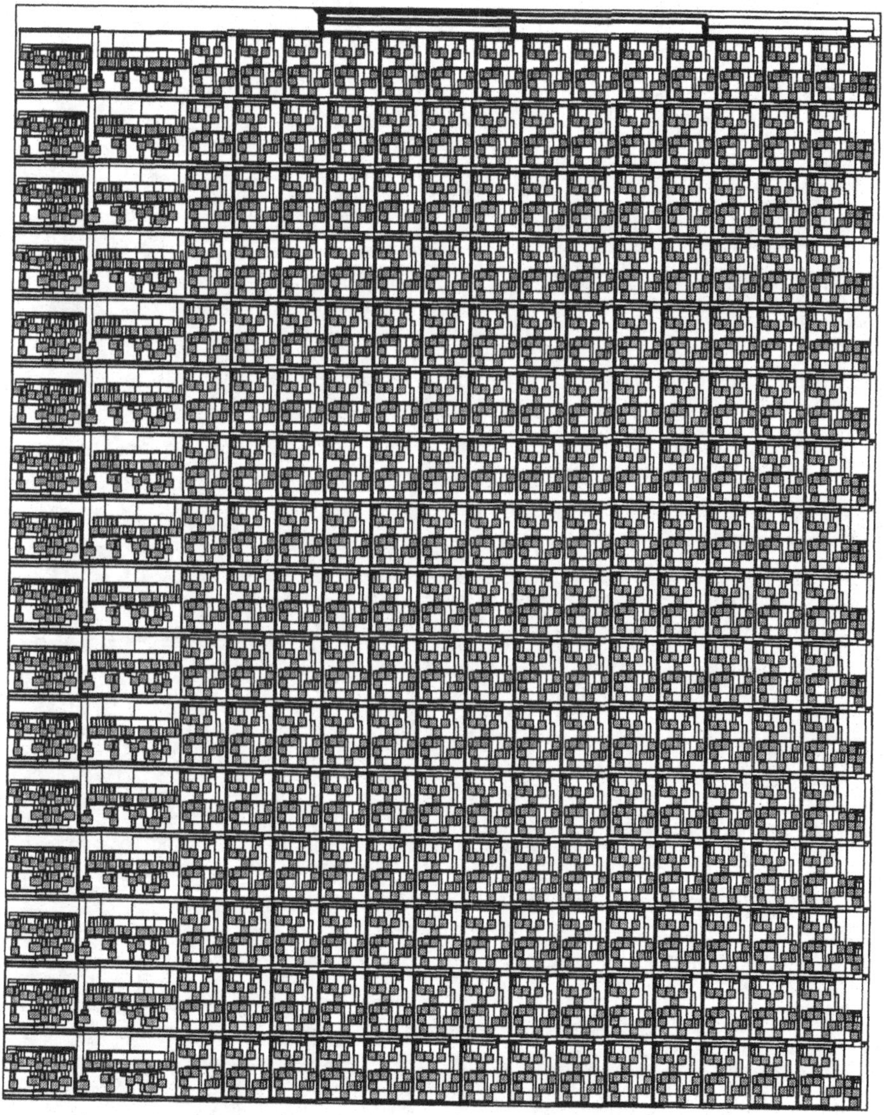

Appendix A
An example for an FPGA device

As an example for an FPGA device we show here some details of the commercially available Altera FLEX 10K chip. Figure A.1 shows the overall structure of the device. It contains a number of *logic array blocks* (LABs), where each LAB contains eight *logic elements* (LEs). Each logic element consists of one 4–input lookup table and one additional flip–flop. In addition to LABs, the chip also contains RAM blocks, which are implemented by *embedded array blocks* (EABs). LABs and EABs are connected by interconnection wires arranged in rows and columns. These wires are also connected to input and output pins of the chip.

Figure A.1. Overall structure of Altera FLEX 10K device (www.altera.com).[1]

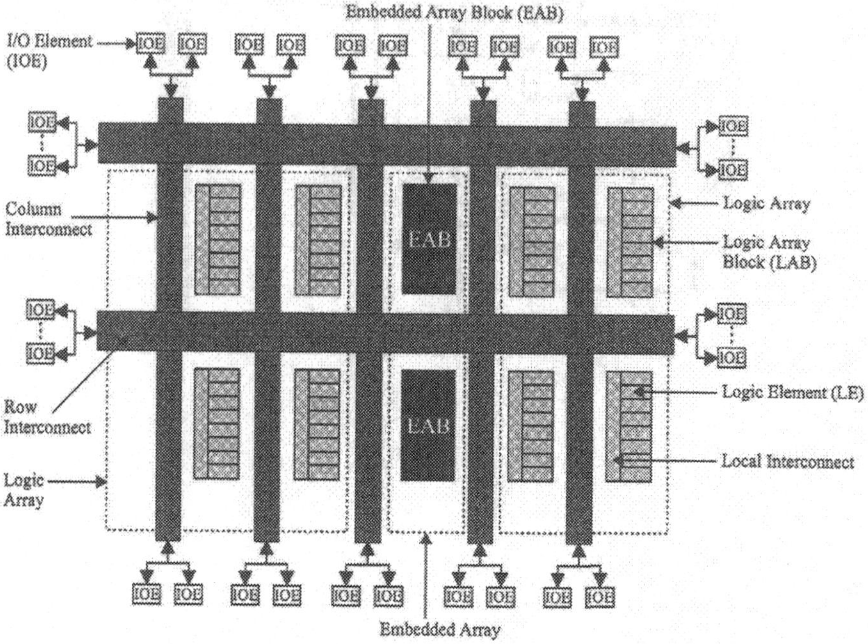

[1]Altera is a trademark and service mark of Altera Corporation in the United States and other countries. Altera products are the intellectual property of Altera Corporation and are protected by copyright laws and one or more U.S. and foreign patents and patent applications.

Figure A.2.   Logic array block of Altera FLEX 10K device (www.altera.com).

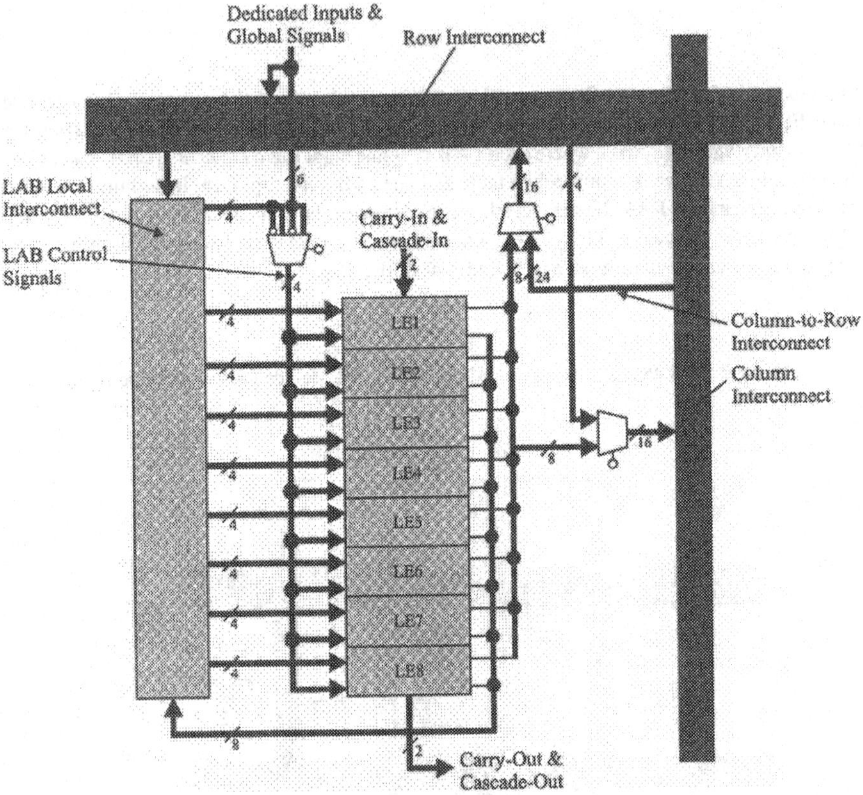

The logic array block is depicted in Figure A.2. An LAB contains eight LEs and a set of local interconnect wires. The local interconnect wires are used to realize feed back connections of the outputs of the LEs to their inputs and to connect the LE inputs to the row interconnect wires. The LE outputs can be connected to row and column interconnect wires.

Figure A.3.　Logic element of Altera FLEX 10K device (www.altera.com).

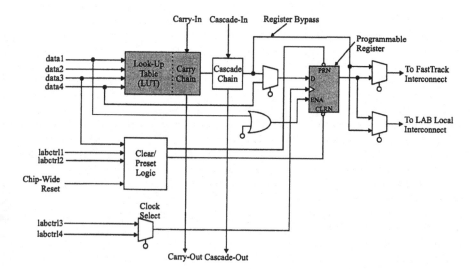

Finally, Figure A.3 shows the architecture of the logic element (LE). An LE consists of one 4–input LUT and one flip–flop. Two programmable multiplexers are used to connect either the LUT output or the flip–flop output to the local and global interconnect wires. There is also extra circuitry to support the implementation of adders: For the implementation of adders, the 4–input LUT can be used to implement both carry and sum functions of a full–adder. The carry bit is passed to the next LE via the carry–out wire.

Appendix B
Complexity of CM

We prove Theorem 2.2 on page 42: The problem CM is NP–hard.

The proof uses a polynomial time transformation from the NP complete problem 'Partition into Cliques' (PC) to CM.

The problem PC is (see (Garey and Johnson, 1979), page 193):

Partition into Cliques (PC)

Given: An undirected graph $G = (V, E)$ and a natural number $K \leq |V|$.

Find: Can V be partitioned into K disjoint sets $V_1, \ldots, V_K$, such that for $1 \leq i \leq K$ the subgraph of all nodes in V_i is a complete graph?

The idea of the proof is based on considering the compatibility relation $\sim$ for rows of the decomposition matrix as a graph. The problem of determining the minimum number of compatibility classes for rows of the decomposition matrix is equivalent to the problem of finding a minimum partition of the graph into complete subgraphs.

PROOF: Let an instance of problem PC consisting of a graph $G = (V, E)$ and a natural number $K \leq |V|$ be given. We can find in polynomial time an incompletely specified function f with $ncc((X^{(1)}, f) \leq K$ if and only if G can be partitioned into K complete subgraphs. f is constructed as follows:

Let p be minimal with $2^p \geq |V|$. Define $f : \{0, 1\}^{2p} \rightsquigarrow \{0, 1\}$ by giving a decomposition matrix $Z(X^{(1)}, f)$ with respect to $X^{(1)} = \{x_1, \ldots, x_p\}$. For simplicity assume w.l.o.g. $V = \{0, \ldots, |V| - 1\}$.
Set

$$Z(X^{(1)}, f)_{ij} = \star \text{ for } i \geq |V| \text{ or } j \geq |V|.$$

For all $0 \leq i, j \leq |V| - 1$ we define:

$$Z(X^{(1)}, f)_{ij} = \begin{cases} \star, & \text{if } i < j \\ 1, & \text{if } i = j \\ 0, & \text{if } i > j \text{ and } \{i, j\} \notin E \\ \star, & \text{if } i > j \text{ and } \{i, j\} \in E \end{cases}$$

223

$Z(X^{(1)}, f)$ has the following structure:

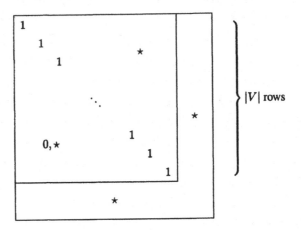

Let $i, j \in \{0, \ldots, |V| - 1\}$ and w.l.o.g. $i > j$.
For all columns $k < j$ we have:

$$Z(X^{(1)}, f)_{jk}, Z(X^{(1)}, f)_{ik} \in \{0, \star\}.$$

For all columns $k > j$ we have:

$$Z(X^{(1)}, f)_{jk} = \star.$$

For column j we have:

$$Z(X^{(1)}, f)_{jj} = 1 \text{ and } Z(X^{(1)}, f)_{ij} = 0 \overset{\text{Def. of } f}{\Longleftrightarrow} \{i, j\} \notin E.$$

Thus rows i and j are compatible ($bin_p(i) \sim bin_p(j)$), iff $\{i, j\} \in E$.
Nodes $i_1, \ldots i_m$ form a complete subgraph, if and only if the rows with indices $i_1, \ldots i_m$ are pairwise compatible.
Rows with indices $i \geq |V|$ are compatible to all other rows. Thus we have:

- If there is a partition of $V = \{0, \ldots, |V| - 1\}$ into K disjoint sets $V_1, \ldots, V_K$, such that the subgraphs of all nodes in V_i ($1 \leq i \leq K$) are complete graphs, then

$$\{V_1, \ldots, V_{K-1}, V_K \cup \{|V|, \ldots, 2^p - 1\}\}$$

is a partition of the row indices, such that all subsets contain indices of pairwise compatible rows. Then $ncc(X^{(1)}, f) \leq K$.

- If $ncc(X^{(1)}, f) \leq K$, then there is a partition $\{PK_1, \ldots, PK_K\}$ of the row indices $\{0, \ldots, 2^p - 1\}$, such that rows with indices in sets PK_i ($1 \leq i \leq K$) are pairwise compatible. Since rows with indices $i \geq |V|$ are compatible to all other rows, we can assume that for all $1 \leq i \leq K$ $PK_i \cap V \neq \emptyset$. Then there is a partition of V into sets

$$V_1 = PK_1 \cap V, \ldots, V_K = PK_K \cap V,$$

such that the subgraphs of all nodes in V_i are complete subgraphs.

$\square$

Appendix C
Characterization of weak and strong symmetry

We prove Lemma 2.6 on p. 54, which characterizes weak and strong symmetry.

PROOF:

1. '$\Longleftarrow$': Suppose that f is not strongly symmetric in P. Then there must be $\lambda_i \in P$, such that f is not strongly symmetric in λ_i and there must be a pair of variables (x_i, x_j) $\in \lambda_i$, such that f is not strongly symmetric in (x_i, x_j). By definition there must be $e_1 = (\epsilon_1, \ldots, \epsilon_i, \ldots, \epsilon_j, \ldots, \epsilon_n)$ and $e_2 = (\epsilon_1, \ldots, \epsilon_j, \ldots, \epsilon_i, \ldots, \epsilon_n)$, such that $e_1 \in D$ and $e_2 \notin D$ or $e_1, e_2 \in D$ with $f(e_1) \neq f(e_2)$. But e_1 and e_2 belong to the same weight class C of P. Both cases lead to a contradiction: In the first case we have $\{dc, 1\} \subseteq f(C)$ or $\{dc, 0\} \subseteq f(C)$, in the second case $\{0, 1\} \subseteq f(C)$.

   '$\Longrightarrow$': If f is strongly symmetric in P, then the following holds for all $\sigma \in \Sigma = \{\sigma_{i,l} \mid \exists \lambda_j \in P$ with $x_i, x_l \in \lambda_j\}$[1]: $\forall e = (\epsilon_1, \ldots, \epsilon_n) \in \{0, 1\}^n$ $f(e) = f(\sigma(e))$ (including the extended interpretation $f(e) = f(\sigma(e)) = dc$). Let e_1 and e_2 be members of an arbitrary weight class C of P. Then there is a sequence of permutations $\sigma_1 \ldots \sigma_k \in \Sigma$ with $e_2 = (\sigma_1 \circ \ldots \circ \sigma_k)(e_1)$. Thus $f(e_1) = f(e_2)$ holds, such that $f(C) = \{0\}$ or $f(C) = \{1\}$ or $f(C) = \{dc\}$.

2. '$\Longleftarrow$': Let $\forall 0 \leq w_i \leq |\lambda_i| (1 \leq i \leq k)$ $\{0, 1\} \not\subseteq f(C^P_{w_1, \ldots, w_k})$.
   We have to prove that there is a completely specified extension f' of f, which is symmetric in P.
   Define f' as follows:
   If $f(C) = \{\epsilon\}$ ($\epsilon \in \{0, 1\}$) for a weight class C, then $f'(C) = f(C)$.
   If $f(C) = \{dc\}$ for a weight class C, then $f'(C) = 0$.
   If $f(C) = \{\epsilon, dc\}$ ($\epsilon \in \{0, 1\}$) for a weight class C, then $f'(C) = \epsilon$.
   Then f' is a completely specified function and because of part 1 of the theorem, f' is strongly symmetric in P and thus symmetric in P according to the symmetry definition for completely specified functions.

---

[1]$\sigma_{i,l} : \{0,1\}^n \rightarrow \{0,1\}^n$, $\sigma_{i,l}(\epsilon_1, \ldots, \epsilon_i, \ldots, \epsilon_l, \ldots, \epsilon_n) = (\epsilon_1, \ldots, \epsilon_l, \ldots, \epsilon_i, \ldots, \epsilon_n)$
$\forall (\epsilon_1, \ldots, \epsilon_n) \in \{0,1\}^n$

'$\Longrightarrow$': Let f be (weakly) symmetric in P. Thus there is a completely specified extension f' of f, which is symmetric in P. If there were a weight class C with $\{0,1\} \subseteq f(C)$, then $\{0,1\} \subseteq f'(C)$, since f' is an extension of f. Since f' is completely specified, we have, for all weight classes C of P, according to part 1 of the theorem: $f'(C) = \{0\}$ or $f'(C) = \{1\}$, which contradicts our assumption.

$\square$

Appendix D
Complexity of MSP

We prove Theorem 2.3 on p. 55: The problem **MSP** is NP–hard.

PROOF: The proof is based on a polynomial–time transformation from the NP–complete problem 'Partition into Cliques' (**PC**) (see (Garey and Johnson, 1979)) to MSP:

Let an instance of PC be given by a graph $G = (V, E)$ and a number $K \leq |V|$. We can determine in polynomial time ROBDDs $f_{G_{on}}$ and $f_{G_{dc}}$ of an incompletely specified function f_G with the property that there is a partition of G into K cliques iff there is a partition P of the variable set X of f with $|P| = K$, such that f is symmetric in P.

W.l.o.g. $V = \{x_1, \ldots, x_n\} = X$.
$f_G \in BP_n$ is defined by

$$
f_G(\epsilon_1, \ldots, \epsilon_n) = \begin{cases} 1 & \text{if} & \begin{aligned} &\epsilon_1 = \ldots = \epsilon_i = 1, \\ &\epsilon_{i+1} = \ldots = \epsilon_n = 0, \\ &\quad 1 \leq i \leq n-1 \end{aligned} \\[2ex] 0 & \text{if} & \begin{aligned} &\epsilon_1 = \ldots = \epsilon_{i-1} = 1, \\ &\epsilon_i = \ldots = \epsilon_{j-1} = 0, \\ &\epsilon_j = 1, \\ &\epsilon_{j+1} = \ldots = \epsilon_n = 0, \\ &\quad 1 \leq i \leq n-1, j > i \\ &\quad \text{and } \{x_i, x_j\} \notin E \end{aligned} \\[2ex] dc & \text{otherwise} \end{cases}
$$

The ON–set $ON(f_G)$ consists of all vectors

$$
\epsilon^{(i)} := (\underbrace{1, \ldots, 1}_{i \text{ times}}, 0, \ldots, 0)
$$

with $1 \leq i \leq n-1$. The OFF–set $OFF(f_G)$ contains exactly the vectors $\sigma_{ij}(\epsilon^{(i)})$ [1] $(1 \leq i \leq n-1)$ with $j > i$ and $\{x_i, x_j\} \notin E$.

---

[1] $\sigma_{ij} : \{0,1\}^n \to \{0,1\}^n, \sigma_{ij}(\epsilon_1, \ldots, \epsilon_i, \ldots, \epsilon_j, \ldots, \epsilon_n) = (\epsilon_1, \ldots, \epsilon_j, \ldots, \epsilon_i, \ldots, \epsilon_n)$

Let f_G be symmetric in a partition $P = \{\lambda_1, \ldots, \lambda_K\}$ of the input variables. Consider a pair of variables $x_i, x_j \in \lambda_l$ ($1 \le l \le K$) with $j > i$. Since $\epsilon^{(i)} \in ON(f_G)$, $\sigma_{ij}(\epsilon^{(i)})$ cannot be in $OFF(f_G)$, since f_G is symmetric in (x_i, x_j). Thus we can conclude $\{x_i, x_j\} \in E$ from the definition of f_G. Consequently, the variables (nodes) from λ_l form a clique in G.

Inversely, let $P = \{\lambda_1, \ldots, \lambda_K\}$ be a partition of the input variables, such that the nodes from λ_i form cliques in G. Suppose that f is *not* symmetric in P.

We prove a contradiction as follows: According to Lemma 2.6 there is a weight class $C^P_{w_1, \ldots, w_K}$ for P with $\{0, 1\} \subseteq f_G(C^P_{w_1, \ldots, w_K})$. According to the definition of f_G the only vertex of $C^P_{w_1, \ldots, w_K}$ with function value 1 is

$$\epsilon^{(w)} = (\underbrace{1, \ldots, 1}_{w \text{ times}}, 0, \ldots, 0) \quad (w = \sum_{i=1}^{k} w_i).$$

Since all vectors of the same weight class have the same 1–weight, there has to be $j > w$ with

$$\sigma_{ij}(\epsilon^{(w)}) = (\underbrace{1, \ldots, 1}_{w-1 \text{ times}}, 0, 0, \ldots, 0, \underbrace{1}_{\epsilon^{(0)}_j}, 0, \ldots, 0) \in C^P_{w_1, \ldots, w_K}$$

and $f(\sigma_{ij}(\epsilon^{(w)})) = 0$, i.e., $(x_w, x_j) \notin E$.

Let $\lambda' \in P$ with $x_w \in \lambda'$. If $x_j \notin \lambda'$, then $w^1_{\lambda'}(\sigma_{ij}(\epsilon^{(w)})) = w^1_{\lambda'}(\epsilon^{(w)}) - 1$. This contradicts the fact that $\epsilon^{(w)}$ and $\sigma_{ij}(\epsilon^{(w)})$ are in the same weight class. If $x_j \in \lambda'$, then we obtain a contradiction to the fact that $(x_w, x_j) \notin E$.

Since $ON(f_G)$ and $OFF(f_G)$ are of polynomial size, the ROBDDs for $f_{G_{on}}$ and $f_{G_{dc}}$ can be computed in polynomial time. $\qquad \square$

Appendix E
Making a function strongly symmetric

We prove Theorem 2.4 found on p. 59, which is needed to make functions strongly symmetric in a partition P.

PROOF: Let $P = \{\lambda_1, \lambda_2, \lambda_3, \ldots, \lambda_l\}$ and w.l.o.g. $\lambda_1 = \{x_i\}$ and $\lambda_2 = \{x_{j_1}, \ldots, x_{j_k}\}$.
f is strongly symmetric in P and we have to show that $f^{(k)}$ is strongly symmetric in $P' = \{\lambda_1 \cup \lambda_2, \lambda_3, \ldots, \lambda_l\}$.
Because of Lemma 2.6 we have to show that for all weight classes $C^{P'}_{w_{1,2}, w_3, \ldots, w_l}$ of P' the equation

$$f^{(k)}(C^{P'}_{w_{1,2}, w_3, \ldots, w_l}) = \begin{cases} \{0\} & \text{or} \\ \{1\} & \text{or} \\ \{dc\} \end{cases}$$

holds.

Case 1: $w_{1,2} = 0$ or $w_{1,2} = k + 1$
  Then the following holds:

$$C^{P'}_{w_{1,2}, w_3, \ldots, w_l} = C^{P}_{0,0, w_3, \ldots, w_l} \quad \text{or} \quad C^{P'}_{w_{1,2}, w_3, \ldots, w_l} = C^{P}_{1,k, w_3, \ldots, w_l}$$

and thus $|f(C^{P'}_{w_{1,2}, w_3, \ldots, w_l})| = 1$ because of the strong symmetry of f in P.
If $f(C^{P'}_{w_{1,2}, w_3, \ldots, w_l}) = \{c\}$, $c \in \{0, 1\}$, then $f^{(p)}(C^{P'}_{w_{1,2}, w_3, \ldots, w_{n+1-k}}) = \{c\}$ for all $1 \leq p \leq k$, since $f^{(p)}$ is an extension of f.
If $f(C^{P'}_{w_{1,2}, w_3, \ldots, w_l}) = \{dc\}$, then $f^{(p)}(C^{P'}_{w_{1,2}, w_3, \ldots, w_l}) = \{dc\}$ for all $1 \leq p \leq k$, since $make\_strongly\_symm(f^{(p-1)}, x_i, x_{j_p})$ provides a minimal extension, which is strongly symmetric in (x_i, x_{j_p}), and $w^1_{\lambda_1}(\epsilon) = w^1_{\lambda_2}(\epsilon) = 0$ or $w^0_{\lambda_1}(\epsilon) = w^0_{\lambda_2}(\epsilon) = 0$ for all $\epsilon \in C^{P'}_{w_{1,2}, w_3, \ldots, w_l}$.

Case 2: $1 \leq w_{1,2} \leq k$
  In this case we have the following disjoint union:

$$C^{P'}_{w_{1,2}, w_3, \ldots, w_l} = C^{P}_{0, w_{1,2}, w_3, \ldots, w_l} \cup C^{P}_{1, w_{1,2}-1, w_3, \ldots, w_l}.$$

It follows from our precondition

$$f(C^P_{0,w_{1,2},w_3,\ldots,w_l}) = \left\{ \begin{array}{l} \{0\} \text{ or} \\ \{1\} \text{ or} \\ \{dc\} \end{array} \right.$$

and

$$f(C^P_{1,w_{1,2}-1,w_3,\ldots,w_l}) = \left\{ \begin{array}{l} \{0\} \text{ or} \\ \{1\} \text{ or} \\ \{dc\} \end{array} \right.$$

Case 2.1: $f(C^P_{0,w_{1,2},w_3,\ldots,w_l}) = f(C^P_{1,w_{1,2}-1,w_3,\ldots,w_l})$

Since the calls of $make\_strongly\_symm(f^{(p-1)}, x_i, x_{j_p})$ give minimal extensions, which are strongly symmetric in (x_i, x_{j_p}), the assignment for $C^P_{0,w_{1,2},w_3,\ldots,w_l}$ and $C^P_{1,w_{1,2}-1,w_3,\ldots,w_l}$ is not changed.

$f^{(p)}(C^P_{0,w_{1,2},w_3,\ldots,w_l}) = f^{(p)}(C^P_{1,w_{1,2}-1,w_3,\ldots,w_l})$ holds and thus

$$f^{(p)}(C^{P'}_{w_{1,2},w_3,\ldots,w_l}) = \left\{ \begin{array}{ll} \{0\} & \text{or} \\ \{1\} & \text{or} \\ \{dc\} \end{array} \right.$$

Case 2.2: $f(C^P_{0,w_{1,2},w_3,\ldots,w_l}) \neq f(C^P_{1,w_{1,2}-1,w_3,\ldots,w_l})$

Since f is symmetric in (x_i, x_{j_1}), there are $c \in \{0,1\}$ and $u \in \{0,1\}$, such that $f(C^P_{u,w_{1,2}-u,w_3,\ldots,w_l}) = \{dc\}$ and $f(C^P_{\overline{u},w_{1,2}-\overline{u},w_3,\ldots,w_l}) = \{c\}$.
In the following we assume $u = 0$ (case $u = 1$ is analogous).

From the definition of $make\_strongly\_symm$ it follows that for all $1 \leq p \leq k$
$f^{(p)}(\epsilon) \in \{c, dc\} \, \forall \epsilon \in C^P_{0,w_{1,2},w_3,\ldots,w_l} \cup C^P_{1,w_{1,2}-1,w_3,\ldots,w_l}$.
A call of $make\_strongly\_symm(f^{(p)}, x_i, x_{j_{p+1}})$ assigns function values to vectors $\epsilon \in C^P_{0,w_{1,2},w_3,\ldots,w_l}$ with $\epsilon_i = 0$ and $\epsilon_{j_{p+1}} = 1$, namely $f^{(p+1)}(\epsilon)$ is set to the value $f^{(p)}(\sigma_{ij_{p+1}}(\epsilon)) = c$ $(\sigma_{ij_{p+1}}(\epsilon) \in C^P_{1,w_{1,2}-1,w_3,\ldots,w_l})$.[1]
It remains to be shown that $f^{(k)}(\epsilon) = c \, \forall \epsilon \in C^P_{0,w_{1,2},w_3,\ldots,w_l}$, i.e., that the sequence of k calls is enough to assign function value c to *all* elements of $C^P_{0,w_{1,2},w_3,\ldots,w_l}$.
The following statement is proven by induction:
$f^{(p)}(\epsilon) = c \, \forall \epsilon \in C^P_{0,w_{1,2},w_3,\ldots,w_l}$ with $\epsilon_{j_1} = 1$ or $\epsilon_{j_2} = 1$ or $\ldots$ or $\epsilon_{j_p} = 1$.

$p = 0$: Trivial.

$p \to p + 1$:
Because of the inductive assumption and since $f^{(p+1)}$ is an extension of $f^{(p)}$, we have:
$f^{(p+1)}(\epsilon) = c \, \forall \epsilon \in C^P_{0,w_{1,2},w_3,\ldots,w_l}$ with $\epsilon_{j_1} = 1$ or $\epsilon_{j_2} = 1$ or $\ldots$ or $\epsilon_{j_p} = 1$.

We have to show that $f^{(p+1)}(\epsilon) = c \, \forall \epsilon \in C^P_{0,w_{1,2},w_3,\ldots,w_l}$ with $\epsilon_{j_{p+1}} = 1$.

Let $\delta \in C^P_{1,w_{1,2}-1,w_3,\ldots,w_l}$ with

$$\delta_i = \overline{\epsilon_i} = 1^{\dagger}, \delta_{j_{p+1}} = \overline{\epsilon_{j_{p+1}}} = 0$$

---

[1] $\sigma_{ij} : \{0,1\}^n \to \{0,1\}^n, \sigma_{ij}(\epsilon_1,\ldots,\epsilon_i,\ldots,\epsilon_j,\ldots,\epsilon_n) = (\epsilon_1,\ldots,\epsilon_j,\ldots,\epsilon_i,\ldots,\epsilon_n)$
[†] For all elements ϵ of the weight class $C^P_{0,w_{1,2},w_3,\ldots,w_l}$ we have $\epsilon_i = 0$.

$$\text{and } \delta_l = \epsilon_l \text{ for } l \neq i, j_{p+1},$$

thus $\delta = \sigma_{i,j_{p+1}}(\epsilon)$.

We have

$$f^{(p)}(\delta) = f(\delta) = c$$

and thus

$$f^{(p+1)}(\epsilon) = f^{(p)}(\sigma_{i,j_{p+1}}(\epsilon)) = f^{(p)}(\delta) = c.$$

It follows from the statement shown by induction:

$$f^{(k)}(\epsilon) = c \; \forall \epsilon \in C^P_{0,w_{1,2},w_3,\ldots,w_l} \text{ with } \epsilon_{j_1} = 1 \text{ or } \ldots \text{ or } \epsilon_{j_k} = 1$$

or equivalently

$$f^{(k)}(\epsilon) = c \; \forall \epsilon \in C^P_{0,w_{1,2},w_3,\ldots,w_l} \text{ with } w^1_{\lambda_2}(\epsilon) \geq 1.$$

But $w^1_{\lambda_2}(\epsilon) = w_{1,2} \geq 1$ holds for *all* $\epsilon \in C^P_{0,w_{1,2},w_3,\ldots,w_l}$ (assumption in Case 2).

$\square$

Appendix F
Compatibility of don't care minimization methods

We present the proof of Lemma 2.9 found on p. 62, which is needed to prove that two don't care assignment methods are compatible: the don't care assignment to obtain strong symmetries and the don't care assignment to minimize the number of linking nodes.

PROOF: W.l.o.g. let $\cup_{j=1}^{i}\lambda_j = \{x_1, \ldots, x_p\}$ and $\cup_{j=i+1}^{k}\lambda_j = \{x_{p+1}, \ldots, x_n\}$.

Suppose we apply the minimization of linking nodes as defined in Section 2.2.1.3 to a cut line between two symmetric groups λ_i and λ_{i+1}. The minimization of linking nodes works as follows:

- The ROBDD nodes below the cut line correspond to all different cofactors of f with respect to the first p variables. Let the set of these cofactors be $COF = \{cof_1, \ldots, cof_l\}$. Note that these cofactors are incompletely specified functions.

- Two cofactors cof_i and cof_j are compatible iff there is no $(\epsilon_{p+1}, \ldots, \epsilon_n) \in \{0,1\}^{n-p}$ such that $cof_i(\epsilon_{p+1}, \ldots, \epsilon_n) = c, cof_j(\epsilon_{p+1}, \ldots, \epsilon_n) = \bar{c}$ for $c \in \{0,1\}$. A partition $P_{COF} = \{COF_1, \ldots, COF_m\}$ of COF is computed such that all pairs $cof_i, cof_j \in COF_q$ $(1 \leq q \leq m)$ are compatible.

- For a set COF_q of compatible cofactors an extension $extension_q$ is computed as follows: $extension_q(\epsilon_{p+1}, \ldots, \epsilon_n) = c$ $(c \in \{0,1\})$ iff $\exists cof_j \in COF_q$ with $cof_j(\epsilon_{p+1}, \ldots, \epsilon_n) = c$ and $extension_q(\epsilon_{p+1}, \ldots, \epsilon_n) = dc$ iff $\forall cof_j \in COF_q$ $cof_j(\epsilon_{p+1}, \ldots, \epsilon_n) = dc$.

- The cofactors $cof_j \in COF_q$ are all replaced by their (common) extension $extension_q$. This leads to an extension f' of f. The number of ROBDD nodes in the representation for f' which are located immediately below the cut line between λ_i and λ_{i+1} equals the size m of the partition P_{COF}.

We have to prove that f' is strongly symmetric in all sets $\lambda_j \in P = \{\lambda_1, \ldots, \lambda_k\}$.

Case 1: $j \leq i$

Let x_{j_1} and $x_{j_2} \in \lambda_j$.
Choose $\epsilon^{(1)}, \epsilon^{(2)} \in \{0,1\}^n$ arbitrarily with

$$\epsilon^{(1)} = (\epsilon_1, \ldots, \epsilon_{j_1}, \ldots, \epsilon_{j_2}, \ldots, \epsilon_n) \text{ and}$$

$$\epsilon^{(2)} = (\epsilon_1, \ldots, \epsilon_{j_2}, \ldots, \epsilon_{j_1}, \ldots, \epsilon_n).$$

Since f is strongly symmetric in x_{j_1}, x_{j_2}, the cofactors $f_{x_1^{\epsilon_1} \ldots x_{j_1}^{\epsilon_{j_1}} \ldots x_{j_2}^{\epsilon_{j_2}} \ldots x_p^{\epsilon_p}}$ and $f_{x_1^{\epsilon_1} \ldots x_{j_2}^{\epsilon_{j_2}} \ldots x_{j_2}^{\epsilon_{j_1}} \ldots x_p^{\epsilon_p}}$ are equal. During the minimization of linking nodes this cofactor is replaced by some extension $extension_q$, and of course the corresponding cofactors $f'_{x_1^{\epsilon_1} \ldots x_{j_1}^{\epsilon_{j_1}} \ldots x_{j_2}^{\epsilon_{j_2}} \ldots x_p^{\epsilon_p}}$ and $f'_{x_1^{\epsilon_1} \ldots x_{j_2}^{\epsilon_{j_2}} \ldots x_{j_2}^{\epsilon_{j_1}} \ldots x_p^{\epsilon_p}}$ of the result f' of this replacement are still equal. Thus, $f'(\epsilon^{(1)}) = f'(\epsilon^{(2)})$ and f' is strongly symmetric in x_{j_1} and x_{j_2}.

Case 2: $j \geq i+1$

Let x_{j_1} and $x_{j_2} \in \lambda_j$.
Choose $\epsilon^{(1)}, \epsilon^{(2)} \in \{0, 1\}^n$ arbitrarily with

$$\epsilon^{(1)} = (\epsilon_1, \ldots, \epsilon_p, \ldots, \epsilon_{j_1}, \ldots, \epsilon_{j_2}, \ldots, \epsilon_n) \text{ and}$$

$$\epsilon^{(2)} = (\epsilon_1, \ldots, \epsilon_p, \ldots, \epsilon_{j_2}, \ldots, \epsilon_{j_1}, \ldots, \epsilon_n).$$

Suppose the cofactor $f_{x_1^{\epsilon_1} \ldots x_p^{\epsilon_p}}$ is in the set COF_q. Since f is strongly symmetric in x_{j_1} and x_{j_2}, *all cofactors* $\in COF_q$ are strongly symmetric in x_{j_1} and x_{j_2}.
If for all cofactors

$$cof_j \in COF_q \quad cof_j(\epsilon_{p+1}, \ldots, \epsilon_{j_1}, \ldots, \epsilon_{j_2}, \ldots, \epsilon_n) = dc$$

then also for all cofactors

$$cof_j \in COF_q \quad cof_j(\epsilon_{p+1}, \ldots, \epsilon_{j_2}, \ldots, \epsilon_{j_1}, \ldots, \epsilon_n) = dc$$

because of strong symmetry and according to the definition of $extension_q$ given above

$$extension_q(\epsilon_{p+1}, \ldots, \epsilon_{j_1}, \ldots, \epsilon_{j_2}, \ldots, \epsilon_n) =$$
$$extension_q(\epsilon_{p+1}, \ldots, \epsilon_{j_2}, \ldots, \epsilon_{j_1}, \ldots, \epsilon_n) = dc.$$

If $cof_j \in COF_q$ exists with

$$cof_j(\epsilon_{p+1}, \ldots, \epsilon_{j_1}, \ldots, \epsilon_{j_2}, \ldots, \epsilon_n) = c \ (c \in \{0, 1\})$$

then

$$cof_j(\epsilon_{p+1}, \ldots, \epsilon_{j_2}, \ldots, \epsilon_{j_1}, \ldots, \epsilon_n) = c$$

because of strong symmetry and according to the definition of $extension_q$

$$extension_q(\epsilon_{p+1}, \ldots, \epsilon_{j_1}, \ldots, \epsilon_{j_2}, \ldots, \epsilon_n) =$$
$$extension_q(\epsilon_{p+1}, \ldots, \epsilon_{j_2}, \ldots, \epsilon_{j_1}, \ldots, \epsilon_n) = c.$$

$f_{x_1^{\epsilon_1} \ldots x_p^{\epsilon_p}}$ is replaced by $extension_q$ and for the result f' we have

$$f'(\epsilon_1, \ldots, \epsilon_p, \ldots, \epsilon_{j_1}, \ldots, \epsilon_{j_2}, \ldots \epsilon_n) =$$
$$f'(\epsilon_1, \ldots, \epsilon_p, \ldots, \epsilon_{j_2}, \ldots, \epsilon_{j_1}, \ldots \epsilon_n).$$

$\square$

Appendix G
Symmetries and decomposability

We prove Theorem 3.3 found on p. 110, which looks into the correlation of symmetries in the bound set of a decomposed function and the number of decomposition functions needed in the decomposition.

PROOF: Let $P = (p_1, \ldots, p_p)$ be a polarity vector, which is constructed as in the proof of Theorem 2.1 (page 36). We assume $\mu_k = \{x_{i_1}, \ldots, x_{i_{l_k}}\}$.

- $p_{i_1}, \ldots, p_{i_{l_k}}$ are chosen arbitrarily for $1 \leq k \leq t$, i.e., if μ_k is a set of multiply symmetric pairs of variables,

- and otherwise $p_{i_1}, \ldots, p_{i_{l_k}}$ are determined according to the following procedure: Choose $p_{i_1} = \epsilon \in \{0, 1\}$ arbitrarily. If f is (nonequivalence) symmetric in (x_{i_1}, x_{i_2}), then choose $p_{i_2} = \epsilon$, if f is equivalence symmetric in (x_{i_1}, x_{i_2}), then choose $p_{i_2} = \bar{\epsilon}$. Continue the procedure with (x_{i_2}, x_{i_3}), etc., until $p_{i_{l_k}}$ is determined.

Now f_P with $f_P(x_1, \ldots, x_n) = f(x_1^{p_1}, \ldots, x_p^{p_p}, x_{p+1}, \ldots, x_n) \; \forall (x_1, \ldots, x_n) \in \{0, 1\}^n$ is (nonequivalence) symmetric in $P_{\sim_{esym}} \cup \{\{x_j\} \mid x_j \in X \setminus X^{(1)}\}$.

For a set $\mu_k = \{x_{i_1}, \ldots, x_{i_{l_k}}\}$ let the P–weight $w_{\mu_k}^P$ with respect to μ_k of $(\epsilon_1, \ldots, \epsilon_p) \in \{0, 1\}^p$ be defined as

$$w_{\mu_k}^P(\epsilon_1, \ldots, \epsilon_p) = |\{\epsilon_{i_j} \mid \epsilon_{i_j} = p_{i_j}, j \in \{1, \ldots, l_k\}\}|.$$

(Note that for $P = (1, \ldots, 1)$ (i.e., when we only have nonequivalence symmetry) $w_{\mu_k}^P$ is exactly the 1–weight $w_{\mu_k}^1$ of the μ_k–part.)

To compute $nrp(X^{(1)}, f)$ we have to determine the number of different cofactors $f_{x_1^{\epsilon_1} \ldots x_p^{\epsilon_p}}$ for all $(\epsilon_1, \ldots, \epsilon_p) \in \{0, 1\}^p$.

Claim 1: For $k \in \{t + 1, \ldots, l\}$ and arbitrary $\epsilon, \delta \in \{0, 1\}^p$ with

- $\epsilon_j = \delta_j$, if $x_j \notin \mu_k$ and
- $w_{\mu_k}^P(\epsilon_1, \ldots, \epsilon_p) = w_{\mu_k}^P(\delta_1, \ldots, \delta_p)$

we have:

$$f_{x_1^{\epsilon_1} \ldots x_p^{\epsilon_p}} = f_{x_1^{\delta_1} \ldots x_p^{\delta_p}}.$$

Remember $\mu_k = \{x_{i_1}, \ldots, x_{i_{l_k}}\}$.

The basic idea of the proof is simple: Assume for the moment that $p_{i_1} = 1, \ldots, p_{i_{l_k}} = 1$, i.e., all pairs of variables in μ_k are nonequivalence symmetric. The conditions of Claim 1 mean that δ_j and ϵ_j are equal for all positions j not corresponding to a variable in μ_k and that the 1–weights $w^1_{\mu_k}(\delta)$ and $w^1_{\mu_k}(\epsilon)$ of the μ_k–parts are equal. Now it is clear that we can transform δ into ϵ by a series of pairwise exchanges of bit positions in $\{i_1, \ldots, i_{l_k}\}$. Because of the nonequivalence symmetry for all pairs $x_i, x_j \in \mu_k = \{x_{i_1}, \ldots, x_{i_{l_k}}\}$, the cofactor with respect to the current bit vector does not change during this transformation and finally $f_{x_1^{\epsilon_1} \ldots x_p^{\epsilon_p}} = f_{x_1^{\delta_1} \ldots x_p^{\delta_p}}$.

When we also have to consider equivalence symmetry, pairwise variable exchanges must be replaced by pairwise variable negations and a few additional technical details are introduced. The detailed proof is as follows:

We construct a series $\delta^{(1)}, \ldots, \delta^{(l_k)}$ of vectors in $\{0, 1\}^p$ with

$$f_{x_1^{\delta_1^{(1)}} \ldots x_p^{\delta_p^{(1)}}} = \cdots = f_{x_1^{\delta_1^{(l_k)}} \ldots x_p^{\delta_p^{(l_k)}}}, \text{ where}$$

- $\delta^{(1)} = \delta$,
- $\delta_j = \delta_j^{(1)} = \cdots = \delta_j^{(l_k)} = \epsilon_j$ for $j \notin \{i_1, \ldots, i_{l_k}\}$,
- $(\delta_{i_1}^{(j+1)}, \ldots, \delta_{i_j}^{(j+1)}) = (\epsilon_{i_1}, \ldots, \epsilon_{i_j})$ for $1 \leq j \leq l_k - 1$,
- $w^P_{\mu_k}(\delta^{(j+1)}) = w^P_{\mu_k}(\delta^{(j)})$ for $1 \leq j \leq l_k - 1$.

Because of $(\delta_{i_1}^{(l_k)}, \ldots, \delta_{i_{l_k}}^{(l_k)}) = (\epsilon_{i_1}, \ldots, \epsilon_{i_{l_k}-1}, \delta_{i_{l_k}}^{(l_k)})$ and $w^P_{\mu_k}(\delta^{(l_k)}) = w^P_{\mu_k}(\delta) = w^P_{\mu_k}(\epsilon)$ we also have $\delta_{i_{l_k}}^{(l_k)} = \epsilon_{i_{l_k}}$ and thus $\delta^{(l_k)} = \epsilon$.

During the construction we make use of nonequivalence and equivalence symmetries in μ_k:

If $\delta_{i_j}^{(j)} = \epsilon_{i_j}$, then choose $\delta^{(j+1)} = \delta^{(j)}$.

If $\delta_{i_j}^{(j)} = \overline{\epsilon_{i_j}}$, then we have to differentiate between two cases:

Case 1: $\epsilon_{i_j} = p_{i_j}$
    Then $\delta_{i_j}^{(j)} \neq p_{i_j}$.

Now we determine m with $j < m \leq l_k$, such that $\delta_{i_m}^{(j)} = p_{i_m}$.

Such an m always exists, since otherwise we would have $w^P_{\mu_k}(\delta_1^{(j)}, \ldots, \delta_p^{(j)}) < w^P_{\mu_k}(\epsilon_1, \ldots, \epsilon_p) = w^P_{\mu_k}(\delta_1, \ldots, \delta_p)$ because of $\delta_{i_j}^{(j)} \neq p_{i_j}$ and $\epsilon_{i_j} = p_{i_j}$.

Define
$$\delta_s^{(j+1)} = \begin{cases} \overline{\delta_s^{(j)}}, & \text{if } s = i_j \text{ or } s = i_m \\ \delta_s^{(j)}, & \text{otherwise.} \end{cases}$$

With this definition we have $(\delta_{i_1}^{(j+1)}, \ldots, \delta_{i_j}^{(j+1)}) = (\epsilon_{i_1}, \ldots, \epsilon_{i_j})$.

- Assuming $p_{i_m} = p_{i_j}$, i.e., f is nonequivalence symmetric in the variables x_{i_j} and x_{i_m}, we can conclude $f_{x_1^{\delta_1^{(j)}} \ldots x_p^{\delta_p^{(j)}}} = f_{x_1^{\delta_1^{(j+1)}} \ldots x_p^{\delta_p^{(j+1)}}}$ because of $\delta_{i_m}^{(j)} \neq \delta_{i_j}^{(j)}$.
- Assuming $p_{i_m} \neq p_{i_j}$, i.e., f is equivalence symmetric in the variables x_{i_j} and x_{i_m}, we can conclude $f_{x_1^{\delta_1^{(j)}} \ldots x_p^{\delta_p^{(j)}}} = f_{x_1^{\delta_1^{(j+1)}} \ldots x_p^{\delta_p^{(j+1)}}}$ because of $\delta_{i_m}^{(j)} = \delta_{i_j}^{(j)}$.

Since $\delta_{i_j}^{(j)} \neq p_{i_j}$, $\delta_{i_m}^{(j)} = p_{i_m}$ and thus $\delta_{i_j}^{(j+1)} = p_{i_j}$, $\delta_{i_m}^{(j+1)} \neq p_{i_m}$, we have

$$w_{\mu_k}^P(\delta_1^{(j)}, \ldots, \delta_p^{(j)}) = w_{\mu_k}^P(\delta_1^{(j+1)}, \ldots, \delta_p^{(j+1)}).$$

__Case 2:__ $\epsilon_{i_j} \neq p_{i_j}$
  Then $\delta_{i_j}^{(j)} = p_{i_j}$.

Now we determine m with $j < m \leq l_k$, such that $\delta_{i_m}^{(j)} \neq p_{i_m}$.

Such an m always exists, since otherwise we would have $w_{\mu_k}^P(\delta_1^{(j)}, \ldots, \delta_p^{(j)}) > w_{\mu_k}^P(\epsilon_1, \ldots, \epsilon_p) = w_{\mu_k}^P(\delta_1, \ldots, \delta_p)$ because of $\delta_{i_j}^{(j)} = p_{i_j}$ and $\epsilon_{i_j} \neq p_{i_j}$.

$\delta_1^{(j+1)}, \ldots, \delta_p^{(j+1)}$ are defined as in Case 1 and the remaining argumentation also follows by analogy to Case 1.

After $l_k - 1$ steps we consequently have

$$f_{x_1^{\delta_1} \ldots x_p^{\delta_p}} = f_{x_1^{\delta_1^{(1)}} \ldots x_p^{\delta_p^{(1)}}} = \ldots f_{x_1^{\delta_1^{(l_k)}} \ldots x_p^{\delta_p^{(l_k)}}} = f_{x_1^{\epsilon_1} \ldots x_p^{\epsilon_p}}$$

and we have proven Claim 1.

__Claim 2:__ For $k \in \{1, \ldots, t\}$ and arbitrary $\epsilon, \delta \in \{0, 1\}^p$ with

* $\epsilon_j = \delta_j$, if $x_j \notin \mu_k$ and
* $w_{\mu_k}^1(\epsilon_1, \ldots, \epsilon_p) \bmod 2 = w_{\mu_k}^1(\delta_1, \ldots, \delta_p) \bmod 2$

we have:

$$f_{x_1^{\epsilon_1} \ldots x_p^{\epsilon_p}} = f_{x_1^{\delta_1} \ldots x_p^{\delta_p}}.$$

Let again $\mu_k = \{x_{i_1}, \ldots, x_{i_{l_k}}\}$.

The construction for variable set μ_k has $|\mu_k| - 1$ steps similarly to the proof of Claim 1:

If $\delta_{i_j}^{(j)} = \epsilon_{i_j}$, then there is nothing to do in step j.

Otherwise we make use of the nonequivalence symmetry in variables x_{i_j} and $x_{i_{j+1}}$ in the case $\delta_{i_j}^{(j)} \neq \delta_{i_{j+1}}^{(j)}$, and in the case $\delta_{i_j}^{(j)} = \delta_{i_{j+1}}^{(j)}$ we make use of the equivalence symmetry in x_{i_j} and $x_{i_{j+1}}$.

Defining

$$\delta_s^{(j+1)} = \begin{cases} \overline{\delta_s^{(j)}}, & \text{if } s = i_j \text{ or } s = i_{j+1} \\ \delta_s^{(j)}, & \text{otherwise.} \end{cases}$$

we have

$$f_{x_1^{\delta_1^{(j)}} \ldots x_p^{\delta_p^{(j)}}} = f_{x_1^{\delta_1^{(j+1)}} \ldots x_p^{\delta_p^{(j+1)}}}.$$

Moreover we have

$$(\delta_{i_1}^{(j+1)}, \ldots, \delta_{i_j}^{(j+1)}) = (\epsilon_{i_1}, \ldots, \epsilon_{i_j})$$

and

$$w_{\mu_k}^1(\delta_1^{(j)}, \ldots, \delta_p^{(j)}) \bmod 2 = w_{\mu_k}^1(\delta_1^{(j+1)}, \ldots, \delta_p^{(j+1)}) \bmod 2.$$

After $l_k - 1$ steps $(\delta_1^{(l_k)}, \ldots, \delta_p^{(l_k)})$ results with

$$(\delta_{i_1}^{(l_k)}, \ldots, \delta_{i_{l_k}}^{(l_k)}) = (\epsilon_{i_1}, \ldots, \epsilon_{i_{l_k-1}}, \delta_{i_{l_k}}^{(l_k)})$$

and because of

$$w^1_{\mu_k}(\delta_1, \ldots, \delta_p) \bmod 2 = w^1_{\mu_k}(\delta_1^{(l_k)}, \ldots, \delta_p^{(l_k)}) \bmod 2 = w^1_{\mu_k}(\epsilon_1, \ldots, \epsilon_p) \bmod 2$$

we also have $\delta^{(l_k)}_{i_{l_k}} = \epsilon_{i_{l_k}}$.

Altogether we have

$$f_{x_1^{\delta_1} \ldots x_p^{\delta_p}} = \ldots = f_{x_1^{\epsilon_1} \ldots x_p^{\epsilon_p}}.$$

Using Claim 1 and 2 we can draw the following conclusion:

Since there are at most $|\mu_k| + 1$ different values for $w^P_{\mu_k}$ and $t + 1 \leq k \leq l$ and since there are at most 2 different values for $w^1_{\mu_k} \bmod 2$ for $1 \leq i \leq t$, the total number of different cofactors $f_{x_1^{\epsilon_1} \ldots x_p^{\epsilon_p}}$ for $(\epsilon_1, \ldots, \epsilon_p) \in \{0, 1\}^p$ can be at most

$$2^t \cdot (|\mu_{t+1}| + 1) \cdot \ldots \cdot (|\mu_l| + 1)$$

and thus Theorem 3.3 is proven. $\qquad\qquad\qquad\qquad\qquad\qquad\qquad\qquad\qquad\square$

Appendix H
Complexity of CDF

We prove Theorem 4.1 found on p. 139: The problem **CDF** is NP–hard.

The Theorem is proven by a polynomial time transformation from the NP–complete problem 3–PARTITION to CDF.

The problem 3–PARTITION is defined as follows (see (Garey and Johnson, 1979) or (Mehlhorn, 1984)):

Problem 3P (3–PARTITION)

Given: Weights $a_1, \ldots, a_{3n} \in \mathbf{Z}^+$, a limit $B \in \mathbf{Z}^+$, such that $B/4 < a_i < B/2 \ \forall i$ and $\sum_{i=1}^{3n} a_i = n \cdot B$.

Find: Is there a partition $S_1, \ldots, S_n$ of $\{1, \ldots, 3n\}$, such that $\sum_{j \in S_i} a_j = B \ \forall i$?

THEOREM H.1   *The problem* **3–PARTITION** *is strongly NP–complete, i.e., it is even NP–complete, when the numbers of its input are given in a unary encoding.*

PROOF: Proof by a transformation from 3DM, see (Garey and Johnson, 1979), p. 96.   □

The polynomial time transformation to prove that CDF is NP–hard proceeds in two steps. First we prove that the problem 2^l–PARTITION is NP–complete. 2^l–PARTITION is defined as follows:

Problem 2^l–PARTITION

Given: Integers $c_1, \ldots, c_k \in \mathbf{Z}^+$, a limit $B' = 2^b$ with $b \in \mathbf{N}$ and a number $M = 2^l$ with $l \in \mathbf{N}$, such that $\sum_{i=1}^{k} c_i = M \cdot B' = 2^{b+l}$ $(k > M)$.

Find: Is there a partition $T_1, \ldots, T_M$ of $\{1, \ldots, k\}$, such that $\sum_{j \in T_i} c_j = B'$ for $1 \le i \le M$?

THEOREM H.2   *The problem 2^l–PARTITION is strongly NP–complete, i.e., it is even NP–complete when the numbers of its input are given in a unary encoding.*

PROOF: We only give the proof that 2^l–PARTITION is NP–hard. The proof uses a polynomial time transformation from 3–PARTITION to 2^l–PARTITION.

Let an instance of 3–PARTITION be given by $a_1, \ldots, a_{3n}, B$. We compute the following instance of 2^l–PARTITION in polynomial time:

$$
\begin{aligned}
B' &= 2^{\lceil \log B \rceil + 2} \quad (\text{i.e., } b = \lceil \log B \rceil + 2) \\
k &= 3n + 2^{\lceil \log n \rceil} \\
c_i &= a_i \quad \forall 1 \le i \le 3n \\
c_i &= B' - B = 2^{\lceil \log B \rceil + 2} - B \quad \forall 3n < i \le 4n \\
c_i &= B' \quad \forall 4n < i \le 3n + 2^{\lceil \log n \rceil} \\
M &= 2^{\lceil \log n \rceil} \quad (\text{i.e., } l = \lceil \log n \rceil)
\end{aligned}
$$

This implies

$$
\begin{aligned}
\sum_{i=1}^{k} c_i &= \sum_{i=1}^{3n} a_i + \sum_{i=3n+1}^{4n} (B' - B) + \sum_{i=4n+1}^{3n+2^{\lceil \log n \rceil}} B' \\
&= n \cdot B + n \cdot (B' - B) + (2^{\lceil \log n \rceil} - n) \cdot B' \\
&= 2^{\lceil \log n \rceil} \cdot B' \\
&= M \cdot B'
\end{aligned}
$$

Moreover we have $k > M$.

We have to show the following:

$$
\begin{aligned}
&\exists \text{ partition } S_1, \ldots, S_n \text{ of } \{1, \ldots, 3n\}, \text{ such that } \sum_{j \in S_i} a_j = B \; \forall 1 \le i \le n \\
&\qquad\qquad\qquad\qquad\qquad\qquad \Longleftrightarrow \\
&\exists \text{ partition } T_1, \ldots, T_M \text{ of } \{1, \ldots, k\}, \text{ such that } \sum_{j \in T_i} c_j = B' \; \forall 1 \le i \le M
\end{aligned}
$$

"$\Longrightarrow$":

Assume that $\exists$ partition $S_1, \ldots, S_n$ of $\{1, \ldots, 3n\}$, such that $\sum_{j \in S_i} a_j = B \; \forall 1 \le i \le n$. Construct $T_1, \ldots, T_M$ as follows:

$$
\begin{aligned}
\text{For } 1 \le i \le n: &\quad T_i = S_i \cup \{3n + i\} \\
\text{For } n < i \le 2^{\lceil \log n \rceil}: &\quad T_i = \{3n + i\}
\end{aligned}
$$

Since $S_1, \ldots, S_n$ is a partition of $\{1, \ldots, 3n\}$, we can conclude that $T_1, \ldots, T_M$ is a partition of $\{1, \ldots, k\}$. It remains to show that $\sum_{j \in T_i} c_j = B'$ for $1 \le i \le M$.

Case 1: $1 \le i \le n$:

$$
\sum_{j \in T_i} c_j = \sum_{j \in S_i} c_j + c_{3n+i} = \sum_{j \in S_i} a_j + (B' - B) = B + B' - B = B'
$$

Case 2: $n < i \le 2^{\lceil \log n \rceil}$:

$$
\sum_{j \in T_i} c_j = c_{3n+i} = B'
$$

"$\Longleftarrow$":

Assume that $\exists$ partition $T_1, \ldots, T_M$ of $\{1, \ldots, k\}$, such that $\sum_{j \in T_i} c_j = B' \; \forall 1 \leq i \leq M$
Then the following holds:

- There are $2^{\lceil \log n \rceil} - n$ sets T_i with $T_i = \{3n + j\}$, where $j \in \{n+1, \ldots, 2^{\lceil \log n \rceil}\}$, since $c_{3n+j} = B'$ and $c_i > 0 \; \forall 1 \leq i \leq k$. Let w.l.o.g. $T_i = \{3n + i\}$ for $n + 1 \leq i \leq 2^{\lceil \log n \rceil}$.

- For the remaining T_i with $1 \leq i \leq n$ there is no T_i with $\{i_1, i_2\} \subseteq T_i$ and $3n < i_1, i_2 \leq 4n$, because otherwise

$$
\begin{aligned}
c_{i_1} + c_{i_2} &= (B' - B) + (B' - B) \\
&= B' + (B' - 2B) \\
&= B' + (2^{\lceil \log B \rceil + 2} - 2B) \\
&= B' + (4 \cdot 2^{\lceil \log B \rceil} - 2B) \\
&\geq B' + (4B - 2B) \\
&= B' + 2B \\
&> B', \quad \text{since } B > 0.
\end{aligned}
$$

$\Longrightarrow \forall 1 \leq i \leq n \quad T_i$ contains at most one i_1 with $3n < i_1 \leq 4n$.
However, since all i_1 with $3n < i_1 \leq 4n$ must be contained in one of the sets T_i $(1 \leq i \leq n)$, it holds that, for all $1 \leq i \leq n$, T_i contains *exactly* one i_1 with $3n < i_1 \leq 4n$. We assume w.l.o.g.: $3n + i \in T_i \quad \forall 1 \leq i \leq n$

Now define

$$
S_i = T_i \setminus \{3n + i\} \quad \forall 1 \leq i \leq n
$$

$$
\Longrightarrow \quad S_i \subseteq \{1, \ldots, 3n\}
$$

$S_1, \ldots S_n$ form a partition of $\{1, \ldots, 3n\}$. For $1 \leq i \leq n$ we have:

$$
\sum_{j \in S_i} a_j = \sum_{j \in S_i} c_j = \sum_{j \in T_i} c_j - c_{3n+i} = B' - (B' - B) = B.
$$

$\square$

Now we can prove Theorem 4.1 based on the fact that 2^l–PARTITION is NP–hard (even if the numbers have a unary encoding):

PROOF: Proof by polynomial transformation from 2^l–Partition to CDF:
Let an instance of 2^l–Partition be given by $c_1, \ldots, c_k \in \mathbf{Z}^+, B' = 2^b, M = 2^l$ with $\sum_{i=1}^k c_i = M \cdot B' = 2^{b+l}, k > M$. All numbers have are unary encoded.

We compute a corresponding instance of CDF in polynomial time:
We compute functions f_1 and $f_2 \in B_n$ with input variables $x_1, \ldots, x_p, y_1, \ldots, y_p, p = l + b + 1$. The functions are defined by their decomposition matrices Z_1 and Z_2 with respect to the bound set $X^{(1)} = \{x_1, \ldots, x_p\}$. The decomposition matrices have 2^{l+b+1} rows and 2^{l+b+1} columns, respectively.

The matrices are constructed with k blocks corresponding to the k weights $c_1, \ldots, c_k$. (These blocks are constructed in such a way that common decomposition functions of f_1 and f_2 have to assign the same function values to the indices of the same block.)

We use the following abbreviation in the definition of Z_1 and Z_2: $un(i)$ is defined as

$$\underbrace{1\ldots1}_{i \text{ times}} \quad \underbrace{0\ldots0}_{2^{l+b+1}-i \text{ times}} .$$

$un(i)$ is a row pattern of the decomposition matrix beginning with i ones followed by zeros. The structure of block i is defined as follows:

<u>Case 1</u>: $c_i = 1$

Block i consists both for Z_1 and for Z_2 of the equal rows:

For Z_1:
$$\begin{array}{|c|} \hline un(\sum_{j=1}^{i-1} c_j) \\ un(\sum_{j=1}^{i-1} c_j) \\ \hline \end{array}$$

For Z_2:
$$\begin{array}{|c|} \hline un(\sum_{j=1}^{i-1} c_j) \\ un(\sum_{j=1}^{i-1} c_j) \\ \hline \end{array}$$

<u>Case 2</u>: $c_i > 1$

Block i consists both for Z_1 and for Z_2 of $2c_i$ rows:

For Z_1:
$$\begin{array}{|c|} \hline un(\sum_{j=1}^{i-1} c_j) \\ un(\sum_{j=1}^{i-1} c_j) \\ un(\sum_{j=1}^{i-1} c_j + 1) \\ un(\sum_{j=1}^{i-1} c_j + 1) \\ \vdots \\ un(\sum_{j=1}^{i-1} c_j + (c_i - 2)) \\ un(\sum_{j=1}^{i-1} c_j + (c_i - 1)) \\ un(\sum_{j=1}^{i-1} c_j + (c_i - 1)) \\ \hline \end{array}$$

For Z_2:
$$\begin{array}{|c|} \hline un(\sum_{j=1}^{i-1} c_j) \\ un(\sum_{j=1}^{i-1} c_j + 1) \\ un(\sum_{j=1}^{i-1} c_j + 1) \\ un(\sum_{j=1}^{i-1} c_j + 2) \\ \vdots \\ un(\sum_{j=1}^{i-1} c_j + (c_i - 1)) \\ un(\sum_{j=1}^{i-1} c_j + (c_i - 1)) \\ un(\sum_{j=1}^{i-1} c_j) \\ \hline \end{array}$$

Both decomposition matrices have $2\sum_{i=1}^{k} c_i = 2MB' = 2^{l+b+1}$ rows and $2MB'$ columns and they can be computed in polynomial time.

Since the rows of different blocks are different and since two rows of the same block are always pairwise equal, the number of *different* row patterns is equal to $\frac{1}{2}(2^{l+b+1}) = 2^{l+b}$.

$\implies$ The minimum number of decomposition functions in a decomposition with respect to $X^{(1)}$ is $\lceil \log(nrp(X^{(1)}, f_1)) \rceil = \lceil \log(nrp(X^{(1)}, f_2)) \rceil = l + b$ both for f_1 and for f_2.

Finally, we choose the natural number of our instance of CDF as $h = l$, i.e., the problem is to decide whether there are functions $\alpha_1, \ldots, \alpha_l \in B_p$ which can be used as common decomposition functions in one–sided communication minimal decompositions of f_1 and f_2 with respect to $X^{(1)}$.

We have to show the following:

$$\exists \text{ partition } T_1, \ldots, T_M \text{ of } \{1, \ldots, k\}, \text{ such that } \sum_{j \in T_i} c_j = B' \; \forall 1 \le i \le M$$
$$\Longleftrightarrow$$
$$\exists h = l \text{ common decomposition functions of } f_1, f_2.$$

Before this equivalence is proven we define the following notion:

The decomposition matrices have been defined by k blocks. $\{0, 1\}^p$ is partitioned into k disjoint sets $block_1, \ldots, block_k$ with the definition

$$(x_1, \ldots, x_p) \in block_i$$
$$\Longleftrightarrow$$
The row with index $int(x_1, \ldots, x_p)$ belongs to block i.

"$\Longrightarrow$":

Assume that $\exists$ partition $T_1, \ldots, T_M$ of $\{1, \ldots, k\}$, such that $\sum_{j \in T_i} c_j = B' \; \forall 1 \le i \le M$.

Now we define l common decomposition functions $\alpha_1, \ldots, \alpha_l$ of f_1 and f_2:

For all $1 \le i \le 2^l$ and all $j \in T_i$ define for all $\mathbf{x} = (x_1, \ldots, x_p) \in block_j$

$$\alpha(\mathbf{x}) = (\alpha_1(\mathbf{x}), \ldots, \alpha_l(\mathbf{x})) = bin_l(i - 1)$$

(That is, if $T_i = \{i_1, \ldots, i_q\}$, then all blocks $block_{i_1}, \ldots block_{i_q}$ will have the function value $bin_l(i - 1)$.)

The number of different rows patterns $mult_{bin_l(i-1)}$ of rows with function value $bin_l(i-1)$ for α is (both for Z_1 and for Z_2):

$$
\begin{aligned}
mult_{bin_l(i-1)} &= \sum_{j \in T_i} (\text{number of different row patterns in block } j) \\
&= \sum_{j \in T_i} c_j \\
&= B' = 2^b
\end{aligned}
$$

$$\Longrightarrow nrp(X^{(1)}, f_1, \alpha) = nrp(X^{(1)}, f_2, \alpha) = 2^b.$$

According to Lemma 4.2 there exist one–sided decompositions of f_1 and f_2 with respect to $X^{(1)}$ of form

$$
\begin{aligned}
f_1(x_1, &\ldots, x_p, y_1, \ldots, y_p) = \\
&g^{(1)}(\alpha_1(\mathbf{x}), \ldots, \alpha_l(\mathbf{x}), \alpha_{l+1}^{(1)}(\mathbf{x}), \ldots, \alpha_{l+b}^{(1)}(\mathbf{x}), y_1, \ldots, y_p) \\
f_2(x_1, &\ldots, x_p, y_1, \ldots, y_p) = \\
&g^{(2)}(\alpha_1(\mathbf{x}), \ldots, \alpha_l(\mathbf{x}), \alpha_{l+1}^{(2)}(\mathbf{x}), \ldots, \alpha_{l+b}^{(2)}(\mathbf{x}), y_1, \ldots, y_p).
\end{aligned}
$$

"$\Longleftarrow$":

Assume that $\exists$ l common decomposition functions $\alpha_1, \ldots, \alpha_l$ of f_1 and f_2.

Since the number of different row patterns of Z_1 and Z_2 is 2^{l+b}, respectively, all decomposition functions have to be strict (cf. Lemma 3.5), including $\alpha_1, \ldots, \alpha_l$.

<u>Claim 1</u>: $\alpha(x_1, \ldots, x_p) = \alpha(x_1', \ldots, x_p')$ for all $(x_1, \ldots, x_p), (x_1', \ldots, x_p') \in block_i$ $(1 \le i \le k)$.

Proof:

$\alpha_1, \ldots, \alpha_l$ are common decomposition functions of f_1 and f_2.

Rows 1 and 2 of block i in Z_1 are equal $\Rightarrow \alpha$ has the same function value for the indices of rows 1 and 2.

Rows 2 and 3 of block i in Z_2 are equal $\Rightarrow \alpha$ has the same function value for indices of rows 2 and 3.

The claim results from considering rows 3 and 4 of block i in Z_1, etc.

<u>Claim 2</u>: Both for f_1 and f_2, α has the same function values for row indices with *exactly* $2^b = B'$ *different* rows patterns.

Proof:

- According to Lemma 4.2 we have $nrp(X^{(1)}, f_1, \alpha) \leq 2^b$ and $nrp(X^{(1)}, f_2, \alpha)$ $\leq 2^b$. For each function value of α, among all rows with this function value, the number of different row patterns is $\leq 2^b$.
- There are $2^l \cdot 2^b$ different row patterns of Z_1 and Z_2, respectively, and α can have exactly 2^l different function values. So for each function value of α, among all rows with this function value the number of different row patterns is *exactly equal* to 2^b.

It can be concluded from Claims 1 and 2 that the blocks of the decomposition matrices can be partitioned into groups where the number of different row patterns in such a group is exactly 2^b.

Now define for $i = 1, \ldots, 2^l$:

$$T_i = \{j \mid \alpha(\mathbf{x}) = bin_l(i-1) \, \forall \, \mathbf{x} \in block_j\}.$$

$T_1, \ldots, T_M$ form a partition of $\{1, \ldots, k\}$ because of Claim 1. The number of different patterns of rows whose index is mapped to $bin_l(i-1)$ by α is

$$2^b = B' = \sum_{j \in T_i} (\text{Number of different row patterns in block } j) = \sum_{j \in T_i} c_j.$$

Thus for the given partition $T_1, \ldots, T_M$ we have:

$$\sum_{j \in T_i} c_j = B'$$

and $T_1, \ldots, T_M$ is a solution to the 2^l–PARTITION–problem.

$\square$

Appendix I
ROBDD based computation of compatibility classes

In Section 4.3.1.3 the classes $E_1, \ldots, E_l$ of $\{0,1\}^p/_\sim$ are computed based on ROBDD representations. Based on ROBDDs $bdd_j^{(i)}$ representing characteristic functions of equivalence classes $K_j^{(i)}$, a graph $G_\sim$ is computed and the connected components of the graph correspond to classes E_j. However, to compute $G_\sim$ explicitly, for each pair $bdd_{j_1}^{(i_1)}$, $bdd_{j_2}^{(i_2)}$ of nodes in the node set of $G_\sim$ we have to check whether $bdd_{j_1}^{(i_1)} \wedge bdd_{j_2}^{(i_2)} \neq 0$. However, it turns out that it is not necessary to construct $G_\sim$ explicitly and it is not necessary to perform this and–operation for all pairs of nodes. Rather, the number of apply operations (and, or, ... of 2 ROBDDs (Bryant, 1986)) needed is linear in the number of edges which really exist in $G_\sim$.

Figure I.1 shows an algorithm which is able to fulfill these requirements. The algorithm provides sets $CLNR_1^{(i)}, \ldots, CLNR_l^{(i)}$ (for all $1 \leq i \leq m$), such that for all equivalence classes $E_1, \ldots, E_l$ of $\{0,1\}^p/_\sim$ we have: For $1 \leq j \leq l$ $E_j = \bigcup_{s \in CLNR_j^{(i)}} K_s^{(i)}$ for arbitrary $i \in \{1, \ldots, m\}$.

Basically the algorithm performs a depth–first search in the (implicitly defined) graph $G_\sim$. The following idea leads to the reduction of apply–operations compared to the naive approach: Whenever the algorithm processes a node $bdd_j^{(i)}$, it is possible to find all nodes of $G_\sim$ which are connected to $bdd_j^{(i)}$ and which have not yet been visited without needing to know $G_\sim$ explicitly. Here we make use of the facts that is easy to check whether the intersection of two ROBDDs is empty or not, and that it is easy to get an element of $ON(f)$ based on the ROBDD for f (Bryant, 1986).

If, during the depth–first search starting from $bdd_j^{(i)}$, we visit vb nodes which are connected to $bdd_j^{(i)}$ and which have not yet been visited before, then we need only $vb+m-1$ apply–operations to identify these nodes (m is the number of functions $f_1, \ldots, f_m$).

This is guaranteed, since during the computation of a connected component we maintain m ROBDDs $cc^{(1)}, \ldots, cc^{(m)}$, where $cc^{(i)}$ represents (for function f_i) all visited nodes $bdd_j^{(i)}$ of the current connected component. If for $1 \leq i \leq m$ the nodes $bdd_{j_1}^{(i)}, \ldots, bdd_{j_q}^{(i)}$ of the current connected component have already been visited, then $cc^{(i)} = \bigvee_{s=1}^{q} bdd_{j_s}^{(i)}$ (invariant of lines 6 and 7 of the algorithm), such that $cc^{(i)}$ is the characteristic function of $\bigcup_{s=1}^{q} K_{j_s}^{(i)}$.

Figure I.1. Algorithm to compute $\{0,1\}^p/\sim$.

Input:

- Functions $f_1, \ldots, f_m$ of B_n with input variables $x_1, \ldots, x_n$.

- For $1 \leq i \leq m$:
  Partitions into equivalence classes $\{0,1\}^p/\equiv_i = \{K_1^{(i)}, \ldots, K_{nrp(X^{(1)}, f_i)}^{(i)}\}$ for a decomposition of f_i with respect to $X^{(1)} = \{x_1, \ldots, x_p\}$.
  The sets $K_j^{(i)}$ are represented by ROBDDs $bdd_j^{(i)}$ for their characteristic functions.

Output: For all $1 \leq i \leq m$ sets $CLNR_1^{(i)}, \ldots, CLNR_l^{(i)}$, such that for all equivalence classes $E_1, \ldots, E_l$ of $\{0,1\}^p/\sim$ the following holds:
$$\forall 1 \leq j \leq l \; E_j = \bigcup_{s \in CLNR_j^{(i)}} K_s^{(i)} \text{ for arbitrary } i \in \{1, \ldots, m\}.$$

Algorithm:

```
1       procedure search(bdd bdd_j^(i), int i)
2       begin
3          if (i = 1) then mark bdd_j^(i) as visited. fi
4          cc^(i) = cc^(i) ∨ bdd_j^(i)
5          CLNR_ccnr^(i) = CLNR_ccnr^(i) ∪ {j}
6          /* If nodes bdd_j1^(i), ..., bdd_jq^(i) are already visited, then
7                cc^(i) = ∨_{s=1}^q bdd_js^(i) and CLNR_ccnr^(i) = {j1, ..., jq} */
8          for k = 1 to m do
9             if (k ≠ i)
10               then
11                  not_covered = bdd_j^(i) ∧ cc^(k)‾
12                  while (not_covered ≠ 0) do
13                     Choose arbitrary v ∈ ON(not_covered).
14                     Let u ∈ {1, ..., nrp(f_k, X^(1))}, such that v ∈ ON(bdd_u^(k)).
15                     search(bdd_u^(k), k)
16                     not_covered = bdd_j^(i) ∧ cc^(k)‾
17                  od
18               fi
19            od
20       end
22
24       begin
25          ∀1 ≤ j ≤ nrp(f_1, X^(1)):  Mark bdd_j^(1) as not visited.
26          ccnr = 0
27          for j = 1 to nrp(f_1, X^(1)) do
28             if (bdd_j^(1) not yet visited)
29                then
30                   ccnr = ccnr + 1
31                   ∀1 ≤ i ≤ m: cc^(i) = 0, CLNR_ccnr^(i) = ∅
32                   search(bdd_j^(1), 1)
33                fi
34          od
35       end
```

Consequently, we have to initialize $cc^{(i)}$ by 0 before processing a new connected component (line 31 of the algorithm), and for each visit to a node $bdd_j^{(i)}$ we perform the operation $cc^{(i)} = cc^{(i)} \vee bdd_j^{(i)}$ (line 4).

During the loop of lines 8–19 for all $k \in \{1, \ldots, m\} \setminus \{i\}$ we visit, step by step, all nodes $bdd_{j_\bullet}^{(k)}$ which are connected to $bdd_j^{(i)}$ in $G_\sim$ and which are still unvisited.

The following considerations show that we need not know $G_\sim$ explicitly to determine the unvisited nodes $bdd_{j_\bullet}^{(k)}$ connected to $bdd_j^{(i)}$: Let U be the set of all indices u, such that nodes $bdd_u^{(k)}$ of f_k have not yet been visited in the current connected component. Then $\overline{cc^{(k)}} = \bigvee_{u \in U} bdd_u^{(k)}$ holds in line 11 of the algorithm. To find an unvisited node $bdd_u^{(k)}$ of $G_\sim$ which is connected to the current node $bdd_j^{(i)}$ in $G_\sim$, we have to find an unvisited node with $bdd_u^{(k)} \wedge bdd_j^{(i)} \neq 0$. Now the operation $\overline{cc^{(k)}} \wedge bdd_j^{(i)} \neq 0$ (lines 11 and 12) checks *in parallel* for all nodes $bdd_u^{(k)}$ with $u \in U$, which are still unvisited, whether there is an edge from the current node $bdd_j^{(i)}$ to $bdd_u^{(k)}$. If the result of this operation is different from 0 (*not_covered* $\neq 0$ in line 12), then there is such a node. In this case an arbitrary unvisited node $bdd_u^{(k)}$ which is connected to $bdd_j^{(i)}$ is selected in lines 13 and 14. The depth–first search is continued with this node (line 15). When the recursive processing is finished, then the check for another (unvisited) node $bdd_u^{(k)}$ is repeated (in lines 16 and 12). This happens as long as there are no longer any such nodes (*not_covered* = 0).

Thus for each edge in the (implicitly defined) graph $G_\sim$ which leads the depth–first search from $bdd_j^{(i)}$ to an unvisited node, we need one *and*–operation for 2 ROBDDs (lines 11 or 16) and for each node $bdd_j^{(i)}$ and each $k \in \{1, \ldots, m\} \setminus \{i\}$ we need one additional *and*–operation (lines 11 or 16), whose result is 0, such that no edge is followed after the operation.

Since the edges of $G_\sim = (V, E)$ which lead the depth–first search to unvisited nodes in a fixed connected component form a spanning tree for this connected component, the number of *and*–operations whose result is different from 0 is bounded by $|V| - 1$. The number of *and*–operations whose result is 0 is equal to $|V| \times (m - 1)$. Thus, the total number of *and*–operations is bounded by $|V| - 1 + |V| \times (m - 1)$. If we also consider one *or*–operation per node (see line 4), the total number of *apply*–operations for 2 ROBDDs is at most

$$|V| \times (m + 1) - 1.$$

Now it is clear that the total number of *apply*–operations is only linear in the number of edges in $G_\sim$, when we also take into account that the degree of a node in $G_\sim$ is always $m - 1$ or larger.

The analysis given above shows that it is really possible to prove a linear upper bound on the number of apply–operations which are performed during the algorithm. More precisely, we have proved the following lemma:

LEMMA I.1 *The number of apply–operations, which are needed by algorithm* search *in Figure I.1 is bounded by* $|V| \times (m + 1) - 1$ *and thus is linear in the number of edges in* $G_\sim$.

Appendix J
Maximal don't care sets for decomposition and composition functions

In Section 5.4 we claim that the don't care sets for composition and decomposition functions which we compute during the decomposition of incompletely specified functions are maximal. The following theorem gives an exact formulation of the definition for these don't care sets and claims that these don't care sets are maximal (or equivalently that the domains of decomposition and composition function are minimal).

THEOREM J.3 *Let $f \in BP_n(D_f)$. f is decomposed with respect to $\{x_1, \ldots, x_p\}$. Let $PK = \{PK_1, \ldots, PK_v\}$ be a partition of $\{0,1\}^p$, where for the sets PK_i ($1 \leq i \leq v$) we have $\epsilon \sim \delta$ for all $\epsilon, \delta \in PK_i$. $\{PK_1, \ldots, PK_v\}$ has the following 'local optimality' property: For all $i \neq j \in \{1, \ldots, v\}$: $\exists \epsilon \in PK_i, \delta \in PK_j$ with $\epsilon \not\sim \delta$. For all $1 \leq i \leq v$ let $cof_{PK_i} = com\_ext(\{f_{x_1^{\epsilon_1} \ldots x_p^{\epsilon_p}} | (\epsilon_1, \ldots, \epsilon_p) \in PK_i\})$. Let $r = \lceil \log(v) \rceil$.*
For $v = 2^r$ let

$$dc\_set = \{\epsilon \in \{0,1\}^p \mid \forall 1 \leq i \leq v \ cof_{PK_i} \ is \ an \ extension \ of \ f_{x_1^{\epsilon_1} \ldots x_p^{\epsilon_p}}\}$$

and for $v < 2^r$ let

$$dc\_set = \{\epsilon \in \{0,1\}^p \mid (f_{x_1^{\epsilon_1} \ldots x_p^{\epsilon_p}})_{on} = (f_{x_1^{\epsilon_1} \ldots x_p^{\epsilon_p}})_{off} = 0.\}.$$

Define $D_\alpha = \{0,1\}^p \setminus dc\_set$.[1] The decomposition function $\alpha : D_\alpha \to \{0,1\}^r$ is defined by

$$\alpha(PK_i \setminus dc\_set) = (a_1^{(i)}, \ldots, a_r^{(i)}) \ (1 \leq i \leq v)$$

with $(a_1^{(i)}, \ldots, a_r^{(i)}) \neq (a_1^{(j)}, \ldots, a_r^{(j)}) \ \forall 1 \leq i, j \leq v \ with \ i \neq j.$

---

[1] If $v > 1$, then for all PK_i we have $PK_i \setminus dc\_set \neq \emptyset$ because of the 'local optimality' property of PK. This can be concluded from $\epsilon \in dc\_set \implies \epsilon \sim \delta$ for all $\delta \in \{0,1\}^p$: Suppose $PK_i \setminus dc\_set = \emptyset$, then $PK_i \subseteq dc\_set$ and for an arbitrary $1 \leq j \leq v$ with $j \neq i$ we would have $\epsilon \sim \delta$ for all $\epsilon \in PK_i$ and $\delta \in PK_j$. Consequently, the local optimality property of PK would be violated.

249

Define $D_g = \{(a_1^{(i)}, \ldots, a_r^{(i)}, y_1, \ldots, y_{n-p}) \mid \exists \epsilon \in \{0,1\}^p \text{ with } \alpha(\epsilon) = (a_1^{(i)}, \ldots, a_r^{(i)}) \text{ and } (y_1, \ldots, y_{n-p}) \in D(co f_{PK_i})\}$. *The composition function g is defined by*

$$g : D_g \to \{0,1\}, \quad g(a_1^{(i)}, \ldots, a_r^{(i)}, y_1, \ldots, y_{n-p}) = co f_{PK_i}(y_1, \ldots, y_{n-p}).$$

Then D_α and D_g are minimal in the following sense: For each proper subset $D_\alpha' \subset D_\alpha$ and each proper subset $D_g' \subset D_g$ we have: For $\alpha' : D_\alpha' \to \{0,1\}^r, \alpha|_{D_\alpha'} = \alpha'$ there is a completely specified extension $\alpha'' : \{0,1\}^p \to \{0,1\}^r$ of α', a completely specified extension $g'' : \{0,1\}^{r+n-p} \to \{0,1\}$ of g and at least one $\epsilon \in D_f$ with

$$f(\epsilon_1, \ldots, \epsilon_n) \neq g''(\alpha''(\epsilon_1, \ldots, \epsilon_p), \epsilon_{p+1}, \ldots, \epsilon_n).$$

For $g' : D_g' \to \{0,1\}, g|_{D_g'} = g'$ there is a completely specified extension $g'' : \{0,1\}^{r+n-p} \to \{0,1\}$ of g', a completely specified extension $\alpha'' : \{0,1\}^p \to \{0,1\}^r$ of α and at least one $\epsilon \in D_f$ with

$$f(\epsilon_1, \ldots, \epsilon_n) \neq g''(\alpha''(\epsilon_1, \ldots, \epsilon_p), \epsilon_{p+1}, \ldots, \epsilon_n).$$

According to Theorem J.3 the don't care sets of decomposition function α and composition function g can*not* be extended.

PROOF:

Composition function:
    Choose an arbitrary $\epsilon = (\epsilon_1, \ldots, \epsilon_{n-p+r}) \in D_g$. (For $D_g = \emptyset$ the proof is trivial.)

    We have to prove the following: There exists $g'' : \{0,1\}^{n-p+r} \to \{0,1\}$ with $g''|_{D_g \setminus \{\epsilon\}} = g|_{D_g \setminus \{\epsilon\}}$, $\alpha'' : \{0,1\}^p \to \{0,1\}^r$ with $\alpha''|_{D_\alpha} = \alpha$, and $(\delta_1, \ldots, \delta_p) \in \{0,1\}^p$, such that $(\delta_1, \ldots, \delta_p, \epsilon_{r+1}, \ldots, \epsilon_{n-p+r}) \in D_f$ and $g''(\epsilon_1, \ldots, \epsilon_r, \epsilon_{r+1}, \ldots, \epsilon_{n-p+r}) = g''(\alpha''(\delta_1, \ldots, \delta_p), \epsilon_{r+1}, \ldots, \epsilon_{n-p+r}) \neq f(\delta_1, \ldots, \delta_p, \epsilon_{r+1}, \ldots, \epsilon_{n-p+r})$.

    $v > 1$: Since $\epsilon \in D_g$, there is $PK_i \in \{PK_1, \ldots, PK_v\}$ with $\alpha^{-1}(\epsilon_1, \ldots, \epsilon_r) = PK_i \setminus dc\_set$. In addition there must be $\delta \in PK_i$, such that $(\delta_1, \ldots, \delta_p, \epsilon_{r+1}, \ldots, \epsilon_{n-p+r}) \in D_f$, because otherwise $(\epsilon_{r+1}, \ldots, \epsilon_{n-p+r}) \notin D(co f_{PK_i})$, which would contradict $\epsilon \in D_g$. If $\delta \notin dc\_set$, then $\alpha(\delta) = (\epsilon_1, \ldots, \epsilon_r)$ and thus $\alpha''(\delta) = (\epsilon_1, \ldots, \epsilon_r)$, otherwise we choose $\alpha''(\delta) = (\epsilon_1, \ldots, \epsilon_r)$.

    Using $g''(\epsilon_1, \ldots, \epsilon_r, \epsilon_{r+1}, \ldots, \epsilon_{n-p+r}) = \overline{f(\delta_1, \ldots, \delta_p, \epsilon_{r+1}, \ldots, \epsilon_{n-p+r})}$ we have

$$
\begin{aligned}
g''(\alpha''(\delta_1, \ldots, \delta_p), \epsilon_{r+1}, \ldots, \epsilon_{n-p+r}) &= g''(\epsilon_1, \ldots, \epsilon_r, \epsilon_{r+1}, \ldots, \epsilon_{n-p+r}) \\
&= \overline{f(\delta_1, \ldots, \delta_p, \epsilon_{r+1}, \ldots, \epsilon_{n-p+r})}.
\end{aligned}
$$

    $v = 1$: We have $r = 0$ and $f(x_1, \ldots, x_p, y_1, \ldots, y_{n-p}) = g(y_1, \ldots, y_{n-p})$ for $(x_1, \ldots, x_p, y_1, \ldots, y_{n-p}) \in D_f$. There has to be $(\delta_1, \ldots, \delta_p)$ with $(\delta_1, \ldots, \delta_p, \epsilon_1, \ldots, \epsilon_{n-p}) \in D_f$, because otherwise $(\epsilon_1, \ldots, \epsilon_{n-p}) \notin D_g$.

    Using $g''(\epsilon_1, \ldots, \epsilon_{n-p}) = \overline{f(\delta_1, \ldots, \delta_p, \epsilon_1, \ldots, \epsilon_{n-p})}$ we obviously have $g''(\epsilon_1, \ldots, \epsilon_{n-p}) \neq f(\delta_1, \ldots, \delta_p, \epsilon_1, \ldots, \epsilon_{n-p})$.

Decomposition function:
    Choose an arbitrary $\epsilon = (\epsilon_1, \ldots, \epsilon_p) \in D_\alpha$. (For $D_\alpha = \emptyset$ the proof is trivial.)

    We have to prove the following: There is $\alpha'' : \{0,1\}^p \to \{0,1\}^r$ with $\alpha''|_{D_\alpha \setminus \{\epsilon\}} = \alpha|_{D_\alpha \setminus \{\epsilon\}}$, $g'' : \{0,1\}^{r+n-p} \to \{0,1\}$ with $g''|_{D_g} = g$ and $(\delta_1, \ldots, \delta_{n-p}) \in \{0,1\}^{n-p}$, such that $(\epsilon_1, \ldots, \epsilon_p, \delta_1, \ldots, \delta_{n-p}) \in D_f$ and $g''(\alpha''(\epsilon_1, \ldots, \epsilon_p), \delta_1, \ldots, \delta_{n-p}) \neq f(\epsilon_1, \ldots, \epsilon_p, \delta_1, \ldots, \delta_{n-p})$.

$v < 2^r$: There must be $(\delta_1, \ldots, \delta_{n-p})$ with $(\epsilon_1, \ldots, \epsilon_p, \delta_1, \ldots, \delta_{n-p}) \in D_f$, because otherwise $(\epsilon_1, \ldots, \epsilon_p) \notin D_\alpha$.

Choose $(a_1, \ldots, a_r)$, such that $\alpha(\epsilon) \neq (a_1, \ldots, a_r)$ for all $\epsilon \in \{0,1\}^p$. Now set $\alpha''(\epsilon) = (a_1, \ldots, a_r)$. $(a_1, \ldots, a_r, \delta_1, \ldots, \delta_{n-p}) \notin D_g$. Set $g''(a_1, \ldots, a_r, \delta_1, \ldots, \delta_{n-p}) = \overline{f(\epsilon_1, \ldots, \epsilon_p, \delta_1, \ldots, \delta_{n-p})}$.

$$
\begin{aligned}
\implies \quad g''(\alpha''(\epsilon_1, \ldots, \epsilon_p), \delta_1, \ldots, \delta_{n-p}) &= g''(a_1, \ldots, a_r, \delta_1, \ldots, \delta_{n-p}) \\
&= \overline{f(\epsilon_1, \ldots, \epsilon_p, \delta_1, \ldots, \delta_{n-p})}.
\end{aligned}
$$

$v = 2^r$: Because of $\epsilon \in D_\alpha$ there is $PK_i \in PK$, such that cof_{PK_i} is not an extension of $f_{x_1^{\epsilon_1}, \ldots, x_p^{\epsilon_p}}$. Consequently, there is a $\delta \in \{0,1\}^{n-p}$, such that $(\epsilon_1, \ldots, \epsilon_p, \delta_1, \ldots, \delta_{n-p}) \in D_f$ and $\delta \notin D(cof_{PK_i})$ or $cof_{PK_i}(\delta_1, \ldots, \delta_{n-p}) \neq f(\epsilon_1, \ldots, \epsilon_p, \delta_1, \ldots, \delta_{n-p})$.

Let $\alpha(PK_i \setminus dc\_set) = (a_1^{(i)}, \ldots, a_r^{(i)})$. If $\delta \notin D(cof_{PK_i})$ define

$$
g''(a_1^{(i)}, \ldots, a_r^{(i)}, \delta_1, \ldots, \delta_{n-p}) = \overline{f(\epsilon_1, \ldots, \epsilon_p, \delta_1, \ldots, \delta_{n-p})},
$$

and otherwise we also have

$$
\begin{aligned}
g''(a_1^{(i)}, \ldots, a_r^{(i)}, \delta_1, \ldots, \delta_{n-p}) &= g(a_1^{(i)}, \ldots, a_r^{(i)}, \delta_1, \ldots, \delta_{n-p}) \\
&= cof_{PK_i}(\delta_1, \ldots, \delta_{n-p}) \\
&= \overline{f(\epsilon_1, \ldots, \epsilon_p, \delta_1, \ldots, \delta_{n-p})}.
\end{aligned}
$$

If we choose $\alpha''(\epsilon_1, \ldots, \epsilon_p) = (a_1^{(i)}, \ldots, a_r^{(i)})$, then we really have

$$
\begin{aligned}
g''(\alpha''(\epsilon_1, \ldots, \epsilon_p), \delta_1, \ldots, \delta_{n-p}) &= g''(a_1^{(i)}, \ldots, a_r^{(i)}, \delta_1, \ldots, \delta_{n-p}) \\
&= \overline{f(\epsilon_1, \ldots, \epsilon_p, \delta_1, \ldots, \delta_{n-p})}.
\end{aligned}
$$

$\square$

References

Akers, S. (1978). Binary decision diagrams. *IEEE Trans. on Comp.*, 27:509–516.

Ashenhurst, R. (1959). The decomposition of switching functions. In *Int'l Symp. on Theory Switching Funct.*, pages 74–116.

Becker, B., Burch, T., Hotz, G., Kiel, D., Kolla, R., Molitor, P., Osthof, H., Pitsch, G., and Sparmann, U. (1990). A graphical system for hierarchical specifications and checkups of VLSI circuits. In *European Conf. on Design Automation*, pages 174–179.

Becker, B., Hotz, G., Kolla, R., Molitor, P., and Osthof, H. (1987a). CADIC - Ein System zum hierarchischen Entwurf integrierter Schaltungen. In *E.I.S.-Workshop*, pages 235–245.

Becker, B., Hotz, G., Kolla, R., Molitor, P., and Osthof, H. (1987b). Hierarchical design based on a calculus of nets. In *Design Automation Conf.*, pages 649–653.

Bollig, B., Löbbing, M., and Wegener, I. (1995). Simulated annealing to improve variable orderings for OBDDs. In *Int'l Workshop on Logic Synth.*, pages 5b:5.1–5.10.

Bollig, B., Savicky, P., and Wegener, I. (1994). On the improvement of variable orderings for OBDDs. *IFIP Workshop on Logic and Architecture Synthesis, Grenoble*, pages 71–80.

Bollig, B. and Wegener, I. (1996). Improving the variable ordering of OBDDs is NP-complete. *IEEE Trans. on Comp.*, 45(9):993–1002.

Brayton, R., Hachtel, G., McMullen, C., and Sangiovanni-Vincentelli, A. (1984). *Logic Minimization Algorithms for VLSI Synthesis*. Kluwer Academic Publishers.

Brayton, R., Rudell, R., Sangiovanni-Vincentelli, A., and Wang, A. (1987). MIS: A multiple - level logic optimization system. *IEEE Trans. on Comp.*, 6(6):1062–1081.

Brélaz, D. (1979). New methods to color vertices of a graph. *Comm. of the ACM*, 22:251–256.

Breuer, M. (1979). Min–cut placement. *Journal of Design Automation and Fault Tolerant Computation*, 1(4):346–362.

Brown, S., Francis, R., Rose, J., and Vranesic, Z. (1992). *Field-Programmable Gate Arrays*. Kluwer Academic Publisher.

Brown, S. and Vranesic, Z. (2000). *Fundamentals of Digital Logic with VHDL Design*. McGraw-Hill.

Bryant, R. (1986). Graph - based algorithms for Boolean function manipulation. *IEEE Trans. on Comp.*, 35(8):677–691.

Bryant, R. (1992). Symbolic Boolean manipulation with ordered binary decision diagrams. *ACM, Comp. Surveys*, 24:293–318.

Buch, P., Narayan, A., Newton, A., and Sangiovanni-Vincentelli, A. (1997). Logic synthesis for large pass transistor circuits. In *Int'l Conf. on CAD*, pages 663–670.

Chang, S., Cheng, D., and Marek-Sadowska, M. (1994). Minimizing ROBDD size of incompletely specified multiple output functions. In *European Design & Test Conf.*, pages 620–624.

Coudert, O., Berthet, C., and Madre, J. (1989a). Verification of sequential machines based on symbolic execution. In *Automatic Verification Methods for Finite State Systems*, volume 407 of *LNCS*, pages 365–373. Springer Verlag.

Coudert, O., Berthet, C., and Madre, J. (1989b). Verification of sequential machines using Boolean functional vectors. In *Proceedings IFIP International Workshop on Applied Formal Methods for Correct VLSI Design*, pages 111–128.

Curtis, H. (1961). A generalized tree circuit. *Journal of the ACM*, 8:484–496.

Darringer, J., Brand, D., Joyner, W., and Trevillyan, L. (1984). LSS: A system for production logic synthesis. *IBM J. Res. and Develop.*, 28(5):537–545.

Darringer, J., Joyner, W., Berman, L., and Trevillyan, L. (1981). LSS: Logic synthesis through local transformations. *IBM J. Res. and Develop.*, 25(4):365–388.

Dietmeyer, D. and Schneider, P. (1967). Identification of symmetry, redundancy and equivalence of Boolean functions. *IEEE Trans. on Electronic Comp.*, 16:804–817.

Drechsler, R., Becker, B., and Göckel, N. (1995). A genetic algorithm for variable ordering of OBDDs. In *Int'l Workshop on Logic Synth.*, pages 5c:5.55–5.64.

Edwards, C. and Hurst, S. (1978). A digital synthesis procedure under function symmetries and mapping methods. *IEEE Trans. on Comp.*, 27:985–997.

Felt, E., York, G., Brayton, R., and Sangiovanni-Vincentelli, A. (1993). Dynamic Variable Reordering for BDD Minimization. In *European Design Automation Conf.*, pages 130–135.

Francis, R., Rose, J., and Chung, K. (1990). Chortle: A technology mapping program for lookup table-based field programmable gate arrays. In *Design Automation Conf.*, pages 613–619.

Francis, R., Rose, J., and Vranesic, Z. (1991). Chortle-crf: Fast technology mapping for lookup table-based FPGAs. In *Design Automation Conf.*, pages 227–233.

Fujii, H., Ootomo, G., and Hori, C. (1993). Interleaving based variable ordering methods for ordered binary decision diagrams. In *Int'l Conf. on CAD*, pages 38–41.

Fujita, M., Fujisawa, H., and Kawato, N. (1988). Evaluation and improvements of Boolean comparison method based on binary decision diagrams. In *Int'l Conf. on CAD*, pages 2–5.

Fujita, M., Matsunaga, Y., and Kakuda, T. (1991). On variable ordering of binary decision diagrams for the application of multi-level synthesis. In *European Conf. on Design Automation*, pages 50–54.

Garey, M. and Johnson, D. (1979). *Computers and Intractability - A Guide to NP-Completeness*. Freeman, San Francisco.

Hotz, G. (1965). Eine Algebraisierung des Syntheseproblems für Schaltkreise. *EIK Journal of Information Processing and Cybernetics*, 1(1):185–231.

Hotz, G. (1974). *Schaltkreistheorie*. Walter de Gruyter.

Huang, J.-D., Jou, J.-Y., and Shen, W.-Z. (1995). Compatible class encoding in Roth-Karp decomposition for two-output LUT architecture. In *Int'l Conf. on CAD*.

Hwang, T., Owens, R., and Irwin, M. (1990). Exploiting communication complexity for multilevel logic synthesis. *IEEE Trans. on CAD*, 9(10):1017–1027.

Ishiura, N., Sawada, H., and Yajima, S. (1991). Minimization of binary decision diagrams based on exchange of variables. In *Int'l Conf. on CAD*, pages 472–475.

Jiang, J.-H., Jou, J.-Y., and Huang, J.-D. (1998). Compatible class encoding in hyper-function decomposition for FPGA synthesis. In *Design Automation Conf.*, pages 712–717.

Kam, T., Villa, T., Brayton, R., and Sangiovanni-Vincentelli, A. (1994). A fully implicit algorithm for exact state minimization. In *Design Automation Conf.*, pages 684–690.

Karp, R. (1963). Functional decomposition and switching circuit design. *J. Soc. Indust. Appl. Math.*, 11(2):291–335.

Karplus, K. (1991). Amap: a technology mapper for selector-based field-programmable gate arrays. In *Design Automation Conf.*, pages 244–247.

Kim, B.-G. and Dietmeyer, D. (1991). Multilevel logic synthesis of symmetric switching functions. *IEEE Trans. on CAD*, 10(4).

Lai, Y.-T., Pan, K.-R., and Pedram, M. (1994a). FPGA synthesis using function decomposition. In *Int'l Conf. on Comp. Design*, pages 30–35.

Lai, Y.-T., Pan, K.-R., and Pedram, M. (1996). OBDD-based function decomposition: Algorithms and implementation. *IEEE Trans. on CAD*, 15(8):977–990.

Lai, Y.-T., Pan, K.-R., Pedram, M., and Sastry, S. (1993a). FGMap: A technology mapping algorithm for look-up table type FPGAs based on function graphs. In *Int'l Workshop on Logic Synth.*, pages 9b1–9b4.

Lai, Y.-T., Pedram, M., and Vrudhula, S. (1993b). BDD based decomposition of logic functions with application to FPGA synthesis. In *Design Automation Conf.*, pages 642–647.

Lai, Y.-T., Pedram, M., and Vrudhula, S. (1994b). EVBDD-based algorithms for integer linear programming, spectral transformation, and function decomposition. *IEEE Trans. on CAD*, 13(8):959–975.

Lee, C. (1959). Representation of switching circuits by binary decision diagrams. *Bell System Technical Jour.*, 38:985–999.

Legl, C., Wurth, B., and Eckl, K. (1996a). A boolean approach to performance-directed technology mapping for LUT-based FPGA designs. In *Design Automation Conf.*, pages 730–733.

Legl, C., Wurth, B., and Eckl, K. (1996b). An implicit algorithm for support minimization during functional decomposition. In *European Conf. on Design Automation*, pages 412–417.

Lupanov, O. (1958). A method of circuit synthesis. *Izv. VUZ Radioviz*, 1:120–140.

Lupanov, O. (1970). On circuits of functional elements with delay. *Probl. Kibern.*, 23:43–81.

Malik, S., Wang, A., Brayton, R., and Sangiovanni-Vincentelli, A. (1988). Logic verification using binary decision diagrams in a logic synthesis environment. In *Int'l Conf. on CAD*, pages 6–9.

Mehlhorn, K. (1984). *Data Structures and Algorithms 2: Graph Algorithms and NP - Completeness*. Springer Verlag.

Molitor, P. (1988). Free net algebras in VLSI-theory. *Fundamenta Informaticae, Annales Societatis Mathematicae Polonae*, XI:117–142.

Molitor, P. and Scholl, C. (1994). Communication Based Multilevel Synthesis for Multi-Output Boolean Functions. In *Great Lakes Symp. VLSI*, pages 101–104.

Möller, D., Mohnke, J., and Weber, M. (1993). Detection of symmetry of Boolean functions represented as ROBDDs. In *Int'l Conf. on CAD*, pages 680–684.

Möller, D., Molitor, P., and Drechsler, R. (1994). Symmetry based variable ordering for ROBDDs. *IFIP Workshop on Logic and Architecture Synthesis, Grenoble*, pages 47–53.

Moret, B. (1982). Decision trees and diagrams. In *Computing Surveys*, volume 14, pages 593–623.

Morgenstern, C. (1992). A new backtracking heuristic for rapidly four-coloring large planar graphs. Technical Report CoSc-1992-2, Texas Christian University, Fort Worth, Texas.

Murgai, R., Brayton, R., and Sangiovanni-Vincentelli, A. (1994). Optimum functional decomposition using encoding. In *Design Automation Conf.*, pages 408–414.

Murgai, R., Brayton, R., and Sangiovanni-Vincentelli, A. (1995). *Logic Synthesis for Field-Programmable Gate Arrays*. Kluwer Academic Publisher.

Murgai, R., Nishizaki, Y., Shenoy, N., Brayton, R., and Sangiovanni-Vincentelli, A. (1990). Logic synthesis for programmable gate arrays. In *Design Automation Conf.*, pages 620–625.

Murgai, R., Shenoy, N., Brayton, R., and Sangiovanni-Vincentelli, A. (1991). Improved logic synthesis algorithms for table look up architectures. In *Int'l Conf. on CAD*, pages 564–567.

Panda, S. and Somenzi, F. (1995). Who are the variables in your neighborhood. In *Int'l Conf. on CAD*, pages 74–77.

Panda, S., Somenzi, F., and Plessier, B. (1994). Symmetry detection and dynamic variable ordering of decision diagrams. In *Int'l Conf. on CAD*, pages 628–631.

Paul, W. (1976). Realizing boolean functions on disjoint sets of variables. *IEEE Trans. on Circ. and Systems*, 2:383–396.

Roth, J. and Karp, R. (1962). Minimization over Boolean graphs. *IBM J. Res. and Develop.*, 6(2):227–238.

Rudell, R. (1989). Logic synthesis for VLSI design. Technical report, UCB/ERL Memorandum M89/49, University of California, Berkeley.

Rudell, R. (1993). Dynamic variable ordering for ordered binary decision diagrams. In *Int'l Conf. on CAD*, pages 42–47.

Saldanha, A., Villa, T., Brayton, R., and Sangiovanni-Vincentelli, A. (1992). A framework for satisfying input and output encoding constraints. In *Design Automation Conf.*

Sangiovanni-Vincentelli, A., Gamal, A., and Rose, J. (1993). Synthesis methods for field programmable gate arrays. *Proc. of the IEEE*, 81:1057–1083.

Sasao, T. (1993). *Logic Synthesis and Optimization*. Kluwer Academic Publisher.

Sauerhoff, M. and Wegener, I. (1996). On the complexity of minimizing the OBDD size for incompletely specified functions. *IEEE Trans. on CAD*, 15(11):1435–1437.

Sawada, H., Suyama, T., and Nagoya, A. (1995). Logic synthesis for look-up table based FPGAs using functional decomposition and support minimization. In *Int'l Conf. on CAD*, pages 353–358.

Schlichtmann, U. (1993). Boolean matching and disjoint decomposition. *IFIP Workshop on Logic and Architecture Synthesis, Grenoble*, pages 83–102.

Scholl, C. (1996). *Mehrstufige Logiksynthese unter Ausnutzung funktionaler Eigenschaften.* PhD thesis, Universität des Saarlandes.

Scholl, C. (1997). Multi-output functional decomposition with exploitation of don't cares. In *Int'l Workshop on Logic Synth.*

Scholl, C. (1998). Multi-output functional decomposition with exploitation of don't cares. In *Design, Automation and Test in Europe*, pages 743–748.

Scholl, C., Melchior, S., Hotz, G., and Molitor, P. (1997). Minimizing ROBDD sizes of incompletely specified functions by exploiting strong symmetries. In *European Design & Test Conf.*, pages 229–234.

Scholl, C. and Molitor, P. (1993). Mehrstufige Logiksynthese unter Ausnutzung von Symmetrien und nichttrivialen Zerlegungen. Technical Report 02/93, Sonderforschungsbereich 124, TP B6.

Scholl, C. and Molitor, P. (1994). Efficient ROBDD Based Computation of Common Decomposition Functions of Multi-Output Boolean Functions. In *IFIP Workshop on Logic and Architecture Synthesis, Grenoble*, pages 61–70.

Scholl, C. and Molitor, P. (1995a). Communication based FPGA synthesis for multi-output Boolean functions. In *ASP Design Automation Conf.*, pages 279–287.

Scholl, C. and Molitor, P. (1995b). Efficient ROBDD based computation of common decomposition functions of multioutput boolean functions. In Saucier, G. and Mignotte, A., editors, *Novel Approaches in Logic and Architecture Synthesis*, pages 57–63. Chapman & Hall.

Scholl, C., Möller, D., Molitor, P., and Drechsler, R. (1999). BDD minimization using symmetries. *IEEE Trans. on CAD*, 18(2):81–100.

Schweikert, D. and Kernighan, B. (1972). A proper model for the partitioning of electrical circuits. In *Design Automation Conf.*, pages 57–62.

Sentovich, E., Singh, K., Lavagno, L., Moon, C., Murgai, R., Saldanha, A., Savoj, H., Stephan, P., Brayton, R., and Sangiovanni-Vincentelli, A. (1992). SIS: A system for sequential circuit synthesis. Technical report, University of California, Berkeley.

Shannon, C. (1938). A symbolic analysis of relay and switching circuits. *Trans. AIEE*, 57:713–723.

Shannon, C. (1949). The synthesis of two–terminal switching circuits. *Bell System Technical Jour.*, 28:59–98.

Shen, W.-Z., Huang, J.-D., and Chao, S.-M. (1995). Lambda set selection in Roth-Karp decomposition for LUT-based FPGA technology mapping. In *Design Automation Conf.*, pages 65–69.

Shiple, T., Hojati, R., Sangiovanni-Vincentelli, A., and Brayton, R. (1994). Heuristic minimization of BDDs using don't cares. In *Design Automation Conf.*, pages 225–231.

Slansky, J. (1960). Conditional-sum addition logic. *IEEE Trans. on Electronic Comp.*, 9:226–231.

Takagi, N. and Yajima, S. (1988). An on-line error-detectable divider with a redundant binary representation and a residue code. In *Int'l Symp. on Fault-Tolerant Comp.*, pages 174–179.

Tani, S., Hamaguchi, K., and Yajima, S. (1993). The complexity of the optimal variable ordering problem of shared binary decision diagrams. In *ISAAC'93*, volume 762 of *LNCS*, pages 389–398. Springer Verlag.

Wallace, C. (1964). A suggestion for a fast multiplier. *IEEE Trans. on Comp.*, 13:14–17.

Wang, K., Hwang, T., and Chen, C. (1993). Restructuring binary decision diagrams based on functional equivalence. In *European Conf. on Design Automation*, pages 261–265.

Wurth, B., Eckl, K., and Antreich, K. (1995). Functional multiple-output decomposition: Theory and implicit algorithm. In *Design Automation Conf.*, pages 54–59.

About the Author

Christoph Scholl studied computer science and electrical engineering at the University of Saarland, Germany, from 1988 to 1993. He received his Dipl.–Inform. and Dr.–Ing. degrees in 1993 and 1997, respectively, from the University of Saarland.

In 1993 he was with the Sonderforschungsbereich 'VLSI Design Methods and Parallelism' at University of Saarland. From 1993 to 1996 he held a scholarship position in the Graduiertenkolleg 'Efficiency and Complexity of Algorithms and Computers' at the University of Saarland.

Since 1996 he has been with the Institute of Computer Science at Albert-Ludwigs-University, Freiburg im Breisgau, Germany.

His research interests include logic synthesis, verification, and testing of VLSI circuits.

Index